Increasing Intelligence

Increasing Intelligence

Norbert Jaušovec
University of Maribor
Department of Psychology
Slovenia

Anja Pahor
University of Maribor
Department of Psychology
Slovenia

ELSEVIER

ACADEMIC PRESS
An imprint of Elsevier
elsevier.com

Academic Press is an imprint of Elsevier
125 London Wall, London EC2Y 5AS, United Kingdom
525 B Street, Suite 1800, San Diego, CA 92101-4495, United States
50 Hampshire Street, 5th Floor, Cambridge, MA 02139, United States
The Boulevard, Langford Lane, Kidlington, Oxford OX5 1GB, United Kingdom

Library of Congress Cataloging-in-Publication Data
A catalog record for this book is available from the Library of Congress

British Library Cataloguing-in-Publication Data
A catalogue record for this book is available from the British Library

ISBN: 978-0-12-809407-5

For information on all Academic Press publications
visit our website at https://www.elsevier.com/books-and-journals

Publisher: Nikki Levy
Acquisition Editor: Emily Ekle
Editorial Project Manager: Barbara Makinster
Production Project Manager: Punithavathy Govindaradjane
Designer: Mark Rogers

Contents

Acknowledgments

Increasing Intelligence is a book about intelligence and the possibility of improving it with interventions ranging from classic cognitive training to directly influencing the brain. The topic has been fiercely discussed with a line of reasoning often driven by beliefs and opinions rather than facts provided by research. Anja and I have tried to provide the reader with an unbiased overview of the state of the art in the research field. We are inclined to be on the side of those who think that intelligence can be increased, probably because our own research has shown positive results.

Here we would like to thank the participants in our experiments who solved endless boring tasks that at first glance made no sense, although as ensured by the experimenter, they presented a crucial part of the study. Not to mention EEG preparations ruining the hairstyle of our female participants, bewilderment caused by the NIRS setup, and tingling scalp sensations during transcranial alternating current stimulation. All in the name of science …

Chapter 1

The Mystery of Intelligence

Few questions have generated such passionate discussions in the scientific community than those related to intelligence and intelligence testing. To give some examples: Arthur Jensen (1969) published an article in the *Harvard Educational Review* (HER) titled "*How much can we boost IQ and scholastic achievement?,*" in which he argued that compensatory education in the US failed to produce lasting beneficial effects on children's intelligence quotient (IQ) and achievement, therefore the premises on which these programs were based should be reexamined. Nine years later, he admitted that the firestorm that was ignited by his writing came as an unpleasant surprise. Less than 10% of the 123 pages long article dealt with ethnical differences suggesting that genetic factors may be the reason for the one standard deviation lower performance of Afro-Americans as compared to whites. The reaction was "swift and severe," with regular near-riotous demonstrations by students at colleges where Jensen was invited to speak. His office was picketed and his classes were regularly disrupted or, when lecturing at other universities, cancelled at the last minute. Students' newspapers were filled for weeks with fierce statements against his research, suggesting that the university should get rid of him. The violence of student activists escalated to the point to which it threatened to get out of control. At that time, the campus police assigned two bodyguards to accompany Jensen around campus (Snyderman and Rothman, 1988). Even 9 years later and, as far as Australia, there was still need for police protection (Jensen, 1978).

The media were among the first to give full coverage on Jensen. The *New York Times*, the *Times Magazine,* and local newspapers throughout the United States reported on the topic. The *Times Magazine*, after publishing a lengthy article entitled: *Jensenism, n. The Theory that IQ Is Determined Largely by the Genes,* received more letters than any other issue since the assassination of President Kennedy (Snyderman and Rothman, 1988).

The reaction of colleagues was mixed—condemning the rude treatment on the one hand, while declaring their own disagreement with Jensen's conclusions, on the other. Few publically agreed with Jensen's statements, albeit, privately and in letters they showed support for his ideas. Shortly after the paper appeared in the HER, the American Anthropological Association held a convention at which they passed a resolution condemning Jensen's position and encouraged members to fight racism with all available means.

Increasing Intelligence. http://dx.doi.org/10.1016/B978-0-12-809407-5.00001-4

Twenty-five years later, *The Bell Curve* by Herrnstein and Murray (1994) sparked almost the same obfuscation as Jensen's (1969) article, although the controversy was less physically violent. Jim Naureckas (1/2/1995) in *Extra!*, a magazine of the national media watch group (FAIR) in an article titled *Racism Resurgent: How Media Let The Bell Curve's Pseudo-Science Define the Agenda on Race*, analyzed the media responses. His conclusion was that *The Bell Curve* received broad media attention, mainly sharp criticism, but also some acceptance of premises and evidence. In Naureckas's opinion and in the view of some of the magazine editors referred to later, the book was part of a campaign favoring and justifying racism.

Nearly all the research that Murray and Herrnstein relied on for their central claims about race and IQ was funded by the Pioneer Fund, described by the London *Sunday Telegraph* (3/12/89) as a "neo-Nazi organization closely integrated with the far right in American politics." The fund's mission is to promote eugenics, a philosophy that maintains that "genetically unfit" individuals or races are a threat to society. [...] Another person whose advice Murray and Herrnstein "benefitted especially from"—and who shows up constantly in their footnotes—is Arthur Jensen, whose very similar claims about blacks having innately lower IQs were widely discredited in the 1970s. The Pioneer Fund has given more than $1 million to this "giant in the profession," as Pioneer chief Weyher describes him (GQ, 11/94). And it's easy to see why: "Eugenics isn't a crime," Jensen has said (*Newsday*, 11/9/94). "Which is worse, to deprive someone of having a child, or to deprive the child of having a decent set of parents?"

My third example is anecdotal, based on my own experience. I started my professional career at the University of Maribor, where I was given a 3-year grant to investigate the relation among academic success, intelligence, and creativity in first year undergraduates. My research report would have probably ended up unnoticed on a library bookshelf, or in the best case, would have been discussed by a university committee that would spend years ruminating about the same issues without ever arriving at a solution. However, at that time, in former Yugoslavia, the Universities organized biannual meetings, which took place at Dubrovnik (now Croatia). The general theme of the conferences was University Today, but each time a different topic, either educational or organizational, was chosen as the conference's main motto. The topic of the 1987 conference was creativity in education and I was invited to give a presentation. The event was picked up by local media with a short report in the education section. The only thing reported about my talk in Dubrovnik was that the average IQ of students from the faculty of agriculture was lower in comparison to students of other faculties. This was a marginal finding used to describe sample characteristics. The difference was expected, because the agriculture program was at that time a 2-year undergraduate program, while the others were 4-year university programs. It is well established that high correlations are found between measures of intelligence and educational attainment.[a] In a

a. For operational purposes, educational attainment is usually measured with respect to the highest education program successfully completed, which is typically certified by a recognized qualification.

review of 16 studies, Ceci (1991) estimated that the correlation ranges between 0.50 and 0.90. Shortly after the appearance of the short note in the newspaper, I received a phone call from the secretary of the faculty of agriculture, informing me that I am not allowed to continue the research at their facility. Soon after the grant was discontinued, after only 1 year of research. The story does not end here. In 1996, 9 years later, I was up for promotion to tenure track. I did not pay much attention to the process, but when it ended our faculty commission representative asked me what I had done to offend the dean of the faculty of agriculture, because he strongly opposed my promotion.

Would the reaction be the same, if instead of differences in IQ scores I reported differences in maximal oxygen uptake between students who aspired to be either teachers of physical education or psychologists? Will the book *Race and Sport*, in which Dutton and Lynn (2015) explicitly state that West Africans and those of West African descent dominate male short distance running, provoke a similar hostile reaction as *The Bell Curve*? Not likely, or perhaps just for the six pages in Chapter 7, in which race and intelligence differences are discussed. Indeed, Jensen (1969) himself reported several examples of Afro-American infant's superior performance on different motor tests, which was nearly one standard deviation above white norms, and a developmental IQ of about half a standard deviation above white norms. To my knowledge, these data were never identified as racist.

Intelligence is a precious asset needed in many contexts, and highly valued in society. Thus it is not surprising that considering intelligence as a constant that is completely dependent on natural selection, which started in Africa some 150–100 kyr ago (kyr = 1000 years), took away all hope that with adequate education, effort, and dedication, ethnic differences in intelligence would disappear like morning mist. These circumstances come rather close to what is known as *learned helplessness,* a phenomenon describing passive acceptance of uncontrollable aversive events (Seligman and Maier, 1967). The fierce reactions and criticism triggered by reports on individual differences in intelligence related to race or sex are not surprising, particularly at the time when Jensen's article appeared in HER, and the memories of the Nazi Holocaust were still vivid and folk knowledge about genetics was limited. It was also the first time after the Second World War that someone had stated that a genetic factor accounts for the difference in black and white intelligence. In a several-hour conversation with Jared Taylor in 1992, reprinted in *American Renaissance* (10/29/2012), Arthur Jensen expressed similar opinions:

> *The view being a very common one today, still, that blacks would be no different from the rest of us — the rest of the population—if they simply had the same education and all of that. [...] This is probably the most highly valued trait. When people are asked what characteristics they want their children most to have, the two things they mention first are good health and good intelligence. If you suggest that people differ in intelligence for reasons that they themselves are not responsible for, because of the particular assortment of genes they happen to get, this seems terribly unfair. I think that World War II was really the main turning point in this. We'd been headed in that direction [egalitarianism], but the turning*

point, I think, was the revulsion against the Nazi Holocaust. People pointed to that as an example of what would happen if we recognized differences.

1.1 INTELLIGENCE—THE MOST VALUED TRAIT

Let's look at an example from folk psychology.[b] When asked about your IQ, being either above or below average, the majority would say that it is above average. This is of course impossible as average by definition means that half of the population is above and half below average. The tendency to overly favorably judge our abilities is called *illusory superiority* (Kruger and Dunning, 1999). The authors found that particularly those whose performance on ability tests was in the bottom quartile grossly overestimated their performance. Even though they scored at the 12th percentile, they estimated their performance being at the 62nd percentile. This was explained as a deficit in metacognitive skill—knowing how well one is performing. Predominantly men have been found to be notorious exaggerators when it comes to identifying their own IQ, which is known as the *hubris-humility* effect. The bias that men significantly overestimate and women significantly underestimate their abilities is strongest for mathematical–logical and spatial IQ subtests, followed by overall (g) and verbal IQ subtests. It was suggested that the observed sex differences might result from implicit theories of intelligence—men are believed to perform better on mathematical-logical and spatial tests (Furnham et al., 2002; Storek and Furnham, 2013). The explanation sounds plausible, although it struggles to explain why the same pattern of male hubris is observed for performance on tests of emotional intelligence (Petrides and Furnham, 2000). The construct of intelligence will be discussed in more detail in the next chapter, therefore only a brief commentary will be presented in order to aid understanding of the argumentation. The construct of intelligence usually consists of a verbal and a performance component. Thorndike (1920) added a third component called social intelligence. Due to a lack of adequate tests, the third component was ignored in mainstream intelligence research in the early years of the 20th century. In recent years however, this has changed, particularly with the introduction of the construct of emotional intelligence (EI)—the ability to recognize emotion, reason with emotion, and emotion-related information, and process emotional information as part of general problem-solving (Mayer et al., 2000). It was found that females surpass males on EI (Amelang and Steinmayr, 2006; Mayer et al., 2000; Van Rooy et al., 2005) and on emotional awareness, which is a similar construct (Boden and Berenbaum, 2007; Ciarrochi et al., 2005; Parker et al., 2003). Even though we are not aware of a study that would report sex differences in implicit theories of EI,[c] I assume that

b. People's conceptions also named implicit theories (R.J. Sternberg, 2000).

c. Dean Burnett in *The Guardian: Calm down, dear: the dark side of 'emotional intelligence'* (21/4/2015), suggested that it is widely believed that women are more emotional than men, backing up the statement with the response to Hillary Clinton's presidential bid.

folk psychology would judge that women score higher on that trait, comparable to verbal fluency. Therefore, in my opinion the reason for the male tendency to overestimate their own intelligence must be seen in a broader, more evolutionary context and not just as a result of the influence of peoples' concepts on judgment. Intelligence is regarded as one of the most highly valued traits, thus, having a high IQ means an advantage, and having a low IQ is a handicap. Hence males' overestimations of their intellectual abilities well be part of their mating strategy—increasing attractiveness to catch the attention of possible mates, similar to courtship displayed in response to precopulatory sexual selection observed in many animal species. Support for our hypothesis lends the finding that males who perceive themselves as more successful tend to opt for short-term mating (Landolt et al., 1995). Our hypothesis is also supported by Miller (2000) proposing that during human evolution, mate choice focused increasingly on intelligence as a major indicator of biological fitness. Miller has even gone one step further, suggesting that many human behaviors like conversation, music production, artistic ability, and humor may have evolved principally to advertise intelligence during courtship.

1.1.1 Once Upon a Time in Africa

The general opinion that it is "smart to be smart" is also shared by nature in applying principles of selection to human evolution, thereby favoring intelligence and brain development more than other traits or organs. The work of Darwin (1871) propelled the idea that human intelligence is related to a correspondingly large brain. According to Wilson (1978), no organ in the history of human evolution has grown faster than the brain. When, roughly 4 million years ago, true men diverged from their ape ancestors the brain added about a tablespoonful (16 cm^3) every 100 kyr. The rate continued until the appearance of modern men some 250 kyr ago. Geary (2005) in his book *The Origin of Mind: Evolution of Brain, Cognition, and General Intelligence*, proposed that the rapid expansion of brain size, which is assumed to covary with intelligence (n-weighted $r = 0.20$; Vernon et al., 2000), resulted from a shift from ecological dominance to social selection pressures. However, there is still some controversy about the weight these different selection pressures had on the evolution of intelligence. Among the most essential pressures identified were: climatic, ecological, social, and cultural/social learning (Geary, 2005; Reader et al., 2011).

The turnover-pulse hypothesis (TPH), put forward by Vrba (1975, 1993), is probably the most known theory on the effects of climatic and geological changes on evolution. The hypothesis makes predictions about the potential response of species to large-scale climatic changes, like glaciations and global cooling, considering them to be a major cause of faunal turnover—extinction, speciation, and migration. The THP has inspired debate and controversial viewpoints either in favor of the hypothesis, or suggesting that climate change had a minor impact on human development (Faith and Behrensmeyer, 2013). *The ecological intelligence hypotheses* suggested that foraging challenges drove human brain evolution (Parker and Gibson, 1977). Among the most essential are shifts to more energy-rich or more easily processed diets, constraints related to the complexity of cognitive maps to cover the size of home range and inertial navigation (day journey length). The main critique of the ecological hypothesis was that the costs of maintaining a large brain would be a waste of energy when used just to increase the capability of searching for food and supplies. The selection factor to develop large brains is insufficient to overcome the steep cost gradient (Dunbar, 1998). Dunbar's argument is based on the well known fact that our brain represents about 2% of the body weight but consumes about 20% of total energy intake (Aiello and Wheele, 1995). The percentage of brain energy consumption in other species is much lower (and remarkably constant) than the one observed in humans. In 42 species, mainly mammals, this percentage was estimated between 2 and 8% (Mink et al., 1981).[d] The question is, why would primates or humans need larger brains than other species in order to accomplish the same job? In the late 1980s, an alternative hypothesis suggested that primates' larger brains reflect the demands of complex social interactions involving processes such as tactical deception and coalition-formation. This was termed the *Machiavellian intelligence hypothesis*, albeit the term *social intelligence hypothesis* is preferred (Dunbar, 1998). Yet another theory dubbed *the cultural intelligence hypothesis* suggested that the superiority of social learning (learning by observation of others) over independent individual exploration provided opportunities to learn vital cultural and routine skills (van Schaik et al., 2012).

Comparisons of these conflicting hypotheses are flawed by serious methodological problems. Variables that separate the theories like group size (*social intelligence hypothesis*) or diet (*ecological intelligence hypothesis*) are intercorrelated and are measured with considerable error. Further, the authors of the conflicting theories used different scaling factors for their brain measures. Advocates of the *social intelligence hypothesis* relied on measures of neocortex size relative to the size of the rest of the brain, rather than on the size of the whole brain relative to body size, as did the supporters of the *ecological intelligence hypotheses*. It was argued that a better understanding of the enlargement of the human brain during evolution could be obtained by research focusing on microscopic and macroscopic brain anatomy (Barton, 2006).

An example of the microscopic level of study is *the nutritional neurotrophic neoteny theory*, put forward by Amen-Ra (2007). Amen-Ra argued that human encephalization is paradoxical. Because brain tissue is in metabolic terms highly expensive, enlarging its volume in periods of limited food resources makes no sense. Two strands of research attempted to explain this paradox. In

---

d. An exception is the electric fish *Gnathonemus petersii*—its brain is responsible for approximately 60% of body energy consumption. The exceptionally high energetic cost of the *G. petersii* brain is a consequence both of the brain being very large and of the fish being ectothermic—energy expenditures are about 1/13 of those of endotherms (Nilsson, 1996).

an intriguing study, Bodoky et al. (1995) compared the levels of energy (ATP concentration) in the brain relative to the liver in three groups of rats on different dietary regimes: *ad libitum* access to food; continuous parenteral infusion of nutrients (100% of the daily intake) and 4 days of complete food deprivation. They found substantially reduced levels of ATP in the liver of the starving group of rats, while there were no differences in brain ATP levels compared to the other two groups of rats that had full access to food. These findings support the theory of preferential supply of nutrients to the brain. The second strand of research investigates the influence of caloric restriction on expression of neurotrophins which facilitate influx of glucose into neurons, and promote the proliferation of neural precursor cells that are similar to stem cells (Burkhalter et al., 2003; Kolb and Whishaw, 2009). Lee et al. (2000) found that keeping adult rats on a dietary restriction resulted in a significant increase in the number of newly produced neural cells in the dentate gyrus of the hippocampus. In humans, the hippocampus has been linked to episodic and autobiographic memory formation and particularly the formation of autonoetic awareness of time, which allows us to time travel to the past and future (Kolb and Whishaw, 2009). It can be therefore concluded that caloric restriction promotes the preservation of progenitor neurons in cortical regions involved in memory. Better memory is not only beneficial in search for food and resources, but particularly important in social interaction and learning. To this day, I am embarrassed when at work somebody greets me, starts a conversation and I have no idea who that person might be.

On a macroscopic level, two competing theories have evolved: *mosaic* versus *global brain evolution*. A simplified description of the difference between the competing views was put forward by Finlay et al. (2001). The mosaic *evolution theory* suggests that brain areas can be discriminated according to their function, hence, evolution must have enabled such development. Parts of the system can change (e.g., auditory or vision systems) while the rest of the system is left mostly unchanged. On the other hand, the *global brain evolution theory* suggests that the whole system changed in response to selection pressures. Finlay et al. (2001) compared multiple regressions on allometric data for 131 mammalian species. The main findings supported the *global evolution theory* indicating that for 9 out of 11 brain structures, taxonomic and body size factors were less important than covariance of these major structures with each other. This explanation seems plausible, especially in the view of contemporary brain theories that describe the brain's connectome architecture as a comprehensive network map of the nervous system of a given organism. A review of empirical and computational studies strongly suggested that brain hubs[e] play important roles in information integration underpinning complex cognitive function (van den Heuvel and Sporns, 2013). This idea can be traced back to Harry Jerison (1991) who viewed the brain as a "mapping machine," in which maps are

---

e. Hub: a node occupying a central position in the overall organization of a network (van den Heuvel and Sporns, 2013).

different representations (sensory and motor) of the external world. The author further proposed that the number of these maps is related to the complexity of viewing and representing the external world. Therefore, the increase in relative brain size can be explained as a need for improving ways of knowing reality. For example, the squirrel has three visual areas, the owl monkey has 14 representations of the visual world, and for humans it is estimated that there are 30 visual areas, therefore we can "see" 27 kinds of things which the squirrel cannot see (Kaas, 1987; Kolb and Whishaw, 2009).

Alternatively, the *mosaique evolution theory* advocates an independent size change in individual brain components. An example is Falk's *radiator theory* (1990)—the author devised it based on two personal, yet unrelated events: a comment of her mechanic that the engine can only be as big as the radiator can cool, and a letter from a colleague informing her that emissary veins contribute to the regulation of brain temperature. Combining these two pieces of information led her to the *radiator theory* suggesting that vascular systems of early bipeds became reoriented allowing for less limited growth of the ape's brain. Falk observed a difference in the vein distribution between robust australopithecines and graciie aystraiopithecine through *Homo sapiens* lineage. The latter had a widespread network of small veins allowing the brain to grow larger (Falk, 1990).

To overcome the aforementioned limitations to explain variations in brain size due to evolutionary pressures, Geary (2005) proposed an evolved human disposition, labeled *motivation to control,* as the major drive of human behavior. Humans are motivated to control social relations, the behavior of other people, and biological and physical resources important for survival and reproduction. The concept is similar to Kelly's (1955) core tendency of human personality which is to predict and control experienced events. Similar processes have been also identified in human ontogenetic development by Piaget and Inhelder (1974). Piaget suggested two complementary processes, assimilation and accommodation, to be involved in the construction of new concepts, that is, schemata. The process is triggered by an inconsistency that brings the equilibrium state of a schemata out of balance. Homeostasis is achieved either by incorporating new information into the already learned schemata (assimilation), or by changing the existing schemata and creating new knowledge (accommodation). A nice example of this process is that of a child having developed a schemata of a cat, which when seen is called "mau-mau." The child applies the same "mau-mau" schemata when he or she sees a squirrel (assimilation). It is likely that parental intervention pointing out that this is not a cat but a squirrel has no immediate effect on the child's concept. Only when the child recognizes an inconsistency in the "mau-mau" concept that is triggered, for example, by the bushy tail of the squirrel that is different from that of a cat, a new concept is formed (accommodation).

Even though there is no consensus on how different selection pressures affected the evolution of human intelligence, there is general agreement that these pressures shaped our ancestors about 1.8 million years ago, most likely *Homo ergaster/erectus* in eastern Africa. *H. erectus* was a globetrotter—its remains were

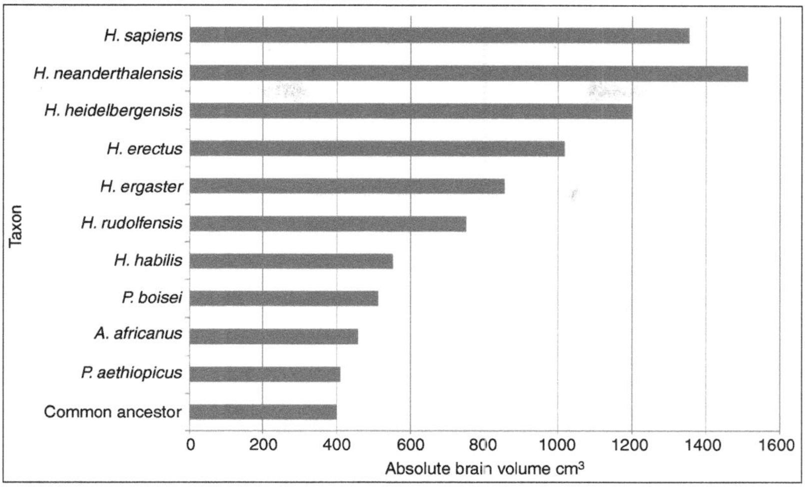

FIGURE 1.1 Estimated brain volumes of the hominid species. The histogram is based on findings by Johanson and Edey (1981) and Wood and Collard (1999).

found in Java, China, and parts of southern Europe (Gabunia et al., 2000). Because the exact relationships among hominid species are not known, they constitute a discontinuous sequence with a *Common ancestor* at the base and *H. sapiens* at the top (Johanson and Edey, 1990). Initially, only fossil remains that showed evidence of stone tool manufacturing were included in the taxonomy of the *Homo* genus. More contemporary classifications grant inclusion based on brain size, inferences about language ability, hand function, and assumptions about expertise in making stone tools (Wood and Collard, 1999). Although there is some disagreement about the inclusion of certain species into the Human genus, the general pattern is clear (Geary, 2005). Fig. 1.1 shows absolute brain volumes of the human family, based on different reports found in the literature (Johanson and Edey, 1990; Wood and Collard, 1999; Geary, 2005). As can be seen in Fig 1.1, brain size in the australopiths increased modestly over a period of 1.5 million years, with almost a "runaway" increase when the first hominids appeared.

The problem in comparing absolute measures of brain volume is that brain size is highly correlated with body mass and that this relationship is not linear (Alba, 2010). Cross-species comparisons can therefore be biased. For example, the sperm whale (*Physeter macrocephalus*), has a brain mass of about 8028 g, which is almost six times the size of a human brain that weighs about 1300 g (Marino et al., 2007). However, we do not expect the sperm whale to be smarter than we are, as already established by Darwin (1871) suggesting that intellect cannot be accurately determined by just looking at the cubic content of a skull.

There are at least two reasons why brain size increased, a cognitive one due to selection pressures for increasing intelligence, and a passive one due to an increase in body mass, which required more brain for the maintenance of a greater body

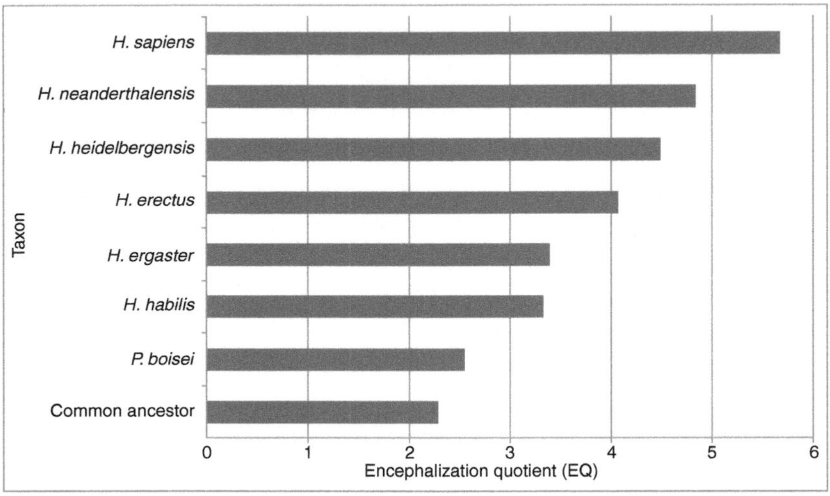

FIGURE 1.2 Estimated encephalization quotients (EQ) of the hominid species. The histogram is based on findings by Johanson and Edey (1981), Ruff et al. (1997), Wood and Collard (1999), and Alba (2010). Brain mass was derived from cranial capacity using the formula: *Brain mass* = 1.147 × *cranial capacity*$^{0.976}$. EQ was derived using Martin's (1981) relation between body mass in (kg) and brain mass in (g): *EQ* = *brain mass*/11.22 × *body mass*$^{0.76}$.

(Deaner et al., 2007). Harry Jerison (1973) solved the problem by proposing an allometric regression measure—the encephalization quotient (EQ), which is the ratio between actual brain size and expected brain size based on body mass. For example, an EQ of 1 has been estimated for the cat. Fig. 1.2 shows EQ values based on the available body and brain masses in the literature (Johanson and Edey, 1990; Wood and Collard, 1999). The EQ values reported in the literature slightly differ in size due to methodological issues, mainly because a different scaling relation between body and brain mass was assumed. For modern humans, EQ values lie between 5.0 and 7.0, with most estimates between 5.0 and 6.0 (Geary, 2005).

As can be seen in Fig. 1.2, the EQ value of the australopithecines is less than half of that of modern humans, while *Homo ergaster's* EQ value is more than half of *Homo sapiens's EQ value.* Ruff et al. (1997) suggested that EQ began reaching its peak with the first anatomically modern humans approximately 90 kyr ago and has since remained fairly constant. However, absolute brain size has decreased by 11% in the last 35 kyr, of which the last 10 kyr were associated with a decrease of 8%. This decrease was accompanied by a corresponding decrease in body mass, thus EQ values remained stable.

1.1.2 Intelligence, Still a Story of Success

The finding that absolute brain size has decreased for 8% in the last 10 kyr has given rise to some speculations that the urge for being intelligent—"brilliant" has lessened because the environmental pressures have become weaker

(Geary, 2005). Is human intelligence still evolving? Hunter (2009) suggested that we might have reached a point where being highly intelligent no longer represents a selective advantage, or this might even be reversed in some societies. For the latter, he provided an example of the genocidal government in Cambodia who slaughtered those deemed as "brilliant," as they would represent a threat to the regime. At the moment, the answer to this question remains elusive because we are dealing with EQ or skull size on the one hand and IQ on the other. EQ values of *Hominid* were derived from estimates of brain size and body mass for a limited number of fossil remains of skulls, based on a time scale of about 4 to 2 million years. Another aspect that has to be considered is the correlation between head size and intelligence. Vernon et al. (2000) analyzed 35 studies with a total sample of 56,793 subjects and obtained an *n*-weighted correlation of 0.20. A rough estimation would be that skull size can explain about 4% of intelligence. In contrast, IQ is based on paper-and-pencil tests that are thought to represent the theoretical construct of intelligence,[f] with a recorded history of data collection of less than 100 years. The difference in the time scale used to predict a change is therefore enormous, even if we consider a longer period of 40–10 kyr, when it is assumed that early modern humans appeared.

I will provide some recent data and observations on the influence of intelligence on success in life suggesting that changes in brain structure and function provide a better explanation for the decrease in absolute brain size, implying that even in modern civilizations, intelligence still represents a selective advantage. A metaanalysis of several correlational studies between intelligence and success in life defined as socioeconomic success (academic and job performance), has shown correlations around (0.50), with academic performance being the highest (0.58), followed by educational attainment (0.56), job performance estimated by supervisory rating, (0.53) and occupational attainment[g] (0.43). The number of individuals included in the metaanalysis was 191,033 (Hunter and Hunter, 1984; Poropat, 2009; Strenze, 2007, 2015). It could be argued that, in modern societies, intelligence explains approximately 30% of success in academic and job performance. Lynn and Vanhanen (2012) used a different approach by correlating national IQs (based on the results of individual testing in countries) with the same standard measures as described earlier, and with other variables, such as political institutions (democracy), health, demographic, and sociological variables (e.g., crime, corruption, religiosity), geographic, and climatic variables. The correlations between health variables and IQ are worth mentioning. Geary (2005) and Hunter (2009) postulated that intelligence affects one's professional career but is less related to mortality or reproduction than it was in the past (Geary, 2005; Hunter, 2009). Based on an

f. An example is the Raven's Matrices test (Raven, 1990), based on Spearman's principles of the eduction of relations and correlates (Spearman, 1923, 1927).

g. Occupational attainment measured in terms of the hourly wage rate associated with an individual's occupation.

analysis of 24 studies, Lynn and Vanhanen (2012) reported negative correlations between national IQ and infant mortality, ranging between −0.34 and −0.84 (6 studies, $N = 792$; n-weighted $r = -0.84$), and positive correlations between national IQ and life expectancy, ranging from 0.51 to 0.84 (8 studies; $N = 1033$; n-weighted $r = 0.78$). Both findings do not confirm the prediction put forward by Geary. However, critics would argue that correlations between ill-defined and measured constructs are not evidence that human intelligence has undergone selection in the modern era.

Ability data (i.e., intelligence test data, and more ecological measures of intellectual success, like achievements in art and science) obtained for Ashkenazi[h] Jews, identified as Jewish individuals of Central and Eastern European ancestry, may provide a better explanation of the issue. The Ashkenazi represent the majority of modern Jews. A recent study showed that Ashkenazi Jews (AJ) can trace their roots back to a group of merely ≈350 individuals (Carmi et al., 2014). Contemporary AJ formed 600–800 years ago as a fusion of two ancestral populations: European and Middle Eastern. It is assumed that the split between these two ancestral populations occurred around the time of the last glacial maximum (20.4–22.1 kyr), with a bottleneck around 85–94 kyr, roughly corresponding to the out-of-Africa event (Carmi et al., 2014).

As reported in several studies carried out mainly on the US and UK populations of Jews, their average IQ is estimated to be between 110–115 IQ points. If adjusted for the secular increase of intelligence (Flynn effect), the value is estimated to be around 110 IQ points (Lynn and Longley, 2006). At first glance, the difference is not impressive, but when mapped to the high end of the normal distribution, with a standard deviation of 15, we end up with 6 times as many Jews with an IQ higher than 140 as compared to other northern Europeans (Cochran et al., 2006). To avoid the IQ trap so often attracted by critiques, Lynn and Longley (2006) also reported data on the number of Jewish Nobel prize laureates in the United Kingdom and the United States, Fellows of the Royal Society (UK), and academic affiliates of elite colleges (US) in relation to their number in the population, comparing data between the first and second half of the 20th century. Jews were overrepresented among Nobel prize winners by factors of 8.0 and 12.3 in Britain and in the United States, respectively, and were also overrepresented among Fellows of the Royal Society in the United Kingdom. The most recent statistics concerning Jewish Nobel laureates, as provided by Heilman (2015), revealed that since 1901 when the first Nobel prize was awarded, 855 individuals have received it and about 193 (22%) of those who received this award were of Jewish origin. Jews, however, constitute a miniscule part of the world's population (less than 0.2%). For comparison, the percentage of Jewish Nobel laureates for the time period between 1901 and 1962 was estimated to be 16% (Weyl and Possony, 1963). The data point toward

h. Ashkenaz in Hebrew refers to Germany, whereas Sephardic Jews are from the areas around the Mediterranean Sea.

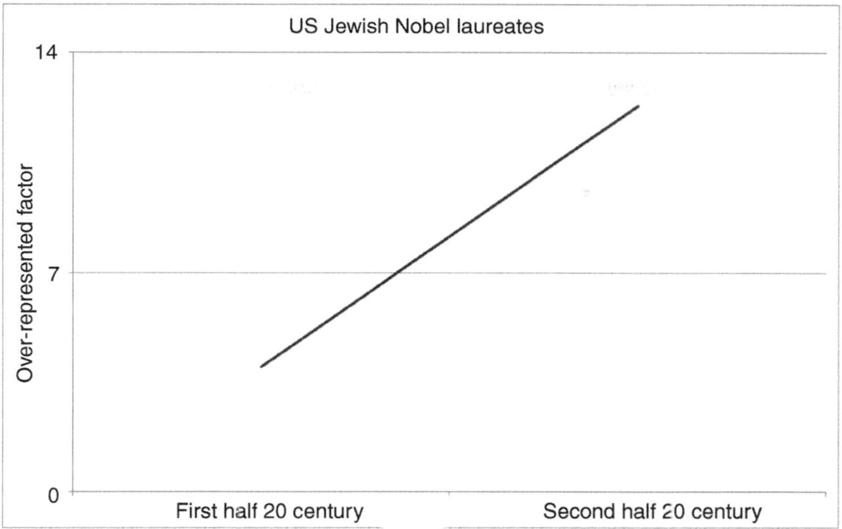

FIGURE 1.3 Estimated over-represented factors for Jewish Nobel prize laureates in United States for the first and second half of the 20th century. The estimates are based on data obtained from the literature (Lynn and Longley, 2006). The factors are calculated as the ratio between the percentages of Jews awarded the Nobel prize and the estimated percentage of Jews in the total population.

an increase in the number of Jewish Nobel laureates in the second half of the 20th century. A similar trend (Fig. 1.3), can be inferred for the data analyzed by Lynn and Longley (2006), suggesting that being highly intelligent represents a selective advantage even in modern developed societies.

Another aspect worth mentioning is that the data are also indicative for a certain amount of malleability of intelligence, not just on a scale of a million years, but also in a shorter time period of one or two generations. This is an important aspect given that the main topic of this book centers on the possibility of increasing intelligence. The mechanisms that might have triggered such a development in the Ashkenazi Jews are diverse, pointing to an interaction between genetic properties and environmental pressures selecting for IQ (Cochran et al., 2006). According to Lynn and Longley (2006), the environmental pressures have been explained by three theories: the eugenic hypothesis, the persecution hypothesis, and the discrimination hypothesis. A fourth theory could be dubbed as educational, albeit not explicitly labeled as such (Heilman, 2015). The eugenic hypothesis proposed that Jews' customs to promote marriage between highly intelligent rabbis and other scholars with the daughters of wealthy merchants can be classified as an eugenic practice that has promoted a greater number of surviving children of the higher intelligent (MacDonald, 1994; Weyl and Possony, 1963). The persecution hypothesis implied that smart Jews had a greater chance to escape or avoid being killed by gentiles. Harassment of Jews is documented throughout history (Weyl and Possony, 1963). The discrimination hypothesis advanced by Cochran et al. (2006) states that in the Middle Ages, Jews in Europe were not allowed to engage in certain kinds of occupation, such as masons, smiths, or arrow makers, and were only permitted to engage in jobs requiring high intelligence, such as moneylenders, tax collectors, import–export merchants, or to deal in second hand goods as traveling vendors. Heilman (2015) suggested that Jewish education, which promotes questioning and allows disobedience in the pursuit of truth, might be the key educational approach that positively affects creativity and intelligence. He backed up the hypothesis via examples from the Midrash commenting and explaining texts of the Torah, but also with the help of texts describing Jewish culture and education. This reminds me of an anecdote I usually tell students when I give a lecture about the concepts of *functional fixedness* and *einstellung—problem solving set*. The story is about a Rabbi trying to enhance what we would today call divergent thinking in his student Joshua. The Rabbi told Joshua about two chimneysweepers who fell down a chimney—the face of one of them was clean while the face of the other was sooty. The Rabbi asked Joshua who of the two will wash his face. Joshua's swift response was that the man with the dirty face will wash it. The Rabbi disagreed, suggesting that the one with the clean face will wash his face, because he will see the other man with the sooty face, assuming that his face is also dirty. The next day, the Rabbi repeats the story and asks the same question. This time, Joshua's answer is that the man with the clean face will wash his face. Yet again, the Rabbi is not satisfied with his answer, stating that the man with the clean face will tell his colleague that he has a sooty face. Joshua gives up, but the Rabbi says: if both have fallen down the chimney, how is it even possible that one would have a clean face and the other a dirty one?

It has been proposed that the combination of the aforementioned environmental constraints and the closure to inward gene flow observed in Askenazi Jews led to an increase in specific genetic variants, in particular the well-known clusters of Ashkenazi genetic diseases, the sphingolipid cluster, and the DNA repair cluster (Cochran et al., 2006). The sphingolipid mutations act as IQ boosters promoting axonal growth and branching (Schwarz et al., 1995). Further support for this hypothesis comes from the finding that cortical expansion in *Homo ergaster/erectus* is assumed to be related to increases in the number of axons and the size of dendrites rather than increases in the number of neurons (Geary, 2005). As discussed earlier, this is in line with the contemporary assumption on brain function and structure defining it as a dense connectome thereby allowing for virtually unconstrained interactions among any pairs of neurons in the brain, either through direct connections or via brain hubs (van den Heuvel and Sporns, 2013).

Another aspect worth mentioning is the fact that most of the UK and US Jews originated from impoverished areas in Eastern Europe and despite the discrimination they faced in their new homes, it can be assumed that their socioeconomic status (SES) improved, which could to some extent explain their increase in IQ. The heritability of intelligence increases from about 30% in early childhood to about 80% in adulthood (Spinath et al., 2003; Edmonds et al., 2008; for a more contemporary viewpoint see Nisbett et al., 2012), and is to some extent related to SES (low versus high) as shown in a study by Turkheimer et al. (2003) and replicated by Harden et al. (2007). The results of both studies point toward a gene–environmental interaction. For children from impoverished families (lower SES), shared environmental influences explained nearly 60% of the variance in IQ, while genetic factors accounted for a negligible variance. A reverse pattern was observed in children from affluent families (high SES) (Turkheimer et al., 2003). The findings of the second study showed a similar though less marked trend identifying income as the main factor predicting the level of heritability (Harden et al., 2007). They proposed a *gene–gene–environmental* model rather than just the *gene–environmental* model put forward in the former study. This means that families with higher income are supposedly more intelligent, which represents the *gene–gene* part of interaction, and they spend more money on better education and training for their offspring, which is the *gene–environmental* part of the model. This model can explain how genetic and environmental factors shaped the IQ of Jews migrating to Western Europe and the US. Support for this explanation also comes from neuroimaging studies. Chiang et al. (2011) conducted a twin study in which they showed that white matter integrity, determined via diffusion tensor imaging[i], was highly heritable but related to SES and the level of intelligence. Higher genetic influence was observed for those who had above average IQ and higher SES.

It can be therefore concluded that intelligence was and still is a central driving force of human development. Yet another piece of evidence for this claim is the growing public demand for cognitive enhancers supporting a billion dollar business promising to raise intelligence.

1.2 CAN WE BOOST INTELLIGENCE?

Yes, we can, but at the moment only in controlled laboratory experiments, for the purpose of verifying or rejecting findings based on correlational studies, thereby widening our understanding of the construct of intelligence. The objective is to combine research methods (correlational and experimental) for the scientific study of human behavior, similar to the one observed in brain imaging where neuroelectric measures of brain activity, such as electroencephalography (EEG), are combined with cardiovascular measures, like magnetic resonance imaging (MRI), or near infrared spectroscopy (NIRS). The point is discussed in more detail in Section 2.2, in which we examine individual alpha frequency (IAF) and its relation to IQ. To date, correlational studies have not been able to

i. Diffusion tensor imaging (DTI) is a MRI method for imaging fiber pathways in the brain by detecting movements of water molecules (Kolb and Whishaw, 2009).

clarify the conflicting results on the IAF–IQ relationship. Influencing IAF by means of neurofeedback or transcranial alternating current stimulation (tACS) might shed light on the controversy (see Section 4.3).

The heated discussion triggered by the study by Jaeggi et al. (2008), which showed that working memory training can increase fluid intelligence, must thus be seen in a broader perspective—in light of the problems Jensen faced after stating that differences in IQ can be, to a certain extent, genetically explained. Nowadays similar viewpoints, although in the opposite direction—intelligence cannot be increased—are often encountered during the peer review process of manuscripts that report positive outcomes of training on the construct level of intelligence. Reviewers' requests often change after each revision, their suggestions are at times contradictory, and they occasionally infer facts that are not present in the study. In contrast, this time public opinion is much more favorable, uncritically claiming that you can *Boost your IQ in 20 minutes*, as reported by Markham Heid in the brain section of *Men's Health* (4/24/2012). Similar claims gave rise to a billion dollar computer-based "brain-games" industry, as recently stressed by Chinthapalli (2015), stating that coffee and crosswords are being replaced with brain training games and so called "smart drugs," such as Adderall, or even edible 24-carat gold flakes for the younger generation entering elite US colleges.

As a consequence in October 2014, a declaration named A *Consensus on the Brain Training Industry from the Scientific Community*, was put forward by the Max Planck Institute for Human Development and the Stanford Center on Longevity signed by several outstanding researchers in the field. The main message was that there is a discrepancy between the state of scientific research on cognitive training, on the one hand, and the claims made by the industry advertising commercial cognitive-training software, on the other. A more moderate opinion was put forward by Papp and Snyder (2012) in their editorial introducing a review article on brain training. Their main suggestion was that the brain training industry has outpaced the scientific evidence for the efficacy of the programs advertised, and that it is not the responsibility of researchers to prove what the industry claims, but that the industry should use more strict scientific methods to support their claims. It seems that this was and still is just wishful thinking. The demand from students, the growing elderly population, and individuals who would like to be smarter is high, allowing the brain-training industry to grow further. Last but not least, this book is also part of it.

The main objective of training studies is to improve fundamental processes that form the basis of intelligent behavior and in that way increase general intelligence (*G*), or fluid intelligence (*Gf*). Test performance can be increased simply by practicing the tests themselves, or by practicing similar tasks. Therefore, training tasks must be substantially different from tests that are used to measure fluid intelligence. For example, in the Jaeggi et al. (2008) study, the training tasks were dual *n*-back tasks, assumed to influence working memory function, hence an increase in working memory capacity, or better performance on more complex *n*-back tasks would point to a near transfer, whereas increased performance on tests of fluid intelligence, like Raven's matrices, would be regarded as a far transfer. Although a more detailed focus on the characteristics and shortcomings of different behavioral training studies will be provided in Chapter 3, at this point I will only discuss the main critiques provoked by the Jaeggi study and studies trying to replicate it. The critiques can be classified into four groups. At extreme ends are those that either reject or accept the possibility of a change in fluid intelligence by means of working memory training. The numerous critiques that lie between the two extremes either state that improvements in the experimental design would make the predictions more valid, or attempt to find a different explanation for the observed increases in tests scores observed after the training has terminated.

Most of those negating that working memory training in healthy adults enhances fluid intelligence, or other measures of cognitive abilities, have come to this conclusion because their studies failed to replicate Jaeggi's findings (Owen et al., 2010; Chooi and Thompson, 2012; Thompson et al., 2013). On the other hand, the possibility to influence fluid intelligence has also been rejected on theoretical grounds. The most rigorous critique was put forward by Haier (2014). In an opinion article, the central objection raised by Haier was that IQ tests are not on a ratio scale (e.g., an increase from 100 IQ points to 105 points is not the same as an increase from 120 IQ points to 125 points), therefore, the interpretation of test score differences is impossible. The suggestion put forward in the paper was that because no IQ test yet exists that would fulfil the requirements of a ratio scale we should direct our efforts into new approaches of IQ measurement. The problem we face next is that intelligence has been defined in many different ways. How can we measure something if we do not know what it is? Hence, the focus should first be on the definition of intelligence. This reminds me of a dispute I saw on television between Joschka Fischer, who was at that time the German Minister for Foreign Affairs, and Donald Rumsfeld, the US Secretary of Defense, about the lack of evidence for nuclear weaponry in Iraq. Joschka was defending the German position of not sending ground forces to fight Saddam. Donald, a little bit angry, replied that in America, there is a saying "don't dig if you are in a hole." From what we know today, Joschka should have said this to Donald.

Among the most often raised critiques of the Jaeggi study were the lack of a placebo (active) control group, the use of just one training task, and only one test measure for the dependent variable of fluid intelligence, there was no control of motivational (expectancy) and repetition effects, nor was the relation between baseline IQ scores and the increment in IQ scores sufficiently analyzed (Shipstead et al., 2010; Thompson et al., 2013). Even stricter was the criticism put forward by Moody (2009). The main objection raised was that different tests of fluid intelligence were used for the so-called control group (who received just 8 days of training), and the experimental group, in which individuals were given an alternative test with a time restriction that may have biased the results. This time restriction made it impossible for the participants to proceed to more demanding items. Since the whole weight of Jaeggi's conclusions rests upon the validity of the measure of fluid intelligence, in Moody's opinion, this brings into question the results and inferences reported in the study.

Support for increasing intelligence by practicing working memory tasks for the most part streams from replication studies that have been designed to overcome the aforementioned shortcomings. In almost all studies that assumed a far transfer effect of working memory training on fluid intelligence, an active control group was introduced and the number of tests to measure fluid intelligence was increased. Some studies also recorded the influence of training on brain activity, used different training tasks, and selected respondents based on their implicit theories on the malleability of intelligence and their level of motivation (Jaeggi et al., 2014; Buschkuehl et al., 2014; Jaušovec and Jaušovec, 2012). On a theoretical level, increasing intelligence was supported by Sternberg (2008) in his comment accompanying the Jaeggi study published in *PNAS*. Sternberg suggested that the recent development in cognitive-based theories of intelligence has provided insight into the relation between working memory processes, roughly divided into storage and maintenance of information, thereby providing a theoretical background for the training design that might be appropriate for influencing intelligence. A limitation that still persists in the aforementioned studies is a rather small number of participants involved in training programs. This shortcoming is related to the metric structure of ordinal-scaled data, like intelligence test scores, suggesting that by increasing n the ordinal scale represents the closest approximation to a truly metric scale (Shepard, 1966), coming close to the ratio scale requested by Haier (2014).

The last group of doubters that intelligence can be raised via training consists of researchers whose replication studies showed training effects on some, but not all tests used to determine fluid intelligence. The conclusion put forward in discussing the results was that fluid intelligence is not increased at the construct level, but that *visuospatial processing* might have been enhanced (Colom et al., 2013). The observed differences were explained by introducing a new construct—*visuospatial processing*, albeit fluid intelligence itself has enough difficulties to be adequately defined. Deary (2000) in his book *Looking Down on Human Intelligence: From Psychometrics to the Brain,* labels such explanatory constructs as "cheap talk" (p. 101). The Colom study I am referring to used an experimental and a passive control group. The two groups were equalized on the main variables studied. The participants (n= 56) were all female. The difficulty with female participants, when tested on three occasions, separated by a time interval of 24 weeks between the pre and posttest measuring fluid intelligence and another 3-week interval when the second posttest for the cognitive tasks was administered, is that the timing of testing should have controlled for their menstrual cycles. Since this is not mentioned in the article, and by taking a close look at the given training and testing schedule reported in the study, I assume that it was not done. With respect to the female menstrual cycle, research has shown that the relative release of sexual hormones in different phases of the menstrual cycle affected cognitive responses of females (Amin et al., 2006; Berman et al., 1997). High levels of estrogen and progesterone were associated with positive affect and higher activity in the prefrontal cortex. The second problem is that the training sessions were interrupted for the Christmas break (2 weeks). One can only speculate about the effects such an interruption might have had on the training outcomes collected on the two posttests. A recent meta-analysis, including 47 studies with 65 group comparisons, showed that factors related to the time configuration of training (e.g., duration of single training sessions, frequency of training per week, and time interval between single sessions) had an influence on far transfer effects (Schwaighofer et al., 2015). Colom et al. (2013) reported changes at the construct level of fluid intelligence with an effect size (*d*) of 0.81 for the training group and 0.46 for the control group concluding that: "there were no significant differences between the training and control groups for any construct, (except for fluid intelligence at a trend level, $p = 0.06$)"; and on the same page (p. 716), when describing changes at the measure level: "There were no significant differences between the training and control groups on the measures of fluid intelligence, although for the RAPM it was at a trend level ($p = 0.06$)". I must admit that I have some difficulties following the argumentation put forward by Colom and coworkers.

In support of the visuospatial interpretation of the observed test increases reported in the Colom study, Hayes et al. (2015) suggested another explanatory construct—*strategy refinement* instead of intelligence gains. The description of the experimental design in the Hayes study is flawed with error. Thirty-five university students completed two short-form Raven's matrices tests separated by 1 week. In the week between pre/posttests, 23 respondents were enrolled in two training sessions practicing Raven-like problems. The remaining group served as a control. During testing the students' eye-movements were recorded with an eye-tracker. No data on the sex of the subjects was provided, nor was equalization with respect to the studied variables reported, except for vision being either normal or corrected-to normal. The authors established that one-third of the variance of score gains could be attributed to test-taking strategy alone, as revealed by eye-fixation patterns. Because the information on the students' gender is missing it can be merely speculated, that if only females were involved, or if the sample consisted of males and females, the recordings were biased, first, because of the influence of the menstrual cycle on brain activity discussed earlier, and second, because several gender related differences in exploratory eye movements were established (Cazzato et al., 2010; Miyahira et al., 2000). However, if only male students participated, than the data obtained have little explanatory value for the Colom study where only female respondents participated. Hayes and coworkers concluded that their data support the possibility that the reported changes in performance on Raven's matrices after working memory training could be attributed to visuospatial processing. The line of argumentation is puzzling. Let me explain: if the experimental design consists of an active control group matched to an experimental group, than both groups would display about the same amount of increase in test scores due to a refinement in test-taking strategies on the posttest session. A factorial design with the within-subjects variables of pre/posttest and the between-subjects variable of group, experimental versus active control, yielding a significant interaction effect, could therefore be evidence enough that beside the test-retest effect also a training effect due to changes in the latent variable of fluid intelligence might be assumed. This kind of analysis is missing in the Colom, as well as the Hayes study. Frankly, it is disappointing that 8 years after the Jaeggi study (2008), so fiercely criticized because of methodological shortcomings, the studies trying to overcome these problems, "with tools from cognitive psychology to open the black box" (Hayes et al., 2015, p. 3), suffer from similar methodological problems. Thus we are witnessing the birth of a new theory based on studies with a flawed experimental design.

To conclude the section in a more optimistic tenor. Analyzing the data from a neuropsychological perspective, at least two aspects are worth mentioning: First, brain-imaging studies showed that exploratory eye movement is related to activity in the frontal cortex (Tsunoda et al., 2005). It was further found that the prefrontal cortex is central in directing attention to novel events (Daffner, 2000). The finding by Hayes et al. (2015) can therefore point to a near transfer effect, being bigger than expected, because there was just 1 week between the two tests[j]. Frontal brain activity is related to attention (Duncan and Owen, 2000), and in terms of a psychological construct attributed to the central executive—a control system of limited attentional capacity that is responsible for the manipulation of information within working memory and for controlling three slave systems (Baddeley et al., 2011). Second, in the light of Jerison's (1991) mapping theory the processes involved are a prerequisite for the formation of different representations of the outer world that are essential for intelligence—a working hypothesis I will elaborate in more detail in the next chapters.

1.3 EDUCATIONAL INTERVENTIONS FOR HIGHER COGNITIVE COMPETENCE

A detailed analysis of different interventions to raise intelligence will be provided in Chapters 3–5, focusing on behavioral training—Chapter 3, transcranial electric stimulation (TES) and transcranial magnetic stimulation (TMS)—Chapter 4, and in the fifth chapter on other approaches, such as nutrition, exercise, meditation, and pharmacological cognitive enhancement. I sometimes refer to the latter as new age methods. Most of them emerged after Jaeggi's influential, but also controversial article in *PNAS* (Jaeggi et al., 2008). The main characteristic of these new age approaches is that they are mainly carried out in controlled laboratory experiments. In contrast, previous attempts to increase intelligence involved

j. In most of the other studies pre/posttest sessions were several months apart. Yet another shortcoming of the experimental design of the Hayes study (Hayes et al., 2015).

educational interventions directed at children from low SES families who exhibited delays in cognitive development and academic achievement. The connection between IQ tests and education has historical roots dating back to the year 1905 when Binet and Simon published their first IQ test with the aim to exactly determine who should enter regular education in France, and to Terman (1916), who provided test norms based on adequate samples.

This section will provide a brief overview of approaches in education and special education designed to help those at risk of being identified as having below average IQ test scores. The results of such interventions were usually labeled as disappointing and were on a regular basis criticized by experts in the field, starting with Jensen's critique of special education in 1969. About 30 years later more or less the same conclusion was put forward by Detterman and Thompson (1997). Because the assessment of the interventions was in most cases based on comparing intelligence test score gains in intervention and control groups, some of the critique (or support) also came from state of the art review papers where intelligence was in the spotlight of attention (Neisser et al., 1996; Nisbett et al., 2012).

The Neisser "Task Force," appointed by the *American Psychological Association (APA),* was supposed to bring order to the chaotic discussion that contained both scientific and political arguments, which were triggered by Herrnstein and Murray's book *The Bell Curve.* The report published in the *Am. Psychol.* was supported by all of the contributors. The report tackled intelligence from almost all viewpoints a scientist could think of, concluding the report with the magic number of seven central questions about intelligence providing the answer: "we do not know." Among these seven unanswered questions are those that refer to genetic and environmental influences on IQ, the difference between Blacks and Whites in IQ test scores, and the relation between IQ test scores and other forms of intelligent behavior like wisdom, creativity, practical, and social sense—to mention just those which have some relevance for the topic of the chapter. The possibility to raise cognitive competence was not explicitly included in these seven questions, but received a 350 words long passage in the environmental section of the report. The answer was not a final one,[k] suggesting that long-term effects of interventions on students' intelligence are elusive, yet pointing out two exceptions: First, the Venezuelan Intelligence Project (Hernstein et al., 1986a) and second, the Carolina Abecedarian Project (Campbell and Ramey, 1994). Both interventions showed substantial gains on a wide range of test scores, with a shortcoming of the first approach having no follow-up. The second approach showed that the group of children participating in the enrichment program had higher test scores (about 5 IQ points) than controls that persisted even 7 years after the intervention had finished and they also outperformed the controls in academic achievement.

k. Several authors cite the passage, relying almost exclusively on the elusive part of it, suggesting that intelligence cannot be changed (Thompson et al., 2013; Hayes et al., 2015).

Fifteen years later, Nisbett et al. (2012) published a similar review article on intelligence in which the issue of intervention was discussed in much more detail—more than ten times the length that was devoted to the topic by the Neisser task-force. Just the section on the possibility to increase fluid intelligence is 490 words long. However, with respect to enrichment programs for low SES children, there is not much new in the text, except some validation data on the Abecedarian Project, and the introduction of a new enrichment program termed "Portfolio" by de Acedo Lizarraga et al. (2003), which showed that gains in intellectual capacity, cognitive flexibility, use of self-regulation skills, and academic achievement were maintained 2 years after the intervention. The follow-up of the Abecedarian Project was performed when the participants were 21-years old. The difference in IQ test scores was 4.4 points in favor of the treated group, who also had higher educational attainment, more likely attended a 4-year college, showed a reduction in teenage pregnancy, and less incidence of marijuana use (Campbell et al., 2002).

The Abecedarian Project consisted of an early childhood intervention lasting from birth to 5 years of age and a school-age intervention lasting for 3 years (Ramey et al., 1976; Campbell et al., 2002, 2001; Campbell and Ramey, 2007). In total, 111 children participated in the program, mainly Afro-Americans (98%), all of low SES, most coming from female-headed households. The children were first randomly assigned to an experimental (57) and a control group. Upon entering public school, they were assigned to four groups based on their 48 months Stanford–Binet test score: a control group with no intervention and an experimental group attending preschool and early school interventions. The other two groups were a combination of a control and an experimental condition, attending either the preschool or the school intervention. In the preschool intervention both groups, the experimental and the control, received family support social work services, medical care, nutritional supplements, transportation, and payment for participation. The experimental group attended the Carolina Infant Curriculum consisting of three steps—determining objectives, developing curriculum products, and finally evaluating the products. The curriculum objectives were based on Piaget's developmental theory, the hopes and aspirations parents had for their children, adaptive sets (defined as behaviors that predict success) and risk indicators (warning signs along the developmental continuum). The second step consisted of the development of curriculum products—ideas for games between children and their parents, toys with special properties and parent teacher training materials with the aim of enhancing perceptual-motor, cognitive, language, and social development. The third step was the evaluation of these products. The curriculum was prescribed on an individual basis and carried out as a full-time child care program from 1972 to 1977. The early school intervention consisted of the services of a home school resource teacher whose main obligations were to provide assistance in child education, like designing learning activities for parents to help their child understand concepts learned at school, and to teach reading and mathematics. In addition they helped the families with their problems in everyday life to better adjust within the community. The influence of the intervention on children's IQ test scores (WISC-R) administered at the age of 6.5, 8, 12, and 15 years showed on average a difference of 5.9, 4.1, 6.0, and 5.8 IQ points (0.3–0.5 standard deviation), respectively. A follow-up at the age of 21 years (WAIS-R test) showed that the control group scored an average of 85.24 IQ points (SD = 8.61) whereas the experimental group achieved an average of 89.66 IQ points (SD = 10.09). At first glance the increase does not seem very impressive, given the intensity and duration of the intervention. So, does it pay off? A benefit cost analysis of the Abecedarian project (Reynolds et al., 2011), using data collected up to respondents' age of 26 years showed for the preschool program a total return of $10.83 per dollar invested (18%) and for the school-age intervention the return was $3.97 per dollar invested (10%). A similar beneficial estimation on a national level was estimated by Herrnstein and Murray (1994), using the *National Longitudinal Study on Youth* (NLSY). They estimated that an increase of IQ for just three points (from an average of IQ = 100 to IQ = 103) would give a 25% drop in poverty rate, a 18% drop in welfare recipiency, a 20% drop in children who live in poverty the first 3 years of their lives, 20% less children would live without their parents; high school dropouts would fall by 28% and so forth.

The main purpose of the Venezuelan Intelligence Project was to develop cognitive skills that can help children to learn more efficiently and to develop their intellectual performance independently of subject matter (Hernstein et al., 1986b). The tasks were designed to enhance elementary cognitive processes, such as observation and classification, and more complex processes like problem solving and decision making. The total number of lessons designed was approximately 100, of which only 56 were used due to time constrains. The class was instructed 4 times a week, with the fourth day used for review or completion of lessons, while in the preceding three days new lessons were introduced. The students enrolled were seventh-graders, 463 were assigned to the experimental group and 432 were controls. The intervention outcomes were evaluated 3 times: pretest, interim test, and posttest via three ability tests. The effect sizes for the gains of the experimental over the control group were for the Otis–Lennon School Ability Test $d = 0.49$, for the Cattell Culture Fair Intelligence Test $d = 0.11$, and for the General Abilities Tests $d = 0.35$. The gain on the Cattell test was significant just for visual series, whereas for the subtests classification, matrices, and inferring rules no significant gains were observed. The highest gain was observed on the Target ability tests ($d = 0.70$). The Target ability tests were designed to measure the cognitive processes taught in the course. A methodological shortcoming was that the experimental group was superior on almost all pretest measures to the control, and that no follow-up of the study benefits was performed.

As mentioned previously, some of the tasks used in the Venezuelan Intelligence Project were included in the portfolio intervention designed by de Acedo Lizarraga et al. (2003). Also included were the tasks of the Instrumental

Enrichment Program (Feuerstein et al., 1980), and the Philosophy for Children Program (Lipman, 1976). The program was carried out over a 2-year period and consisted of 120 units (hours). Forty students from a private secondary school in Navarre (Spain) participated in the program. The students were tested on a pretest, a posttest when the intervention finished, and on a second posttest after a 2-year interval. The outcome was measured with an IQ test (the Cattell Culture Fair Intelligence Test), with tests measuring cognitive flexibility, metacognition, and learning strategies, and with academic achievement. The latter was defined as mean school grades in the curricular subjects of mathematics, language, and natural sciences. The posttest evaluation showed significant differences between the control and experimental group especially on the intelligence tests and for the variable academic achievement, showing that the gains were maintained for at least 2 years after the intervention.

The Milwaukee Project: Preventing Mental Retardation in Children at Risk has been referred to as the "miracle in Milwaukee" probably because of the reported 32 IQ points increase of the experimental group of children after 3 years of intervention (Garber and Begab, 1988; Jensen, 1989). The scientific community raised doubts with respect to the project's results on several grounds. Probably the most severe, although not directly related to the study outcome, was the fact that the project's director Rick Heber, who received a grant of reportedly $14 million, was convicted for the misuse of funds and sentenced to 9 years of prison. On the same ground of evidence, the project's associate director Patrick Flanigan was also convicted (Jensen, 1989). A second problem more related to the research itself, was the fact that the project started in 1966 and was funded for 15 years till 1981, however, the final report was published seven years later. The third problem is the sample size—only 20 families (3 dropouts) were assigned to the experimental group and 20 families (2 dropouts) were assigned to the control group. An ironic estimation was that $23,000 per IQ point per child was spent. The project design is similar to the Abecedarian project, involving low SES Afro-American families whose children were at risk for identification as cultural-familial mentally retarded, because of the low IQ of their mothers (IQ = 75 or below). The intervention had two components: early childhood stimulation and maternal rehabilitation. Both were only administered to the experimental group. Maternal rehabilitation included basic academic preparation for job adjustment; on-the-job vocational training and the provision of help in finding a job. The idea behind the early childhood stimulation was that retarded mothers create an intellectually and psychologically unfavorable environment, therefore the children spent most of the day in the stimulation center. The designed materials and training aimed at increasing social-emotional, perceptual-motor, and cognitive-language areas of development and were extremely complex. In Garber's report the description of the intervention covers 24 pages. The assessment of the result was also rather complex involving a lot of testing (e.g., The Stanford–Binet test was used 13 times), therefore the test results can only be explained in relation to the control group. The WISC test scores obtained when the children were 7 years old were for the experimental group 103 and for the controls 81, while at the last testing they were 101 for the experimental group and 91 for the controls, with respect to SD (age 7 years 2.24 SD; age 14 years 0.87 SD). However, there were no significant between group differences in school achievement and there was no beneficial effect of the maternal rehabilitation intervention. The dispute about the benefits and results of the intervention program was carried out in the journal *Intelligence* (Garber and Hodge, 1991; Locurto, 1991; Schweinhart and Weikart, 1991). A reassessment of the Milwaukee Project by Besharov et al. (2011) came to a similar conclusion as Jensen (1989), namely that there is inconsistency in the findings showing big gains on IQ tests not being accompanied by academic achievement, further there seems to be some overlap between trained skills and skills found on IQ tests. Moreover, a considerable amount of data is missing in the report, adding fuel to the discussion that the data may be falsified.

The Head Start Program is one of the longest-running programs in the United States with similar goals as the Abecedarian project assuming that environmental enrichment could make poor children more successful at school. The project started in the year 1965 and is still active. In the beginning of the project, the main objective was to raise children's IQ and in that way promote academic achievement, although the credo of the program as stated in the Head Start regulations was to promote the child's social competence (Zigler et al., 1994). This objective was achieved through four program components: education, parent involvement, social services, and health. Because of the critique that usually accompanies enrichment programs which attempt to raise children's IQ (Fryer and Levitt, 2004), the focus of the program moved more toward health issues in the nineties of the 20th century. The main difference between Head Start and previously mentioned programs is also in the number of involved children: as of 2014, Head Start has served more than 32 million children, from birth to age 5, and their families—about one million children per fiscal year. A recent evaluation report determining Head Start's impact on child development, which focused on children participating in the program from 2010 to 2014, was based on outcome measures in the cognitive and emotional domain. The cognitive outcomes were determined with the Peabody Picture Vocabulary Test, the Woodcock-Johnson Letter Word Identification, and Applied Problems variables. No IQ tests were administered in the evaluation. The report came to the following conclusion (Peck and Bell, 2014; p. 23):

> *Applying these analytic innovations to the experimental HSIS evaluation data, we find little evidence that Head Start's impact varies systematically by the level of quality in the program for the available, limited quality measures. The frequency of statistically significant differences in impacts by quality levels is no greater than one would expect to observe by chance alone when no true differences exist. The one exception to this pattern is the discovery that, for 3-year-olds, lower exposure to academic activities is associated with more favorable short-run impacts on*

social development. There is almost no indication that either high or low quality Head Start in any dimension leads to Head Start impacts that last into third grade for either age cohort, consistent with the overall findings of the Head Start Impact Study not disaggregated by quality level.

It is actually amazing that an intervention that has no positive effect on student IQ test scores and a negligible one on children's cognitive competences has managed to survive for such a long time, while on the other hand programs showing substantial influence on IQ test and academic achievement have been abandoned. Speaking from experience, it is difficult to keep teachers' motivation high enough to ensure that the rather work intensive units are implemented. The objective of my PhD thesis was to enhance children's creativity in the first and second grades of elementary school (Jaušovec, 1987). In the program, 300 units per year were designed for the subject mathematics, Slovene language, fine art and a subject called understanding nature and society. However, after the program finished, albeit the units were still available, they were not used by the teachers.

REFERENCES

Alba, D.M., 2010. Cognitive inferences in fossil apes (Primates, Hominoidea): does encephalization reflect intelligence? J. Anthropolog. Sci. 88, 11–48.

Aiello, L., Wheele, P., 1995. The expensive-tissue hypothesis. Curr. Anthropol. 36 (2), 199–220.

Amelang, M., Steinmayr, R., 2006. Is there a validity increment for tests of emotional intelligence in explaining the variance of performance criteria? Intelligence 34 (5), 459–468.

Amen-Ra, N., 2007. How dietary restriction catalyzed the evolution of the human brain: an exposition of the nutritional neurotrophic neoteny theory. Med. Hypotheses 69 (5), 1147–1153.

Amin, Z., Epperson, C.N., Constable, R.T., Canli, T., 2006. Effects of estrogen variation on neural correlates of emotional response inhibition. NeuroImage 32 (1), 457–464.

Baddeley, A.D., Allen, R.J., Hitch, G.J., 2011. Binding in visual working memory: the role of the episodic buffer. Neuropsychologia 49 (6), 1393–1400.

Barton, R.A., 2006. Primate brain evolution: integrating comparative, neurophysiological, and ethological data. Evol. Anthropol. Issues News Rev. 15 (6), 224–236.

Berman, K.F., Schmidt, P.J., Rubinow, D.R., Danaceau, M.A., Van Horn, J.D., Esposito, G., Weinberger, D.R., 1997. Modulation of cognition-specific cortical activity by gonadal steroids: a positron-emission tomography study in women. Proc. Natl. Acad. Sci. USA 94 (16), 8836–8841.

Besharov, D.J., Germanis, P., Caeli, A.H., Call, D.M., 2011. Assessments of Twenty-Six Early Childhood Evaluations. Maryland School of Public Policy Welfare Reform Academy, College Park.

Boden, M.T., Berenbaum, H., 2007. Emotional awareness, gender, and suspiciousness. Cognit. Emot. 21, 268–280.

Bodoky, G., Meguid, M.M., Yang, Z.J., Laviano, A., 1995. Effects of different types of isocaloric parenteral nutrients on food intake and metabolic concomitants. Physiol. Behav. 58 (1), 75–79.

Burkhalter, J., Fiumelli, H., Allaman, I., Chatton, J.-Y., Martin, J.-L., 2003. Brain-derived neurotrophic factor stimulates energy metabolism in developing cortical neurons. J. Neurosci. 23 (23), 8212–8220.

Buschkuehl, M., Hernandez-Garcia, L., Jaeggi, S.M., Bernard, J.A., Jonides, J., 2014. Neural effects of short-term training on working memory. Cogn. Affect. Behav. Neurosci. 14 (1), 147–160, https://doi.org/10.3758/s13415-013-0244-9.

Campbell, F.A., Pungello, E.P., Miller-Johnson, S., Burchinal, M., Ramey, C.T., 2001. The development of cognitive and academic abilities: Growth curves from an early childhood educational experiment. Dev. Psychol. 37 (2), 231–242.

Campbell, F.A., Ramey, C.T., 1994. Effects of early intervention on intellectual and academic achievement: A follow-up study of children from low-income families. Child Dev. 65, 684–698.

Campbell, F.A., Ramey, C.T., 2007. Carolina Abecedarian Project. Human Capital Conference Series on Early Childhood Development. University of Minnesota Center for Early Education and Development and the Federal Reserve Bank of Minneapolis, St. Paul.

Campbell, F.A., Ramey, C.T, Pungello, E., Sparling, J., Miller-Johnson, S., 2002. Early Childhood Education: Young Adult Outcomes From the Abecedarian Project. Appl. Dev. Sci. 6 (1), 42–57.

Carmi, S., Hui, K.Y., Kochav, E., Liu, X., Xue, J., Grady, F., Pe'er, I., 2014. Sequencing an Ashkenazi reference panel supports population-targeted personal genomics and illuminates Jewish and European origins. Nature Commun. 5, 4835.

Cazzato, V., Basso, D., Cutini, S., Bisiacchi, P., 2010. Gender differences in visuospatial planning: an eye movements study. Behav. Brain Res. 206 (2), 177–183.

Ceci, S.J., 1991. How much does schooling influence general intelligence and its cognitive components? A reassessment of the evidence. Dev. Psychol. 27 (5), 703–722.

Chiang, M.-C., McMahon, K.L., de Zubicaray, G.I., Martin, N.G., Hickie, I., Toga, A.W., Thompson, P.M., 2011. Genetics of white matter development: a DTI study of 705 twins and their siblings aged 12 to 29. NeuroImage 54 (3), 2308–2317.

Chinthapalli, K., 2015. The billion dollar business of being smart. BMJ, h4829.

Chooi, W.-T., Thompson, L.A., 2012. Working memory training does not improve intelligence in healthy young adults. Intelligence 40 (6), 531–542.

Ciarrochi, J., Hynes, K., Crittenden, N., 2005. Can men do better if they try harder: sex and motivational effects on emotional awareness. Cogn. Emot. 19 (1), 133–141.

Cochran, G., Hardy, J., Harpending, H., 2006. Natural history of Ashkenazi intelligence. J. Biosoc. Sci. 38 (05), 659–693.

Colom, R., Román, F.J., Abad, F.J., Shih, P.C., Privado, J., Froufe, M., Jaeggi, S.M., 2013. Adaptive n-back training does not improve fluid intelligence at the construct level: gains on individual tests suggest that training may enhance visuospatial processing. Intelligence 41 (5), 712–727.

Daffner, K.R., 2000. The central role of the prefrontal cortex in directing attention to novel events. Brain 123 (5), 927–939.

Darwin, C., 1871. The Descent of Man, and Selection in Relation to Sex. John Murray, London.

de Acedo Lizarraga, M.L.S., Ugarte, M.D., Iriarte, M.D., de Acedo Baquedano, M.T.S., 2003. Immediate and long-term effects of a cognitive intervention on intelligence, self-regulation, and academic achievement. Eur. J. Psychol. Educ. 18 (1), 59–74.

Deaner, R.O., Isler, K., Burkart, J., van Schaik, C., 2007. Overall brain size, and not encephalization quotient, best predicts cognitive ability across non-human primates. Brain Behav. Evol. 70 (2), 115–124.

Deary, I., 2000. Looking Down on Human Intelligence. Oxford University Press, Oxford, Available from: http://www.oxfordscholarship.com/view/10.1093/acprof:oso/9780198524175.001.0001/acprof-9780198524175.

Detterman, D.K., Thompson, L.A., 1997. What is so special about special education? Am. Psychol. 52 (10), 1082.

Dunbar, R.I.M., 1998. The social brain hypothesis. Evol. Anthropol. Issues News Rev. 6 (5), 178–190.

Duncan, J., Owen, A.M., 2000. Common regions of the human frontal lobe recruited by diverse cognitive demands. Trends Neurosci. 23 (10), 475–483.

Dutton, E., Lynn, R., 2015. Race and Sport: Evolution and Racial Differences in Sporting Ability. Ulster Institute for Social Research. London.

Edmonds, C.J., Isaacs, E.B., Visscher, P.M., Rogers, M., Lanigan, J., Singhal, A., Deary, I.J., 2008. Inspection time and cognitive abilities in twins aged 7 to 17 years: age-related changes, heritability and genetic covariance. Intelligence 36 (3), 210–225.

Faith, J.T., Behrensmeyer, A.K., 2013. Climate change and faunal turnover: testing the mechanics of the turnover-pulse hypothesis with South African fossil data. Paleobiology 39 (4), 609–627.

Falk, D., 1990. Brain evolution in Homo: the "radiator" theory. Behav. Brain Sci. 13 (02), 333–344.

Feuerstein, R.F., Rand, Y., Hoffman, M., Miller, R., 1980. Instrumental Enrichment: an intervention program of cognitive modifiability. University Press, Baltimore, MD.

Finlay, B.L., Darlington, R.B., Nicastro, N., 2001. Developmental structure in brain evolution. Behav. Brain Sci. 24 (02), 263–278.

Fryer, R.G., Levitt, S.D., 2004. Understanding the black–white test score gap in the first two years of school. Rev. Econ. Stats. 86 (2), 447–464.

Furnham, A., Hosoe, T., Tang, T.L.-P., 2002. Male hubris and female humility? A crosscultural study of ratings of self, parental, and sibling multiple intelligence in America, Britain, and Japan. Intelligence 30 (1), 101–115.

Gabunia, L., Vekua, A., Lordkipanidze, D., Swisher, C.C., Ferring, R., Justus, A., Mouskhelishvili, A., 2000. Earliest Pleistocene hominid cranial remains from Dmanisi, Republic of Georgia: taxonomy, geological setting, and age. Science 288 (5468), 1019–1025.

Garber, H.L., Begab, M.J., 1988. The Milwaukee project: preventing mental retardation in children at risk. American Association on Mental Retardation, Washington, DC.

Garber, H.L., Hodge, J.D., 1991. Understanding intervention, inoculation, and risk for intellectual deceleration: a reply to Locurto. Intelligence 15 (3), 317–325.

Geary, D.C., 2005. The origin of mind: evolution of brain, cognition, and general intelligence, first ed. American Psychological Association, Washington, DC.

Haier, R.J., 2014. Increased intelligence is a myth (so far). Front. Syst. Neurosci. 8, 34.

Harden, K.P., Turkheimer, E., Loehlin, J.C., 2007. Genotype by environment interaction in adolescents' cognitive aptitude. Behav. Genet. 37 (2), 273–283.

Hayes, T.R., Petrov, A.A., Sederberg, P.B., 2015. Do we really become smarter when our fluid-intelligence test scores improve? Intelligence 48, 1–14.

Heilman, K.M., 2015. Jews, creativity and the genius of disobedience. J. Relig. Health 55, 341–349, http://doi.org/10.1007/s10943-015-0139-x.

Hernstein, R.J., Nickerson, R.S., de Sanchez, M., Swets, J.A., 1986a. Teaching thinking skills. Am. Psychol. 41 (11), 1279.

Hernstein, R.J., Nickerson, R.S., de Sánchez, M., Swets, J.A., 1986b. Teaching thinking skills. Am. Psychol. 41 (11), 1279–1289.

Herrnstein, R.J., Murray, C.A., 1994. The Bell Curve: Intelligence and Class Structure in American Life. Free Press, New York.

Hunter, J.E., Hunter, R.F., 1984. Validity and utility of alternative predictors of job performance. Psychol. Bull. 96 (1), 72–98.

Hunter, P., 2009. How smart is smart? EMBO Reports 10 (11), 1196–1198.

Jaeggi, S.M., Buschkuehl, M., Jonides, J., Perrig, W.J., 2008. From the cover: improving fluid intelligence with training on working memory. Proc. Nat. Acad. Sci. 105 (19), 6829–6833.

Jaeggi, S.M., Buschkuehl, M., Shah, P., Jonides, J., 2014. The role of individual differences in cognitive training and transfer. Memory Cogn. 42 (3), 464–480.

Jaušovec, N., 1987. Spodbujanje otrokove ustvarjalnosti [Enhancing child's creativity]. Državna založba Slovrnije, Ljubljana.

Jaušovec, N., Jaušovec, K., 2012. Can training change individual differences in ability? Int. J. Psychophysiol. 85 (3), 306–307.

Jensen, A., 1969. How much can we boost IQ and scholastic achievement. Harvard Educ. Rev. 39 (1), 1–123.
Jensen, A.R., 1989. Raising IQ without increasing g? A review of The Milwaukee Project: Preventing Mental Retardation in Children at Risk. Dev. Rev. 9, 234–258.
Jensen, A.R., 1978. Citation Classics October 9, 1978. Available from: http://www.garfield.library.upenn.edu/classics1978/A1978FQ53600002.pdf
Jerison, H., 1991. Brain Size and the Evolution of Mind. American Museum of Natural History, New York.
Jerison, H.J., 1973. Evolution of the Brain and Intelligence. Academic Press, New York.
Johanson, D.C., Edey, M.A., 1990. Lucy: the Beginnings of Humankind. Penguin, London.
Johanson, D.C., Edey, M.A., 1981. Lucy, the beginnings of humankind. Simon and Schuster, New York.
Kaas, J.H., 1987. The organization of neocortex in mammals: implications for theories of brain function. Annu. Rev. Psychol. 38 (1), 129–151.
Kelly, E.L., 1955. Consistency of the adult personality. Am. Psychol. 10, 659–681.
Kolb, B., Whishaw, I.Q., 2009. Fundamentals of Human Neuropsychology. Worth Publishers, New York, NY.
Kruger, J., Dunning, D., 1999. Unskilled and unaware of it: how difficulties in recognizing one's own incompetence lead to inflated self-assessments. J. Pers. Soc. Psychol. 77 (6), 1121–1134.
Landolt, M.A., Lalumière, M.L., Quinsey, V.L., 1995. Sex differences in intra-sex variations in human mating tactics: an evolutionary approach. Ethol. Sociobiol. 16 (1), 3–23.
Lee, J., Duan, W., Long, J.M., Ingram, D.K., Mattson, M.P., 2000. Dietary restriction increases the number of newly generated neural cells, and induces BDNF expression, in the dentate gyrus of rats. J. Mol. Neurosci. 15 (2), 99–108.
Lipman, M., 1976. Lisa. Institute for the Advancement of Philosophy for Children (Montclair State College), Upper Montclair, NJ.
Locurto, C., 1991. Beyond IQ in preschool programs? Intelligence 15 (3), 295–312.
Lynn, R., Longley, D., 2006. On the high intelligence and cognitive achievements of Jews in Britain. Intelligence 34 (6), 541–547.
Lynn, R., Vanhanen, T., 2012. National IQs: a review of their educational, cognitive, economic, political, demographic, sociological, epidemiological, geographic and climatic correlates. Intelligence 40 (2), 226–234.
MacDonald, K., 1994. A People that Shall Dwell Alone. Praeger, Westport, CT.
Marino, L., Connor, R.C., Fordyce, R.E., Herman, L.M., Hof, P.R., Lefebvre, L., Whitehead, H., 2007. Cetaceans have complex brains for complex cognition. PLoS Biol. 5 (5), e139.
Martin, R.D., 1981. Relative brain size and basal metabolic rate in terrestrial vertebrates. Nature 293, 57–60.
Mayer, J.D., Caruso, D.R., Salovey, P., 2000. Emotional intelligence meets traditional standards for an intelligence. Intelligence 27, 267–298.
Miller, G., 2000. Sexual selection for indicators of intelligence. The Nature of Intelligence: Novartis Foundation Symposium 233, Vol. 233, Novartis Foundation, John Wiley & Sons, LTD, Chichester, pp. 260–275.
Mink, J.W., Blumenschine, R.J., Adams, D.B., 1981. Ratio of central nervous system to body metabolism in vertebrates: its constancy and functional basis. Am. J. Physiol. Regul. Integr. Comp. Physiol. 241 (3), R203–R212.
Miyahira, A., Morita, K., Yamaguchi, H., Morita, Y., Maeda, H., 2000. Gender differences and reproducibility in exploratory eye movements of normal subjects. Psych. Clin. Neurosci. 54 (1), 31–36.

Moody, D.E., 2009. Can intelligence be increased by training on a task of working memory? Intelligence 37 (4), 327–328.

Neisser, U., Bouchard, T.J., Boykin, A.W., Brody, N., Ceci, S.J., Halpern, D.F., et al., 1996. Intelligence: knowns and unknowns. Am. Psychol. 51 (2), 77–101.

Nilsson, G., 1996. Brain and body oxygen requirements of *Gnathonemus petersii*, a fish with an exceptionally large brain. J. Exp. Biol. 199 (3), 603–607.

Nisbett, R.E., Aronson, J., Blair, C., Dickens, W., Flynn, J., Halpern, D.F., Turkheimer, E., 2012. Intelligence: new findings and theoretical developments. Am. Psychol. 67 (2), 130–159.

Owen, A.M., Hampshire, A., Grahn, J.A., Stenton, R., Dajani, S., Burns, A.S., Ballard, C.G., 2010. Putting brain training to the test. Nature 465 (7299), 775–778.

Papp, K.V., Snyder, P.J., 2012. Editorial to accompany—training the brain: fact and fad in cognitive and behavioral remediation. Brain Cogn. 79 (2), 158, https://doi.org/10.1016/j.bandc.2012.03.005.

Parker, C.P., Baltes, B.B., Young, S.A., Huff, J.W., Altmann, R.A., LaCost, H.A., Roberts, J.E., 2003. Relationships between psychological climate perceptions and work outcomes: a meta-analytic review. J. Org. Behav. 24 (4), 389–416.

Parker, S.T., Gibson, K.R., 1977. Object manipulation, tool use and sensorimotor intelligence as feeding adaptations in cebus monkeys and great apes. J. Hum. Evol. 6 (7), 623–641.

Peck, L.R., Bell, S.H., 2014. The Role of Program Quality in Determining Head Start's Impact on Child Development. OPRE Report #2014-10, Washington, DC: Office of Planning, Research and Evaluation, Administration for Children and Families, US Department of Health and Human Services.

Petrides, K.V., Furnham, A., 2000. Gender differences in measured and self-estimated trait emotional intelligence. Sex Roles 42 (5–6), 449–461.

Piaget, J., Inhelder, B., 1974. Gedaechtnis und Inteligenz. Walter, Olten.

Poropat, A.E., 2009. A meta-analysis of the five-factor model of personality and academic performance. Psychol. Bull. 135 (2), 322–338.

Ramey, C.T., Collier, A.M., Sparling, J.J., Loda, F.A., Campbell, F.A., Ingram, D.L., et al., 1976. The Carolina Abecedarian Project: A longitudinal and multidisciplinary approach to the prevention of developmental retardation. In: Tjossem, T. (Ed.), Intervention strategies for high-risk infants and young children. University Park Press, Baltimore, MD, pp. 629–665.

Raven, J.C., 1990. Advanced progressive matrices: sets I, II. Oxford Univ. Press, Oxford.

Reader, S.M., Hager, Y., Laland, K.N., 2011. The evolution of primate general and cultural intelligence. Phil. Trans. Royal Soc. B 366 (1567), 1017–1027.

Reynolds, A.J., Temple, J.A., White, B.A.B., Ou, S.-R., Robertson, D.L., 2011. Age 26 cost–benefit analysis of the child–parent center early education program: cost–benefit analysis. Child Dev. 82 (1), 379–404.

Ruff, C.B., Trinkaus, E., Holliday, T.W., 1997. Body mass and encephalization in Pleistocene Homo. Nature 387 (6629), 173–176.

Schwaighofer, M., Fischer, F., Bühner, M., 2015. Does working memory training transfer? A meta-analysis including training conditions as moderators. Educ. Psychol. 50 (2), 138–166.

Schwarz, A., Rapaport, E., Hirschberg, K., Futerman, A.H., 1995. A regulatory role for sphingolipids in neuronal growth. Inhibition of sphingolipid synthesis and degradation have opposite effects on axonal branching. J. Biol. Chem. 270 (18), 10990–10998.

Schweinhart, L.J., Weikart, D.P., 1991. Response to "beyond IQ in preschool programs?". Intelligence 15 (3), 313–315.

Seligman, M.E., Maier, S.F., 1967. Failure to escape traumatic shock. J. Exp. Psychol. 74 (1), 1–9.

Shipstead, Z., Redick, T., Engle, R., 2010. Does working memory training generalize? Psychol. Belg. 50 (3–4), 245–276, Retrieved from http://www.psychologicabelgica.com/articles/10.5334/pb-50-3-4-245/toggle_hypothesis=on.

Shepard, R.N., 1966. Metric structures in ordinal data. J. Math. Psychol. 3, 315–387.

Snyderman, M., Rothman, S., 1988. The IQ controversy, the media and public policy. Transaction Books, New Brunswick, NJ.

Spearman, C., 1923. The Nature of "Intelligence" and the Principles of Cognition, second ed. Macmillan, London.

Spearman, C., 1927. The Abilities of Man. Macmillan, London.

Spinath, F.M., Ronald, A., Harlaar, N., Price, T.S., Plomin, R., 2003. Phenotypic g early in life—On the etiology of general cognitive ability in a large population sample of twin children aged 2–4 years. Intelligence 31 (2), 195–210.

Sternberg, R.J., 2000. The concept of intelligence. Handbook of Intelligence. Cambridge University Press, Cambridge, pp. 3–15.

Sternberg, R.J., 2008. Increasing fluid intelligence is possible after all. Proc. Natl. Acad Sci. 105 (19), 6791–6792.

Storek, J., Furnham, A., 2013. Gender, *g*, gender identity concepts, and self-constructs as predictors of the self-estimated IQ. J. Genetic Psychol. 174 (6), 664–676.

Strenze, T., 2007. Intelligence and socioeconomic success: a meta-analytic review of longitudinal research. Intelligence 35 (5), 401–426.

Strenze, T., 2015. Intelligence and socioeconomic success: a study of correlations, causes and consequences (dissertationes sociologicae universitatis tartuensis). University of Tartu Press. Available from: http://dspace.ut.ee/dspace;handle/bitstream/handle/10062/47823/strenze_tarmo.pdf?sequence=5andisAllowed=y

Terman, L.M., 1916. The Measurement Of Intelligence. Houghton Mifflin Company, Boston.

Thompson, T.W., Waskom, M.L., Garel, K.-L.A., Cardenas-Iniguez, C., Reynolds, G.O., Winter, R., Gabrieli, J.D.E., 2013. Failure of working memory training to enhance cognition or intelligence. PLoS one 8 (5), e63614.

Thorndike, E.L., 1920. Intelligence and its use. Harper's New Monthly Magazine, 140, 227– 235.

Turkheimer, E., Haley, A., Waldron, M., D'Onofrio, B.M., Gottesman, I.I., 2003. Socioeconomic status modifies heritability of IQ in young children. Psychol. Sci. 14, 623–628.

Tsunoda, M., Kawasaki, Y., Matsui, M., Tonoya, Y., Hagino, H., Suzuki, M., Kurachi, M., 2005. Relationship between exploratory eye movements and brain morphology in schizophrenia spectrum patients: Voxel? based morphometry of three? dimensional magnetic resonance imaging. Eur. Arch. Psych. Clin. Neurosci. 255 (2), 104–110.

van den Heuvel, M.P., Sporns, O., 2013. Network hubs in the human brain. Trends Cogn. Sci. 17 (12), 683–696.

Van Rooy, D.L., Alonso, A., Viswesvaran, C., 2005. Group differences in emotional intelligence scores: theoretical and practical implications. Pers. Individ. Dif. 38 (3), 689–700.

van Schaik, C.P., Isler, K., Burkart, J.M., 2012. Explaining brain size variation from social to cultural brain. Trends Cogn. Sci. 16 (5), 277–284.

Vernon, P.A., Wickett, J.C., Bazana, P.C., Stelmack, R.M., 2000. The Neuropsychology and Psychophysiology of Human Intelligence. In: Sternberg, R.J. (Ed.), Handbook of Intelligence. Cambridge University Press, Cambridge, pp. 245–264, Available from: http://ebooks.cambridge.org/ref/id/CBO9780511807947A022.

Vrba, E.S., 1975. Some evidence of chronology and palaeoecology of Sterkfontein, Swartkrans and Kromdraai from the fossil bovidae. Nature 254 (27), 301–304.

Vrba, E.S., 1993. Turnover-pulses, the Red Queen, and related topics. Am. J. Sci. 293 (A), 418–452.

Weyl, N., Possony, S., 1963. The geography of intellect. Henry Regnary, Chicago.
Wilson, E.O., 1978. On human nature. Harvard University, Cambridge, MA.
Wood, B., Collard, M., 1999. The human genus. Science 284 (5411), 65–71.
Zigler, E., Piotrkowski, C.S., Collins, R., 1994. Health services in head start. Annu. Rev. Public Health 15 (1), 511–534.

Chapter 2

The Biological Background of Intelligence

Unraveling the neurobiological underpinning of intelligence has attracted cognitive neuroscientists for decades—a feat long thought impossible. On the one hand, we are dealing with a construct: *g*, without knowing its exact meaning and, in the best case, attempting to mark out its limits (Deary, 2000). On the other hand, our brain is regarded as the most complex organ or system we know off. To give a numerical example: neocortical neurons are arranged in overlapping modular columns of different sizes. The macrocolumn has a diameter of 0.5–3.0 mm, and contains about 10^3–10^4 neurons. If we assume a simple binary (on/off) state of the neocortex, we obtain an unimaginable number of 10^{3162} possible states at the level of the macrocolumn. For comparison, the number of electrons that could fit into the volume of the known universe is approximately 10^{120} (Nunez, 1995). Therefore, connecting intelligence and the brain requires a great deal of reductionism, offering ample space for speculation and scientific controversy.

On the construct level, test scores have been used to represent intelligence that, in my opinion, approximate the often repeated definition proposed by Boring (1923), that intelligence is what the test is testing. The theoretical frame for psychometric intelligence was provided by Spearman's (1927) *g-factor*, an empirical outcome of factor analysis derived from the first unrotated principal component that accounts for a large proportion of variance in individuals' mental test scores. In other words, individuals who perform well on one mental task tend to perform well on most others with almost no regard to task content, being either verbal or spatial, simple or complex, symbolic or figural, time-limited or not. Correlations obtained between mental test scores tend to be positive and are often moderately large. The higher the *g*-loading of a test is, the more it is considered to include the essence of intelligence. For example, Colom et al. (2006) used hierarchical factor analysis of the Wechsler Adult Intelligence Scale (WAIS) showing that among the eight WAIS subtests, block design and vocabulary had the highest g-loadings. In a second step, they correlated the test scores with voxel-based morphometry data (gray matter intensities for each voxel) obtained from 48 subjects. Based on the results they concluded (Colom et al., 2006; p. 1359):

Increasing Intelligence. http://dx.doi.org/10.1016/B978-0-12-809407-5.00002-6

[...] it is shown that two prototypical measures of verbal and nonverbal g (i.e., vocabulary and block design) correlate with the amount of regional gray matter across frontal, parietal, temporal, and occipital lobes, suggesting that the general factor of intelligence relates to areas distributed across the brain as opposed to the view that g derives exclusively from the frontal lobes.

The opposing viewpoint that Colom and coworkers alluded to was put forward by Duncan et al. (2000). Although the main idea in both studies was the same, namely to compare high-g tasks with low-g tasks in relation to brain structure or function, the conclusions differed substantially. The Duncan study used positron emission tomography (PET) while 13 respondents solved high/low-g tasks with verbal and spatial content. Regions of significantly greater blood flow in high-g tasks compared to low-g tasks were observed for the spatial tasks bilaterally in the lateral prefrontal cortex, and in a discrete region of the medial frontal gyrus. For the verbal tasks, the only significant difference was visible in the lateral frontal cortex of the left hemisphere, corresponding to the activation observed for the spatial tasks. The conclusion put forward was (Duncan et al., 2000; p. 458):

Such results argue strongly against the possibility that high-g tasks are associated with diffuse neural recruitment, as predicted by broad sampling of the brain's major cognitive functions.

Even though the conclusions put forward in the two exemplary studies differ, the approach they used was the same, namely trying to find evidence in brain structure or function that corresponds to the construct of intelligence, which I call *the construct perspective*. Few studies have adopted an opposite approach: investigating how the brain functions in order to see if this corresponds to the construct of intelligence—*the brain perspective*.

The roots of the idea to examine brain organization in order to explain behavior can be traced back to the article *A default mode of brain function* by Raichle and MacLeod (2001). The default mode that was also given the name *the brain's dark energy*, a term taken from astronomy, refers to the vast amount of energy consumption of the brain that cannot be attributed to observed cognitive involvement. The brain consumes about 20% of the body's energy, 10 times that predicted by its weight alone, but only a fraction of it (estimated between 0.5% and 1.0%) can be explained by momentary demands of the environment (Raichle, 2006). It has recently been suggested that cognition is the consequence of dynamic interactions among distributed brain areas operating in large-scale networks (Bressler and Menon, 2010; Singer, 2013; van den Heuvel and Sporns, 2013). These networks can be defined based on structural or functional relations. Although there is no consensus on the number of large-scale networks, research findings suggest that at least three such networks exist (Fig. 2.1)[a] (Bressler and Menon, 2010; Raichle, 2006; Raichle and MacLeod, 2001; Seeley et al., 2007; Sridharan et al., 2008):

a. Author: Katya Krieger-Redwood.

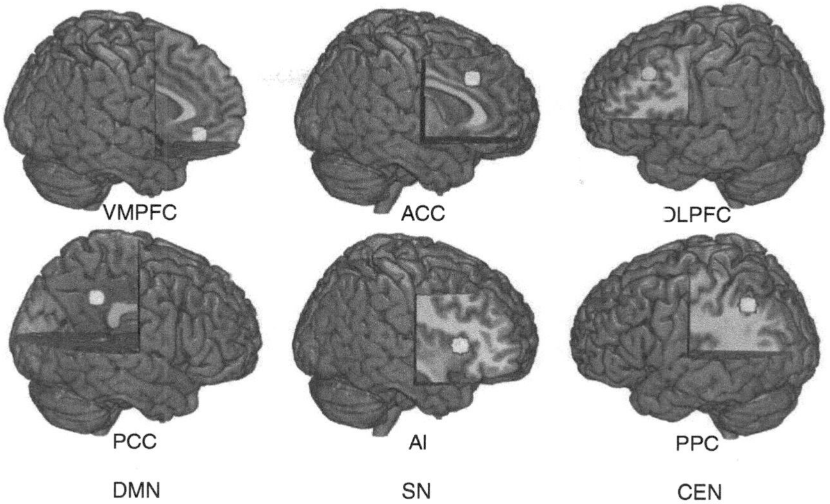

FIGURE 2.1 **Three core large-scale brain maps.** Default mode network *(DMN)* showing involvement in autobiographical, self-monitoring and social cognitive functions. Salience network *(SN)* initiating dynamic switching between DMN and CEN based on AI processed sensory and limbic information. The central-executive network *(CEN)* involved in decision making in goal directed behavior and problem solving. Key nodes of DMN include: ventromedial prefrontal cortex *(VMPFC)* and posterior cingulate cortex *(PCC)*. Key nodes of SN include anterior insula *(AI)* and anterior cingulate cortex *(ACC)*. Key nodes of CEN include dorsolateral prefrontal cortex *(DLPFC)* and posterior parietal cortex *(PPC)*.

1. The default mode network (DMN) shows deactivation during performance on demanding cognitive tasks and increased activity during performance on social cognitive tasks, suggesting its involvement in autobiographical, self-monitoring and social cognitive functions. The network is anchored in the ventromedial prefrontal cortex and the posterior cingulate cortex.
2. The salience network (SN) includes the ventrolateral prefrontal cortex, anterior insula referred to as the fronto-insular cortex, the anterior cingulate cortex, and subcortical areas including the amygdala, substantia nigra, and thalamus. Particularly, the nodes of the SN respond to the degree of subjective salience—cognitive, homeostatic, or emotional, and initiate switching between the DMN and the central-executive network (CEN). In that way, the focus of attention is either on internal events or the external world.
3. The central-executive network (CEN), the key nodes of which include the dorsolateral prefrontal cortex and posterior parietal cortex, is involved in the maintenance and manipulation of information in WM, and plays a role in judgment and decision making in goal-directed behavior and problem solving.

Recent research tried to clarify the mixed results of functional MRI (fMRI) studies that correlated intelligence with brain activity in relation to large-scale brain networks, mainly the DMN and the CEN. In a study by Lipp et al. (2012), less visuospatial intelligent individuals showed stronger deactivation of parts of the default mode network when solving a mental rotation task as compared to highly intelligent individuals. The more demanding the task was, the higher the deactivation was in DMN, suggesting that less intelligent individuals were confronted with greater cognitive demands, causing a more pronounced deactivation of the DMN, and in that way increasing the resources needed to solve the task at hand. In contrast, more intelligent individuals were faced with lower cognitive demands that resulted in weaker DMN deactivation. A similar finding was reported for performance on memory tasks by Basten et al. (2013). The high intelligent group (based on Raven's matrices test) showed stronger activation increases in the central executive network that were accompanied by weaker activation decreases in the default network as compared to the low intelligent group.

The most far reaching attempt in the so-called brain perspective is a study conducted by Hampshire and coworkers (Hampshire and Owen, 2014; Hampshire et al., 2012). The authors compared factor models of brain functional organization with factor models of individual differences in performance. Sixteen individuals solved 12 cognitive tasks while they were in a MRI scanner. The activation levels of each voxel relative to the resting state were calculated for each task and used as an input for exploratory factor analysis. The activation patterns resulted in two principal components in the multiple-demand cortex (MD)[b]: a WM component (MDwm) that included the insula/frontal operculum, the superior frontal sulcus, and the ventral portion of the anterior cingulate cortex, and a reasoning component (MDr) that was derived from performance on tasks that required transformation of information according to logical rules, and included the inferior frontal sulcus, inferior parietal cortex, and the dorsal portion of the anterior cingulate cortex. Factor analysis of the behavioral data resulted in three factors, two of which corresponded to the components extracted based on brain activity, while the third included tasks that used verbal information. The authors conducted an additional analysis in order to identify areas outside the MD cortex that corresponded to the verbal component. The results showed that the verbal component was associated with significant activations in a left lateralized network, including the left inferior frontal gyrus and the bilateral temporal lobes. The interpretation of these findings was that intelligence is not unitary, but is formed of multiple cognitive components—this means there is no *g*, at least when a comparison between data based on the brain/construct perspectives is performed (Hampshire et al., 2014).

In an exploratory study conducted in our lab (Jaušovec and Jaušovec, 2010), an even broader question was put forward, namely: is there a typology of neuroelectric brain activity that could explain individual differences in human behavior? To answer the question, an analysis of electroencephalography (EEG) data based on resting eyes-closed brain activity of 331 right-handed individuals was performed. For each person, levels of verbal and performance

b. The system (MD) refers to frontal and parietal brain area activation patterns related to cognitive performance, including performance on tests assumed to measure fluid intelligence (for review, see Duncan, 2010).

intelligence based on the WAIS-R test; experiential and strategic emotional intelligence, based on the MSCEIT (Mayer et al., 2002), and the Big 5 personality factors based on BFQ (Caprara et al., 2002) were determined. Power and coherence measures of resting EEG were obtained for seven individually determined frequency bands. To obtain a neuroelectric typology of brain activity a two-step approach was used. The data were factorized and Bartlett factor scores were entered in a hierarchical cluster analysis. This analysis revealed a three-cluster solution as optimal. The characteristics that mainly contributed to the between-cluster differences could be summarized as (1) local and long range coding of information, and (2) the excitability dynamics of the cortical network, the way the nervous system entrains to rhythmic environmental stimulation and reacts when the environmental stimuli lack regularity. These differences tentatively correspond to the switching between the default mode network and the central executive network. Although the between cluster differences were significant with respect to the behavioral variables, more pronounced differences could only be observed for the third cluster that was less numerous ($n = 36$). The main characteristic of the third cluster was higher power and brain coupling in the lower frequency bands up to lower-2 alpha (8–10 Hz) between frontal, central, and parietal brain areas (see Fig. 2.2). Further analysis showed that the difference was mainly due to individuals scoring high on performance IQ ($d = 0.40$),

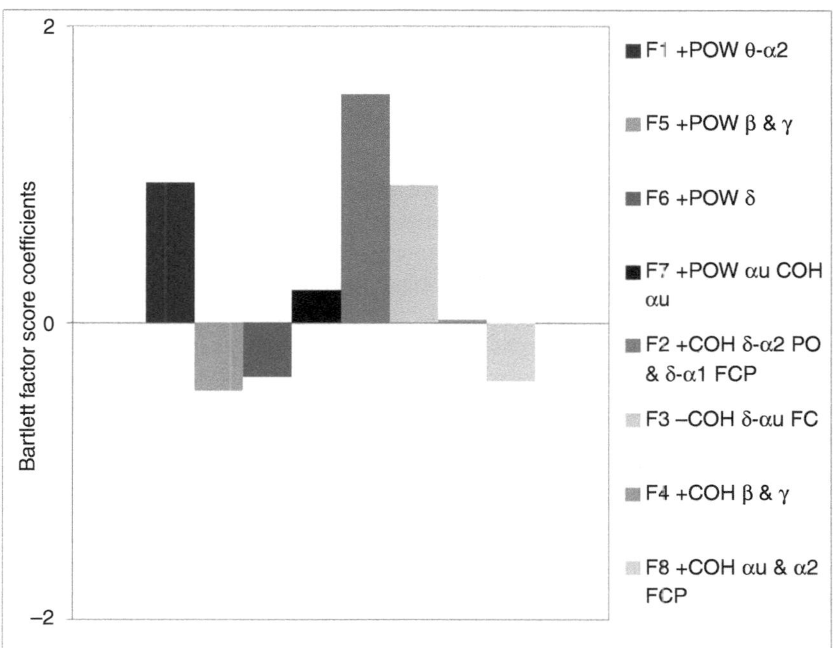

FIGURE 2.2 Centroids of neuroelectric characteristics (Bartlett factor score coefficients) for the third cluster (C3). *POW*, Power; *COH*, coherence; *F*, frontal; *C*, central; *P*, parietal; *O*, occipital.

strategic emotional intelligence ($d = 0.71$), and the personality factors of conscientiousness ($d = 0.48$) and openness ($d = 0.35$). Interestingly, this group showed the lowest scores on verbal intelligence as compared to the other two. This finding is similar to the results reported by Hampshire et al. (2012), and further questions the assumption that our brain organization supports a single *g* solution. However, as recently pointed out by Haier et al. (2014), so far none of the imaging studies have found evidence for a "neuro-g"—yet this does not mean that there is none.

It is evident that an approach that would entirely rely on the "brain" has some difficulties. This was also admonished by Hampshire et al. (2014; p. 17):

> *[...] no neuroimaging method can accurately measure the capacity of a functional brain network; consequently, that approach for estimating 'g' is entirely intractable.*

One way of tackling this problem is to study the brains of individuals who exhibit specific characteristics—either in vivo or postmortem. This method has often been used for the study of individuals with brain lesions or impairments. Henry Gustav Molaison, widely known as patient H.M., became almost an iconic case in neuroscience research. As stressed by Corkin (2002), a lot of what we know about the cognitive and neural organization of memory has to do with Henry Molaison. In 1962, many scientists believed that memory is distributed throughout the brain. Elliott (1969), for example, described the hippocampus as an archaic, vestigial structure that might have been involved in primitive feeding reflexes no longer observed in modern man. This view was revised when Brenda Milner published her findings showing that Henry's implicit memory was fully functioning, even though he had retrograde and anterograde amnesia that manifested in his inability to acquire new semantic or episodic knowledge.

Findings about memory obtained from research done on H.M. suggest that studying brains of exceptional scientists or mathematicians might also advance our understanding of the intelligence construct. The idea is not new and can be traced back to the 19th and the beginnings of the 20th century. For example, Spitzka (1907) provided an extensive review of brain data of 137 distinguished individuals in art, philosophy, mathematics, science, and politics including Beethoven, Turgenev, Gauss, Napoleon, and Lincoln, to name a few. The report consisted mainly of brain morphometric data like weight, head length, width and circumference, and descriptions of brain structure provided by autopsy (gyral convolution, fissuration, and symmetry). As stressed by Witelson et al. (1999), this early work suffers from several limitations, nonetheless, it is worth mentioning that some findings are in line with the results obtained by contemporary research. Spitzka reported that the average brain weight of 108 eminent was 1473 g, exceeding the average brain weight of controls by 75–100 g (not adjusted for age). While no standard deviation was provided, Spitzka used a frequency histogram to conclude that there is a relatively greater number of heavier brains among eminent individuals (Spitzka, 1907; p. 213) and that they display a delay in age-related cerebral atrophy for almost 10 years, compared to the group of average individuals, suggesting a positive correlation between longevity and high intelligence. A second interesting finding was that in the group of

TABLE 2.1 Brain Autopsy Reports of Eight Famous Mathematicians and One Chess Player

Name	Weight (g)	Description
Gauss, Karl Friedrich (1777–1855)	1492	Cerebrum characterized by complexity of convolution, predominantly in frontal brain areas, with enormously expanded parietal areas. The estimated total length of fissures in the frontal lobe was 2145 mm (8%–15% more than in compared average brains).
Dirichlet, Peter Gustave Lejeune (1805–59)	1520	Extreme length and width of the frontal lobes; extreme convolution, more pronounced in the right hemisphere, extreme convolution of the parietal lobes.
Babbage, Charles (1792–1871),	1403	Richness of sulci of the anterior part of the inferior frontal gyrus.
De Morgan, Augustus (1798–1871)	1494	Blind on the right eye since infancy, smaller left hemisphere (interpreted as a consequence of the blindness).
Curtice, Hosea (1825–93)	1612	Extreme fissuration of the cerebrum.
Oliver, James Edward (1830–95)	1416	Cerebrum characterized by complexity of convolution.
Gylden, Hugo (1841–96)	1452	Deaf on the left ear since infancy. The cerebral convolutions are not complex; frontal area excessively traversed by secondary fissures; the subfrontal gyrus and posterior parietal operculum were more developed on the right side.
Kovalevsky, Sonya (1850–91)	1385 (estimated)	Pronounced development of the right supramarginal gyrus.
Steinitz Wilhelm (1836–1900)	1462	Dwarf appearance (150 cm height). Extreme convolution of the frontal area (numerous gyri of reduced size).

Source: Spitzka, E.A., 1907. A study of the brains of six eminent scientists and scholars belonging to the American Anthropometric Society, together with a description of the skull of Professor E. D. Cope. Trans. Am. Philos. Soc. 21 (4), 175; Wagner, R., 1860. Vorstudien zu einer wissenschaftlichen Morphologie und Physiologie des menschlichen Gehirns als Seelenorgan I. Verlag der Dieterichschen Buchhandlung, Göttingen; Wagner, R., 1862. Vorstudien zu einer wissenschaftlichen Morphologie und Physiologie des menschlichen Gehirns als Seelenorgan II. Verlag der Dieterichschen Buchhandlung, Göttingen

individuals of superior intellect, mathematicians still had heavier brains (average weight = 1542 g). Table 2.1 summarizes the findings of the autopsy reports performed on eight mathematicians (one female) and the famous chess player Wilhelm Steinitz, who was unbeaten in match play for 32 years. The most often observed characteristics were increased gyrification, detected mainly in frontal brain areas and, to some extent, also in parietal areas, lateralized in some cases to the right hemisphere. These characteristics could tentatively point to a more developed central executive network in the brains of outstanding mathematicians.

Probably the most popular scientist in recent history is Albert Einstein (1879–1959). Isaacson (2007) provided a vivid account of the amazing and sometimes bizarre stories surrounding Einstein's brain. The decision to preserve Einstein's brain was made by Thomas Harvey, a pathologist who performed the autopsy. Einstein himself wanted to be cremated and his ashes to be scattered as he did not want to become the subject of morbid veneration. Harvey withstood all pressures to hand over his precious possession, even though the requests originated from Einstein's family, the US Army's pathology unit, and researchers who wanted a piece of the brain. He kept Einstein's brain till he was 86 years old, occasionally sending pieces of it to researchers who struck his attention. Finally, he gave the two glass cookie jars where he kept the pieces of AE's brain to his successor at Princeton Hospital.

According to Falk et al. (2013), AE's brain was photographed with a black and white camera, and dissected into 240 blocks (180 of which are at the University Medical Center at Princeton)—5 to 12 sets of 100 or 200 histological slides were stained, of which 567 are at the National Museum of Health and Medicine. The autopsy report has been missing for the last 21 years. Einstein's eyes have also been preserved, and are in private possession. For comparison, after H.M.'s death in 2008, his frozen brain was cut into 2401 slices, each 0.07 mm thick. Each slice was photographed and used to create a 3D model of his brain (Annese et al., 2014).

Published research on the characteristics of AE's brain can be divided into two categories: (1) examining the brain microstructure and (2) evaluating its patterns of gyri and sulci–surface structure (Hines, 2014). The three studies that investigated the microscopic structure showed that the neuron to glia ratio in AE's fronto-parietal areas was smaller (more glia cells) compared to controls (Diamond et al., 1985), and that AE had a higher density of neurons in the right prefrontal area (Anderson and Harvey, 1996). The third study was qualitative pointing to the tentative possibility of a greater number of glia channels and receptors in AE's brain, although of unknown significance as this is also a common feature of Alzheimer's disease (Colombo et al., 2006). The first two studies were heavily criticized, mainly because of methodological shortcomings and inadequate statistical analysis (Colombo et al., 2006; Hines, 2014), hence the conclusion:

> *In summary, the three histological studies of Einstein's brain have, in spite of claims to the contrary, found essentially no differences between his brain and that of controls (Hines, 2014; p. 22).*

Morphological studies based on photographs of AE's brain have shown that in his brain, no parietal operculum could be observed. This feature was unique and not found in any of the control brains. It was further found that AE's parietal areas were 15% larger than that of controls giving the brain a more spherical appearance (Witelson et al., 1999). A recent analysis based on 14 newly discovered photographs revealed that AE's brain did not lack parietal opercula nor was it spherical (Falk et al., 2013). On the other hand, most differences were observed in the frontal and parietal areas. The frontal lobes were highly developed and convoluted suggesting that Einstein's frontal association areas may represent the high end of human brain development. In the left parietal area, an unusually enlarged inferior parietal lobule was observed, while the right superior parietal lobule was wider than those observed in controls. The authors suggested that Einstein's visuospatial and mathematical thinking may have been facilitated by the remarkable parietal organization. In contrast, Hines (2014; p. 24) stated that the most likely explanation of the differences could simply be random variation in cortical morphology. Yet another study based on the same set of newly discovered photographs focused on the structure of the corpus callosum (Men et al., 2014). Their main finding was that AE's corpus callosum was larger (9 out of 10 comparison measures) than the average corpus callosum in the elder control group, while for the younger control group (26.6 years), the ratio was 6 out of 10 in favor of Einstein. The corpus callosum (CC) is the largest bundle of fibers (estimated 200–800 million nerve fibers) that connects the two hemispheres in three ways: (1) connecting identical (homologous) points, (2) areas, and (3) diffuse terminal connections. Therefore the corpus callosum may be involved in any aspect of hemispheric specialization (Aboitiz et al., 1992; Bloom and Hynd, 2005; Kolb and Whishaw, 2009). Men et al. (2014) suggested that their findings about AE's enlarged corpus callosum point to yet another aspect of brain structure that might explain Einstein's intellectual superiority, and his exceptional mathematical and spatial abilities.

Hines (2014) criticized the study by Men et al. (2014) in three points: first, suggesting that although Einstein was a genius, there is no evidence for his superior mathematical ability. His second criticism was based on a study by Luders et al. (2007), who found that intelligence correlated only with the thickness of posterior parts of the corpus callosum, whereas Men et al. (2014) reported that AE's corpus callosum was larger as a whole, showing no topographic differentiation. If AE's callosum were related to his above average IQ, the same pattern would be observed: an enlargement only in the posterior parts of the corpus callosum. The third critical point put forward by Hines was that if one part of the corpus callosum is larger/smaller, all parts will show the same tendency, and that the analyses should have controlled for intercorrelations between the 10 callosal measurements. In my opinion, the critique is a bit far-fetched, concocted only to fit the general objective of the paper: to show that studies of single brains of eminent individuals are unproductive.

The first point of the critique by Hines (2014) can be refuted on two grounds. First, although the quoted section of Men's study is correct, the inference made by Hines is deceptive. Hines implied that Men et al. (2014) explained the larger callosal thickness not as evidence for Einstein's overall greater intelligence, but just for his outstanding mathematical ability. However, on the same page, two lines below the section quoted by Hines, Men et al. (2014; p. 7) explicitly noted that Einstein's extraordinary cognition is related to enhanced communication routes between hemispheres. Moreover, suggesting that Einstein was an average mathematician based on a biographic record stating that he required help from other mathematicians, seems ridiculous. This piece of information is not verified evidence for his superior, or as suggested by Hines (2014), average mathematical ability, at least not in the rigor of analysis demonstrated throughout Hines's paper. Another plausible explanation could well be that asking for advice in a specific mathematical domain that would help solve a problem Einstein was working on, was just a metacognitive decision. It would require too much time and involvement to attain expertise in the specific domain, so a more time efficient approach adopted by Einstein was to ask a colleague for advice. A rather common strategy, especially nowadays, when most of research is to some extent interdisciplinary. While this is not the only possible explanation, there seems to be insufficient evidence to support Hines's statement that Einstein was not an exceptional mathematician, but just a genius. One step further and we open Pandora's box—how much variance in mathematical ability is shared with *g*?

The second critic is related to the study of Luders et al. (2007). In my view, the inference provided by Hines is false. Luders et al. (2007) also found a positive correlation between IQ and CC thickness in the anterior part of CC, significant only in the male subsample. Given that Einstein was male, this comparison seems more adequate than the one provided for the whole sample. In yet another study employing diffusion tensor imaging, it was observed that white matter integrity in the corpus callosum (genu, body, and splenium) positively correlated with intelligence $r = 0.48$ (Navas-Sánchez et al., 2014). Dunst et al. (2014) reported a similar trend for males, and just for the genu of the corpus callosum, corresponding with the findings reported in Luders's study and refutes the critique by Hines (2014). Further, several recent studies have shown that the relation between callosal size and intelligence is age-related (e.g., Hutchinson et al., 2009; Luders et al., 2011). Moreover, in a recent training study, it was shown that long-term transfer of fluid intelligence-related training gains were associated with increased structural integrity in corpus and genu (anterior) of the corpus callosum in normal aging individuals (Wolf et al., 2014). The corpus callosum connects homologous cortical areas of both hemispheres. The corpus and genu connect bilateral premotor, ventral–frontal and dorsolateral prefrontal areas, whereas more posterior parts connect the temporal and parieto-occipital areas (Bloom and Hynd, 2005). In the light of the two most influential theories of neural underpinnings of intelligence, the parieto-frontal integration theory (P-FIT) (Colom et al., 2009; Haier et al., 2009; Jung and Haier, 2007), and the multidemand system theory (Duncan, 2010, 2013; Duncan and Owen, 2000; Duncan et al., 2000), the frontal cortex and its dorsolateral prefrontal part is regarded as a key structure for intelligent performance in humans. Therefore, putting the inconclusive empirical evidence aside, the anterior enlargement observed in Einstein's CC is in agreement with theoretical expectations.

The third critical point proposed by Hines, namely that the 10 CC measures are interconnected, does not make much sense. The finding that the majority of measures taken in AE's CC were larger than those of controls indicates that AE's callosum was larger as a whole, and this was actually acknowledged by Hines, and identified as a shortcoming by incorrectly referring to the Luders study (Luders et al., 2007). Albeit Hines's critique of earlier studies is relevant, the critique of the more recent studies could be considered as unjustified.

Altogether the postmortem morphometric brain studies of exceptional mathematicians certainly do not resolve the issue of the neurobiological underpinnings of intelligence, but as stressed by Witelson et al. (1999), have heuristic value in providing directions for future research and in validating correlational studies where intellectual superiority is almost exclusively determined by a single IQ-test (e.g., Raven's matrices) or batteries of IQ-tests (e.g., WAIS).

Research on the neurobiological background of intelligence has adopted two approaches: (1) individuals are examined at rest, and correlation analyses are used to examine associations between different measures of brain structure (obtained with MRI, PET, DTI) and IQ test measures. In the case of neuroelectric procedures, such as EEG, the measures obtained indicate characteristics of the resting brain, mainly patterns of power and frequency while the participants' eyes are closed or open. (2) In the second approach that we term functional, brains of individuals who are under cognitive load are scanned, and the observed differences between rest and activity are examined in relation to IQ scores. The following sections will present a brief review of studies linking brain structure and function with intelligence. The findings will be discussed in relation to neurobiological theories of intelligence. Some focus will also be devoted to neurobiological underpinnings of speed of processing and working memory, which are considered to be key components of intelligence (Deary, 2012). With respect to the fact that several structural and functional characteristics of brain functioning related to intelligence are sex specific (Jaušovec and Jaušovec, 2008, 2009, 2010, 2012), this topic will be examined in a separate section of the present chapter.

2.1 BRAIN STRUCTURE

The section will rely mainly on metaanalyses and on selected studies introducing novel perspectives important for advancing our understanding of the brain–intelligence relation. Most studies that correlated intelligence with brain structure have done this on a macroscopic level. The development of brain mapping methods, such as diffusion tensor imaging (DTI), caused a shift in research interest toward the microscopic level: analyzing the density and integrity of fibers, and the degree of myelination.

2.1.1 Brain Volume

The best neuroanatomical predictor of intelligence is total brain volume. A positive correlation between different IQ tests and brain volume has been reported in several studies and metaanalyses (Pietschnig et al., 2015; Ritchie et al., 2015a;

TABLE 2.2 Correlations Between Brain Volume (CT or MRI) and Intelligence in Explicitly Normal Populations Reported in Three Metaanalyses

Metaanalysis	Year	Number of samples	*N*	Unweighted mean *r*	*n*-Weighted mean *r*
Vernon et al. (2000)	1987–99	11	432	0.40	0.38
Rushton and Ankney (2009)	1998–2007	28	1389	0.40	0.38
Pietschnig et al. (2015)	1991–2012	38	3254	0.30	–

Included are only studies that have been published in peer reviewed journals.

Rushton and Ankney, 2009; Vernon et al., 2000). The work of Darwin (1871) propelled the idea that human intelligence is related to a correspondingly large brain. Modern brain imaging techniques, such as computerized tomography (CT) and magnetic resonance imaging (MRI), have allowed for in vivo non-invasive measurements of brain volume replacing crude markers like head circumference. For example, Ritchie et al. (2015a) conducted a large-scale study ($n = 672$) in which structural modeling was used on six different brain variables among which were brain volume, cortical thickness, and different measures of white matter structure. In relation to *g* and fluid intelligence, brain volume accounted for the largest portion of variance (12%), followed by cortical thickness (5%). As can be seen in Table 2.2 the correlations between brain volume and intelligence reported in three metaanalyses show robust positive correlations, explaining between 9% and 16% of variance.

The positive correlation between brain volume and intelligence has been explained by a positive association between brain volume and overall number of neurons. It is reasonable to assume that larger brains contain more neurons, have greater computational power, and in that way allow more complex cognitive processing (Ritchie et al., 2015a). The relation between brain volume and number of neurons is based on an analysis of 94 brains of individuals aged between 20 and 90 years (Pakkenberg and Gundersen, 1997). The average numbers of neocortical neurons were estimated to be 19 billion in female brains and 23 billion in male brains. Pakkenberg and Gundersen (1997) further found that about 10% of neurons are lost over the life span. An interesting finding was also that body size had no influence on neuron number.

A recent study revealed that the relation between brain volume and intelligence is more complex than previously assumed (Brouwer et al., 2014). During the lifespan, the brain undergoes several changes, not just the age-related 10% loss of neurons and the 28% volume reduction of white matter reported by Pakkenberg and Gundersen (1997). Changes also occur during childhood, adolescence, and adulthood, as revealed by a review of 56 longitudinal MRI studies (Hedman et al., 2012). The review included a total number of 2211 participants with an age range of 4–88 years. Main changes in brain volume could be summarized as follows: (1) between 9 and 13 years of age an annual increase of 1% was observed, (2) followed by a gradual decrease in brain volume starting at age 13, (3) between 18 and 35 years of age another growth occurred, (4) while after 35 years of age a steady annual volume loss of 0.2% was observed, increasing to 0.5% at age 60, further increasing thereafter (Hedman et al., 2012). The interesting finding in relation to intelligence was that more intelligent individuals showed more pronounced total brain volume increases and attenuated decreases (Brouwer et al., 2014), which is in line with the *brain reserve hypothesis* (Christensen et al., 2007). Briefly, this hypothesis proposes that high intelligence provides a reserve capacity or a buffer against the effects of aging and disease in brain function. A bigger brain represents a reserve that attenuates IQ decrease in normal aging.

These findings, together with changes in cortical thickness described below, may also explain why some individuals show significant changes in IQ over time (Waber et al., 2012), even though IQ is developmentally[c] highly stable. The interim results of a follow up study of normal brain development performed by the National Institutes of Health demonstrated that individual children displayed considerable variability on the same IQ measures across visits, some of which changed by almost 2/3 of a standard deviation (Waber et al., 2012). It was further found that changes in full-scale IQ, as well as in verbal and performance IQ, were related to rates of cortical thinning, mainly in left frontal areas. Individuals with IQ gains showed no changes in cortical thickness, whereas in individuals who showed decreases in IQ, these declines were accompanied by cortical thinning that was most pronounced in those who had the highest IQ decreases (Burgaleta et al., 2014a).

Based on the limited number of studies, it would be premature to make any conclusions, however it seems that the correlation between brain volume and IQ is not straightforward, but that this relation is moderated by brain volume dynamics in the form of volume increases and decreases. The Hedman study further revealed that these dynamic brain changes are to some extent heritable (brain volume h^2 = 43%; cerebellum volume h^2 = 52%), nevertheless leaving room for environmental interplay and in that way opening the possibility of shaping individual brains over time. The latter assumption is crucial for the main question addressed in this book—can we raise intelligence?

2.1.2 Brain Tissue and Surface

Our brain consists of two cell types, glia and neurons, which give the tissue its characteristic color. Gray matter has a brown–gray color due to a high density

---

c. For instance, Deary et al. (2000) reported a mean test–retest correlation between ages 11 and 77 years of $r = 0.73$.

of capillary blood vessels and neuron cell bodies. White matter mainly consists of axons that connect with other neurons in different brain areas. Axons appear white because they are enveloped by the fatty tissue of oligodendroglial cells that provide an insulating myelin layer. Reticular[d] formation consist of a mixture of cell bodies and axons giving the tissue a net-like appearance (for a more detailed description, see Kolb and Whishaw, 2009). With the development of sophisticated imaging techniques like MRI, DTI, and PET it is possible to study brain tissue in more detail therefore there has been an increase of studies correlating brain tissue characteristics in different brain areas with intelligence in the last decades. Gray matter characteristics have been studied with voxel-based and surface-based morphometry methods (MRI). The former has been used to examine the amount of gray matter in the brain, whereas the latter has been used to inspect certain properties of the cortex, such as cortical thickness, and surface area. Properties of white matter have been studied with voxel-based morphometry, diffusion tensor imaging and magnetic resonance spectroscopy (Basten et al., 2015). In a seminal review, Jung and Haier (2007) analyzed 37 modern neuroimaging studies concluding that variations in a distributed network, mainly involving frontal and parietal brain areas, are responsible for individual differences in intelligence. The theory was termed the parieto-frontal integration theory (P-FIT), and will be described in more detail in Section 2.3. In Jung and Haier's analysis, only 11 studies were structural, whereas the others were functional. In the functional studies, fMRI data were recorded while participants solved a series of tasks that are assumed to be highly saturated with *g*, however, two different approaches were used. One approach involved analyzing brain activation patterns during task performance contrasted against a baseline no-task situation. The other approach used the same technique yet took into account the individual intelligence level of the participants. Basten et al. (2015) labeled the former the task approach, and the latter the individual differences approach. In their recent metaanalysis, a more fine-grained selection was applied compared to the one used by Jung and Haier (2007), as they focused only on studies that adopted the individual differences approach in relation to functional or structural measures (Basten et al., 2015). Twelve papers containing data from seven independent subject groups were included in the structural analysis, amounting to 457 participants and 415 foci. The analysis resulted in 12 significant clusters of IQ gray matter correlations distributed over frontal, temporal, and occipital brain areas, whereas no correlation with local amounts of gray matter was observed in parietal areas (see Fig. 2.3). This finding is to some extent surprising given that one of the most influential theories of the brain–intelligence relationship, P-FIT, described the parietal lobe as a key region of intelligence. Basten et al. (2015) attributed their finding to a higher spatial resolution that was more fine-grained than the Brodmann areas used by Jung and Haier (2007). In contrast, an overlap with most of the frontal Brodmann areas reported by Jung

d. *Rete* in Latin means net.

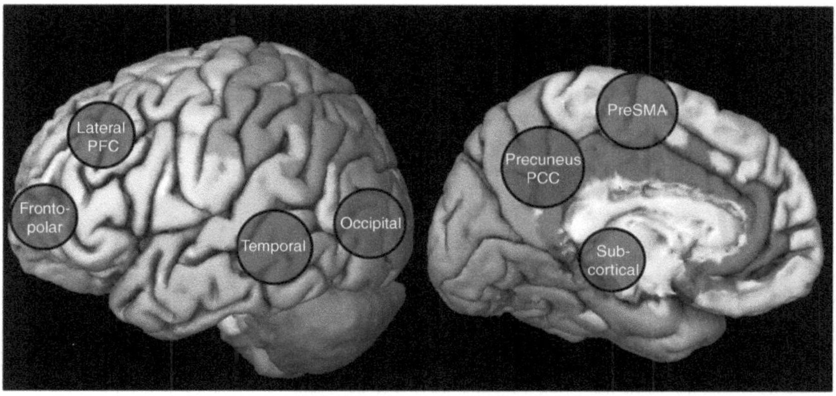

FIGURE 2.3 Structural metaanalysis of voxel based morphometric studies (Basten et al., 2015). Shown are significant correlations of intelligence test data with local amounts of gray matter. *PPC*, Posterior cingulate cortex; *SMA*, supplementary motor area; *PFC*, prefrontal cortex. The brain images were created with Allen Human Brain Atlas, available from: http://human.brain-map.org.

and Haier was observed (e.g., BA10, BA6, BA9). These findings to some extent question the validity of the P-FIT.

The human neocortex has a thickness of only 1.5–4.5 mm. It consists of 4–6 layers that are to some extent functionally specialized. For example, layer IV is involved in sensory analysis, whereas layers V and VI send efferents to other parts of the brain (Kolb and Whishaw, 2009; Luders et al., 2009). Measures of cortical thickness are thought to represent the density and arrangement of neurons per column, as well as glial support and dendritic arborization. In contrast, cortical surface area is related to the number and spacing of minicolumnar units of cells (e.g., Burgaleta et al., 2014a; Luders et al., 2009; Narr et al., 2007; Román et al., 2014). All in all, the few studies that have been conducted show a modest positive correlation ($r = 0.24$) between cortical thickness and intelligence in adults (Ritchie et al., 2015a). The topography of significant correlations was mainly observed in frontal and to a lesser extent in temporal brain areas (Narr et al., 2007; Schnack et al., 2015; Shaw et al., 2006). On the other hand, Karama et al. (2009) found positive correlations between intelligence and cortical thickness in different locations of the brain scattered over the frontal, temporal, parietal, and occipital areas. The correlations were mostly symmetrical although slightly lateralized to the left hemisphere. Yet in another study by Román et al. (2014), significant correlations between IQ and cortical thickness were reported for the right parahippocampal gyrus, right occipital, and left parietal lobe.

Even less numerous are studies correlating cortical surface area with IQ. Román et al. (2014) reported a positive correlation between cortical surface area and the higher order factor of *g* bilaterally in the dorsolateral prefrontal cortex. In a second study by Luders et al. (2008), a similar relation was observed when full-scale IQ was correlated with cortical convolution. Specifically, intelligence scores were positively associated with the degree of gyrification only in the temporo-occipital lobe, particularly in the posterior cingulate gyrus. A second finding was that no negative correlations were observed. This finding corresponds with the previously described analysis of autopsy reports performed on eight famous mathematicians and one chess player, whose brains also showed more convoluted brain surfaces, although mainly in frontal areas.

A second characteristic of the correlations between cortical surface area and thickness with intelligence is that they are not stable, and show considerable developmental fluctuations (Burgaleta et al., 2014a; Schnack et al., 2015; Shaw et al., 2006). In a seminal paper, Shaw et al. (2006) demonstrated that correlations between cortical thickness and intelligence shift from negative in early childhood to positive in late childhood. It was further shown that the trajectories of cortical thickening differed in relation to the level of intelligence. Children classified as superior intelligent (IQ 121–141) displayed a marked increase in cortical thickening peaking at around 11 years of age, even though they started from a relatively thinner cortex. On the other hand, the average intelligent (IQ 83–108) group tended to show a steady decline in cortical thickness. Changes in cortical thickness in the high intelligent (109–120) group followed an intermediate pattern, approximating the average intelligence group. A second characteristic of the latter two groups was also a much shorter period of cortex thickening that peaked at 7–8 years of age.

In a recent study, Schnack et al. (2015) examined the relationship between intelligence and changes in cortical thickness and surface area in 504 individuals aged 9–60 years. The relationship was rather complex with significant relationships mainly observed in the frontal areas of the left hemisphere (inferior frontal gyrus and planum temporale, Broca's speech area). The results further showed that, with increasing age, higher IQ was related to faster thinning of the cortex. In individuals with an IQ of <110, thinning of the left cortex continued after the age of 21, whereas in more intelligent individuals (IQ $>$ 110), thickening of the left cortex was related to higher IQ: by the age of 42, a thicker cortex was associated with higher intelligence. Likewise, a complex pattern of cortical surface expansion and contraction in relation to IQ was observed, leading the authors to conclude that intelligence is more related to the magnitude and timing of changes in brain structure than to brain structure by itself.

In summary, the studies correlating cortical thickness, surface area, and convolution with intelligence in adults showed predominantly positive correlations located mainly in frontal areas. A second finding was that these relations are not fixed, but show developmental changes depending on the individual level of intelligence. One could say that the development of the cortex is never completed, opening up the possibility for intervention.

2.1.3 Corpus Callosum and Subcortical Structures

As already described, the corpus callosum is the largest bundle of fibers in our brain that connects homologous areas of the two hemispheres, therefore it may be involved in any aspect of cognitive functioning, especially in those that require complex problem solving (Aboitiz et al., 1992; Bloom and Hynd, 2005; Kolb and Whishaw, 2009). This assumption is in line with theoretical expectations based on the two most influential theories of neural underpinnings of intelligence, the P-FIT theory (e.g., Jung and Haier, 2007), and the multiple-demand system theory (Duncan, 2010; Duncan et al., 2000). However, empirical evidence for the relationship is rather inconclusive. The only study reporting a positive relation between callosal thickness and intelligence was the study by Luders et al. (2007). For the whole sample (males and females), a positive correlation was observed only in the posterior parts of the CC. On the other hand, for males positive correlations were also observed in the genu of CC, while females displayed mainly negative correlations in the anterior part that did not reach significance. The sample analyzed in the study had a mean age of about 28 years. In contrast, when the CC of younger individuals was analyzed the correlations with full-scale IQ, as well as verbal and performance IQ were mostly negative—especially for children, and the relation also showed sex specific patterns (Ganjavi et al., 2011; Hutchinson et al., 2009; Luders et al., 2011; Moreno et al., 2014). In general, males show negative correlations between CC and IQ scores, while females show positive correlations (Ganjavi et al., 2011; Luders et al., 2011; Moreno et al., 2014), however, negative correlations have also been reported for both sexes (Hutchinson et al., 2009). Some studies (Ganjavi et al., 2011; Luders et al., 2011; Moreno et al., 2014) reported complex interactions between age, sex, type of IQ measure (verbal or performance), and CC locations (anterior–posterior). The problem is that some of the trends observed differ substantially and even point in opposite directions (e.g., Ganjavi et al., 2011; Luders et al., 2011). A somewhat different approach was put forward by Martín-Loeches et al. (2013), who instead of analyzing CC thickness focused on the shape of CC and its relation to 21 specific cognitive variables merged into seven composite variables (fluid and crystallized IQ, working memory, spatial IQ, mental speed, executive functioning, and attentional control). The mean age of the sample was around 20 years. The only significant correlation observed was for attentional control that was positively (higher attentional control values = slower responses) associated with a thinner and more curved CC shape (rotation in the splenium and genu). Overall, based on the studies discussed, no generalization about the relation of callosal thickness and shape with intelligence can be made.

Correlational research on the brain structure–IQ relation has also addressed subcortical areas, such as basal ganglia and the thalamus. The basal ganglia are situated beneath the anterior regions of the cortex, are composed of three structures (the putamen and caudate nucleus also referred to as the dorsal striatum, and the globus pallidus), and are associated with various functions, such as motor control, and simple forms of learning. It has been suggested that the caudate nucleus plays a central role in implicit memory and expressive language (Kolb and Whishaw, 2009). Few studies have examined this topic. In one study, mainly positive correlations were reported between basal ganglia (caudate nucleus and pallidum) volumes and figural and numerical components of intelligence, the former also being sex-specific (Rhein et al., 2014). In a second study, the volume of the caudate nucleus, but not the volumes of other subcortical structures, showed a significant positive association with IQ (Grazioplene et al., 2015). A third study found positive correlations between the shape of right striatal structures and thalamus with spatial and fluid intelligence but not with crystallized intelligence (Burgaleta et al., 2014b). The positive correlations were observed in samples consisting of young adults of about 20–25 years of age. On the other hand, reverse patterns have been observed for children. In 6–10 year olds, increased gray matter volume of the putamen was associated with poor performance on IQ tests and tests of working memory and response speed (Sandman et al., 2014). A shape analysis further revealed that an outward deformation of the left putamen was also related to poor performance on several cognitive tests. These results to some extent resemble those reported for cortical thickness and volume, as well as those reported for corpus callosum thickness in children and young adolescents (e.g., Luders et al., 2011; Shaw et al., 2006).

2.1.4 Microstructure of White Matter

The development of diffusion tensor imaging (DTI) has allowed researchers to explore the microstructure of white matter brain tissue. DTI is an MRI-based imaging technique that became available in the mid-1990s. It measures the random motion (diffusion) of water molecules that have the tendency to move in the direction of the lowest resistance presented by obstacles, primarily in the direction of main fiber orientation (Jones and Leemans, 2011). Mori and Zhang (2006) provided the analogy of a drop of ink on paper making a stain that uncovers the structure of the paper. The shape of diffusion can be round, that is, nondirectional (isotropic) or elongated, that is, directional (anisotropic). This anisotropy, called *fractional anisotropy* (FA) is used as a measure of axonal organization, a marker of the structural integrity of white matter (Clayden et al., 2012). FA is higher in heavily myelinated fiber tracts, and points potentially to more efficient transfer of information (Chiang et al., 2009; Ritchie et al., 2015b). In addition to FA, the average magnitude of water diffusion, mean diffusivity (MD), as well as axial diffusivity (AD; parallel with the main direction of diffusivity, assumed as an indicator of axonal integrity) and radial diffusivity (RD; orthogonal to the main direction of diffusivity; an indirect measure of myelination), can be measured (Clayden et al., 2012; Dunst et al., 2014; Tamnes et al., 2010).

A synthesis of the results of nine studies that examined the IQ–white matter microstructure relationship will be presented hereafter. The total sample size consisted of 2537 healthy individuals aged between 8 and 80 years (Chiang et al., 2009, 2011; Clayden et al., 2012; Dunst et al., 2014; Haász et al., 2013; Li et al., 2009; Navas-Sánchez et al., 2014; Ritchie et al., 2015a; Tamnes et al., 2010). In all studies, white matter integrity was indexed with fractional anisotropy (FA); some studies also used axial radial and mean diffusivity measures to indicate specific relations between myelination and the number of axons (e.g., Dunst et al., 2014; Haász et al., 2013). A finding consistently observed in all of the studies is a positive correlation between FA magnitude and intelligence. Given the considerable sample size, the correlation is as robust as the one reported for brain volume. To some extent this might be explained by the fact that white matter volume is strongly associated with whole brain volume (e.g., Clayden et al., 2012). It has been suggested that the positive correlation between FA and IQ supports the idea that efficient information transfer between hemispheres and brain areas is crucial for higher intellectual capabilities—a finding that is in line with recent trends in cognitive neuroscience describing brain structure and function as a dense connectome allowing for virtually unconstrained interactions among any pair of neurons in the brain, either through direct connections or via a few brain hubs (van den Heuvel and Sporns, 2013). In addition, this finding lends support to the P-FIT theory of Jung and Haier (2007) that proposes that intelligent behavior is related to the fidelity of underlying white matter structures necessary to facilitate rapid and error-free transmission between brain areas.

With respect to the brain areas that showed the most pronounced correlations between IQ and white matter integrity, the analysis showed that most of the correlations were observed in the structures of the corpus callosum (e.g., Chiang et al., 2009, 2011; Clayden et al., 2012; Dunst et al., 2014; Navas-Sánchez et al., 2014; Ritchie et al., 2015b). Some of the studies found widespread significant correlations in almost all major white matter tracts (e.g., Haász et al., 2013), but also in more limited brain areas like the prefrontal lobe and limbic brain areas like the hippocampus (Li et al., 2009).

With regard to gender, some of the studies reported no sex differences (e.g., Clayden et al., 2012; Li et al., 2009; Tamnes et al., 2010). In contrast, Dunst et al. (2014) found a significant correlation between intelligence and increased fractional anisotropy accompanied by decreased radial diffusivity in men, while in women no significant relations were observed. This led the authors to conclude that particularly in men, higher myelination and a higher number of axons is related to a higher level of intelligence. Even more complex findings in relation to white matter integrity, intelligence, socio-economic-status (SES), and genetic factors were reported by Chiang et al. (2009, 2011). The authors found that genetic factors explained 75%–90% of variance in white matter integrity in almost all regions inspected. However, genetic influences were greater in young individuals (adolescents) than in adults, and were more pronounced in males than in females. Additionally, this relation was strongly influenced by SES and the level of intelligence, suggesting that the influence of genetic factors was higher in individuals coming from high SES environments. Moreover, in high intelligent individuals, genetic factors explained almost 80% of the observed FA variability, on the contrary, in individuals of below-average IQ, genetic factors accounted for only 40% of FA variability.

Although in all studies a significant positive correlation between fractional anisotropy and full-scale IQ was observed, some studies also reported specific interactions related to measures of different components of intelligence, such as processing speed, and working memory. Ritchie et al. (2015b) reported that tests measuring the cognitive trait of processing speed had the strongest correlation with white matter FA and showed the most pronounced decline with age. Interestingly, individuals with higher baseline white matter integrity displayed a more attenuated age-related decline in processing speed. This finding is similar to the one observed for age-related decreases in brain volume (Brouwer et al., 2014) and lends further support to the brain reserve hypothesis (Christensen et al., 2007), suggesting that individuals with higher brain white matter integrity are able to maintain their processing speed more efficiently over time. The study by Haász et al. (2013) revealed a positive correlation between FA (negative with MD, RD, and AD), and full-scale intelligence, as for the subtests, significant correlations were reported for processing speed and reasoning components, but not for episodic memory. The results further suggested that information processing speed was closely related to reasoning ability with respect to white matter structure. Significant correlations between processing speed and white matter integrity observed in the two studies (Haász et al., 2013; Ritchie et al., 2015b) are of high importance as they explain the relation between speed of information processing and intelligence on a neurobiological level (Deary, 2012). The cognitive trait component of reasoning ability was to a lesser extent related to white matter integrity, although sharing similar white matter tracts as processing speed, implying that information processing speed is also closely related to reasoning ability at least on the level of white matter structure (Haász et al., 2013).

2.2 NEUROELECTRIC MEASURES OF THE BRAIN AT REST

The alpha rhythm dominates the EEG power spectrum recorded from the scalp during relaxed wakefulness and may tap into general central nervous system functioning, as well as the status of mental health and cognitive performance. It has been suggested that alpha-band oscillations have two roles: inhibition and timing of cortical processing, both of which are related to two fundamental functions of attention (suppression and selection) that enable controlled knowledge access and semantic orientation (Klimesch, 2012; Klimesch et al., 2007; Mazaheri and Jensen, 2010). Scalp alpha rhythms result from sequences of inhibitory and excitatory postsynaptic potentials at the dendrites of cortical pyramidal neurons. These potentials depend mainly on the influence of near and distant cortical modules (Nunez et al., 2001), and on the interactions of excitatory cortico-thalamo-cortical relay fibers and inhibitory thalamic reticular fibers (Lopes da Silva et al., 1980). Alpha power is thought to reflect the number of neurons that discharge synchronously in these integrated cortico-cortical and cortico-thalamocortical systems (Hindriks and van Putten, 2013;

Klimesch, 1999). On the other hand, peak alpha frequency, or individual alpha frequency (IAF) is assumed to reflect a delay in the intracortical circuit at alpha baseline frequency (Hindriks and van Putten, 2013). A study by Smit et al. (2006) showed that genetic influences on IAF and alpha power are largely independent. Therefore it has been suggested that each of these measures, IAF and alpha power, may capture different neural processes. Moretti (2004) suggested that alpha power during wakeful rest is related to the full integrity of the cholinergic systems and cortico-cortical connections more than to integrity of the white matter. Several recent studies lend support to the suggestion that IAF is related to characteristics of the white matter structure like fiber density, axonal diameter, and myelination (Jann et al., 2010, 2012; Valdés-Hernández et al., 2010). IAF is hypothesized to be an index of cognitive capacity termed cognitive preparedness (Angelakis et al., 2004a; Shaw, 2004) that refers to the brain's capacity for higher-level cognitive functions, and is not a synonym for alertness or arousal that modulate lower-level functions.

Within groups of healthy individuals, there are substantial and highly stable between-person differences in alpha frequency and power over test–retest intervals up to several years (Kondacs and Szabó, 1999; Salinsky et al., 1991); showing remarkably high heritability and thus are likely under strong genetic control (Anokhin et al., 2006; Smit et al., 2006). Even though alpha band oscillatory activity may be seen as a relatively stable individual characteristic, it undergoes developmental changes. The frequency of the alpha rhythm increases from approximately 3 to 5 Hz at birth until an adult value near 10 Hz around the age of 20 years (Marshall et al., 2002), followed by an age-related decrease in adults independent of pathology (Aurlien et al., 2004). A similar pattern has also been observed for alpha power (Chiang et al., 2011; Gmehlin et al., 2011).

There have been mixed reports with respect to sex differences in the alpha rhythm in the healthy population (Benninger et al., 1984; Cohn et al., 1985; Gmehlin et al., 2011; Harmony et al., 1990). Sex differences were observed mainly in adolescents. Girls were found to have a maturational lag in the increase of alpha frequency in EEG compared with boys; this lag disappears during early adulthood (Chiang et al., 2011; Clark et al., 2004). Mixed findings have also been reported for sex differences in alpha power, ranging from no differences (Gmehlin et al., 2011; Nikulin and Brismar, 2005), to higher alpha power in males than females (Clarke et al., 2001; Zappasodi et al., 2006), as well as higher alpha power in females as compared to males (Aurlien et al., 2004; Jaušovec and Jaušovec, 2010).

Individual variation in the alpha rhythm has been hypothesized to reflect individual differences in cognition and mental health. Support for this assumption was given by several studies reporting that IAF differs between healthy and clinical individuals. Reduced IAF was observed in patients with Alzheimer's disease (Garcés et al., 2013; Huang et al., 2009), for mild cognitive impairment (Fernández et al., 2006; Garcés et al., 2013), demented patients (Samson-Dollfus et al., 1997), patients with neurogenic pain (Sarnthein, 2005), schizophrenia

(Boutros et al., 2008; Galderisi et al., 2009), and attention-deficit/hyperactivity disorder in boys (Lansbergen et al., 2011). Given that IAF is thought to be related to white matter integrity and as discussed in the previous section, different characteristics of white matter structure were found to be positively related to intelligence, a similar relation would be also expected for individual alpha frequency (IAF). However, empirical research has mostly failed to find such a relation.

Some early research conducted before 1960 reported positive correlations. Taking into consideration the technical possibilities at that time, it is questionable whether the findings can be compared to those obtained with EEG acquisition systems available nowadays. For instance, Mundy-Castle and Nelson (1960) reported a positive correlation (r = 0.34) between what he called alpha frequency and IQ. The data were obtained on 94 men and 2 women in a socially isolated white community in South Africa. The average intelligence score (WAIS) was 75. Another problem reported by Mundy-Castle was that about half of the EEG recordings were characterized with abnormal recording features. No information on EEG acquisition and the determination of alpha frequency, whether obtained in eyes closed or open conditions, was provided. Based on the figure showing an example of two EEG recordings and the description, the analysis was probably done just by visual inspection of one rater[e], and obtained in an eyes-open condition. Another problem observed in many studies is that when the analysis focuses on endophenotypic qualities of the alpha frequency, the eyes closed approach should be employed because only such a method reflects the aggregate generation of alpha (Hooper, 2005).

Studies performed on more sophisticated EEG acquisition systems and software provided mixed findings (Anokhin and Vogel, 1996; Grandy et al., 2013; Jaušovec and Jaušovec, 2000; Pahor and Jaušovec, 2016; Posthuma et al., 2001). The study by Anokhin and Vogel (1996) showed positive correlations between peak alpha frequency (eyes closed) and test scores obtained with Raven's matrices (correlations ranged between 0.20 and 0.36; n = 101 males). In a recent study by Grandy et al. (2013), structural equation models were used to estimate the correlation between composite cognitive performance (Berlin Intelligence Structure test, BIS), and composite eyes open/closed IAF (r = 0.40; n = 85). However, the correlations reported ranged between 0.06 and 0.37; the highest correlations were obtained for memory tests whereas no significant correlations were reported for the reasoning part of BIS. Among the 12 correlations determined (eyes: open/closed; subtests: speed, memory, and reasoning; respondents: young/old), only 5 were significant at a $p < 0.05$ level. Therefore, none would have survived a correction for multiple comparison. On the other hand, two large-scale studies conducted by Jaušovec and Jaušovec (2000) and Posthuma et al. (2001) showed no significant correlations between IAF and IQ obtained with the Wechsler Adult Intelligence Test.

e. For a detailed review of research on the relation between EEG measures and intelligence done before 1964, see Vogel and Broverman (1964).

In a recent large-scale study conducted in our lab ($n = 417$) we attempted to further elucidate the IQ-IAF relationship (Pahor and Jaušovec, 2016). First, correlations between IQ (WAIS) and anterior and posterior measures of IAF were calculated. The data revealed no significant correlations, neither for the whole sample nor for male/female subsamples. One aspect that might have influenced the observed inconsistencies between studies could be test selection. Both large-scale studies reporting no correlations determined intelligence based on the Wechsler test (Jaušovec and Jaušovec, 2000; Posthuma et al., 2001), whereas the studies reporting low positive correlations used Raven's matrices and BIS (Anokhin and Vogel, 1996; Grandy et al., 2013).

To analyze the influence of test type on the relation between intelligence and IAF we adopted the verbal-perceptual-image rotation (VPR) model of human intelligence (Johnson and Bouchard, 2005, 2007a,b). Johnson and Bouchard (2005) constructed a factor model of the residual abilities (after removing the effects of general intelligence) that extracted three factors: one representing a rotation–verbal dimension, one representing focus–diffusion of attention, and one representing content memory. The first two are negatively correlated, that is, individuals relatively strong in abilities lying at one pole of each dimension tend to be relatively weak in those lying at the other pole, and vice-versa. The verbal ability is related to strong performance on vocabulary, spelling, and verbal interpretation tests, while relatively strong performance on mental rotation and shape manipulation tests dominates the rotation dimension. The other dimension captures the negatively correlated focus and diffusion of attention, ranging from problem solving orientations involving attentional focus on details one at a time (verbal, scholastic, spatial, and image rotation abilities), to the application of diffuse attention to a variety of cues simultaneously (fluency, content memory, and perceptual speed abilities). Johnson and Bouchard (2007a) further found that gender differences on these two dimensions were strong. Both men and women are located throughout the ranges of both dimensions, but women are concentrated toward the diffusion and verbal poles while men are concentrated toward the focus and rotation poles. Johnson et al. (2008) tested their model in relation to brain functioning. They correlated the rotation–verbal and focus–diffusion dimensions with brain volumes of gray and white matter. The correlations involving the rotation–verbal dimension were localized relatively to areas of the brain that have been implicated in tasks involving the relevant content areas in prior studies. The correlations observed involving the focus–diffusion dimension were primarily with white matter volumes and were distributed more widely across areas of the brain involved in sensory processing and attention. This division seems plausible since, as discussed in the previous sections, brain gray matter reflects information processing centers in the brain, and brain white matter handles communication between these centers.

The rotation-verbal and focus-diffusion factors were calculated based on the following WAIS subtests: vocabulary, information, digit symbol, digit span, and block design. A positive score indicates orientation toward image rotation ability and orientation toward attentional focus on details one at a time. The correlations with the two factors showed no significant correlations for the whole sample. In contrast, for males significant positive correlations between the rotation-verbal factor and posterior IAF ($r = 0.43$) and between the focus-diffusion factor and anterior IAF ($r = 0.25$) were obtained. For the female sample, both correlations were negative yet they did not reach significance. This distribution of correlations would suggest that a higher peak alpha frequency (IAF) in males is related to spatial ability, as well as to attentional focus on details one at a time. On the other hand, for females higher IAF appears to be related to verbal ability and to a more diffuse mode of attention to a variety of cues simultaneously. This division fits well with the often-reported sex difference in spatial and verbal performance that will be discussed in more detail in the last section of this chapter. Yet another characteristic worth mentioning is the topography of the significant relations between IAF and *g*-factor displayed in males that was posterior for strong performance on mental rotation and shape manipulation tests (verbal-rotation factor), and anterior for problem solving orientations involving attentional focus on details one at a time (focus-diffusion factor). The topography corresponds with the major division of cortical function in which frontal areas play a role in attentional control and parietal areas are involved in spatial processing (Kolb and Whishaw, 2009).

Several studies have reported significant positive correlations between IAF and performance on other cognitive tests, such as reading and memory performance (Angelakis et al., 2004a,b; Clark et al., 2004; Klimesch et al., 1993; Lebedev, 1994). In a large-scale study by Clark et al. (2004) it was established that only frontal IAF was a significant predictor of reverse digit span. Specifically, 1 Hz increase in frequency corresponded to a 0.21 increase in reverse digit span. Still another aspect worth mentioning is that the positive correlation between performance on cognitive tests and IAF was obtained on IAF measures calculated on EEG either during test performance (Klimesch et al., 1993), or immediately after performing the task (Angelakis et al., 2004a,b), while correlations with IAF based on an eyes-closed condition obtained before the task showed no significant correlations (Angelakis et al., 2004a,b). These findings are difficult to explain, but might point to either arousal or cognitive preparedness, or after effects of these processes, considering the timeline of the observed significant correlations.

Research on the relationship between intelligence and different power measures has provided even less consistent results compared to IAF. Using a nonlinear dynamical analysis of multichannel EEG, Lutzenberger et al. (1992a) showed that during resting conditions subjects with high IQs demonstrated higher dimensional complexity in the EEG pattern than subjects with low IQs. However, in a replication study this finding was not confirmed (Anokhin et al., 1999). In a series of studies conducted in our lab, mixed findings were reported. In one study, intelligent students showed lower alpha power (greater mental activity) in a relaxed state than average students did (Jaušovec, 1997). In a second study, this finding could not be replicated (Jaušovec, 1996). Highly intelligent individuals (relaxed with eyes open) displayed an overall higher alpha activity than average individuals. On the other hand, in a relaxed state with eyes closed, no differences were observed. Yet another two studies using alpha power and second order Kolmogorov entropy nonlinear measures showed no significant differences related to the level of intelligence (Jaušovec, 1998, 2000).

Recently, research has focused on functional brain network characteristics, the so-called small-world network attributes (high clustering and short path length) and their relation to psychometric intelligence (Langer et al., 2012). The idea can be traced back to coherence analysis of resting EEG activity. Coherence is the normalized cross-correlation that provides information about cooperation between various brain areas (Petsche, 1996). In an analysis of EEG data obtained during resting eyes-closed conditions, we were able to show that the correlations between intelligence scores and coherence measures in three narrow alpha bands were positive and distributed mostly over the right hemisphere. Highly intelligent individuals showed more coupling between brain areas predominantly between a right temporal site and other sites, in right frontal and parietal brain areas. In the left hemisphere, highly intelligent individuals also displayed more coupling, mostly between parietal and fronto-central sites (Jaušovec and Jaušovec, 2000). A similar finding for small-world network organization was reported by Langer et al. (2012). Higher psychometric intelligence (measured with Raven's matrices) was associated with increased small-world organization of the upper alpha band in the right parietal cortex, whereas negative correlations were observed in the frontal cortex. On the other hand, in a magnetoencephalographic (MEG) resting-state connectivity analysis, in which the relationship between working memory capacity and resting (eyes-closed) connectivity in different frequency bands was assessed, an inverse relationship between functional connectivity across right anterior and left lateral/posterior MEG sensors in the low alpha band and verbal working memory capacity was reported (del Río et al., 2012). At first glance, Langer et al. (2012) and del Río et al. (2012) reported opposite findings, however, they can be viewed as complementary. Activity in the upper alpha band is related to abstract thinking and reasoning, whereas the lower alpha bands are related to attentional demands—active inhibition of task-unrelated areas (Jensen and Mazaheri, 2010). In yet another study, albeit based on resting-state fMRI data, individuals of average intelligence exhibited stronger voxel-mirrored homotopic connectivity than individuals of above average intelligence. The differences were mainly observed in the primary somatosensory cortex (BA 3) and supplementary motor area (BA 6), the visual cortex (BA 23, 31), rolandic operculum (BA 43), and the middle temporal gyrus (BA 21). The authors suggested that a downgrading of interhemispheric connectivity at rest could be associated with higher cognitive functioning (Santarnecchi et al., 2015).

Another characteristic in the resting EEG that has recently attracted the interest of research is cross-frequency coupling and its relation to memory processes.

There is converging evidence that cross-frequency coupling between theta and gamma oscillations plays an important role in working memory processing (for a review, see Roux and Uhlhaas, 2014). Research from rodent studies suggests that in the hippocampus, gamma frequency oscillations are nested within theta oscillations, enabling the representation of multiitem messages that are relevant for working and long-term memory (Belluscio et al., 2012; Colgin et al., 2009; Lisman and Jensen, 2013). Similar findings have been demonstrated in human studies. Axmacher et al. (2010) showed that maintenance of multiple items is associated with hippocampal theta–gamma cross-frequency coupling and that the modulating theta frequency depends on memory load. Further, Kamiński et al. (2011) demonstrated that short-term memory capacity can be predicted by theta-to-gamma cycle length ratio. Long-term memory has also been associated with cross-frequency coupling between theta band activity in frontal brain areas and gamma band activity in parietal and occipital areas (Friese et al., 2013). Moreover, theta and gamma oscillations during encoding predict subsequent memory recall and retrieval (Osipova et al., 2006; Sederberg et al., 2003).

On the other hand, psychometric research has revealed a positive relationship between performance on tasks of working memory and fluid intelligence, with correlations ranking from 0.60 to 0.90 (for a review see Buehner et al., 2005). It has also been demonstrated that WM capacity largely accounts for the relationship between WM and intelligence, and that processing components like updating and control of attention (related to frontal brain processing) have a negligible or nonexistent relationship with intelligence (for a review, see Ackerman et al., 2005). Hence, one could assume that there is also a relationship between intelligence and theta–gamma cross-frequency coupling.

In a recent study conducted in our lab, we explored this possibility by analyzing the resting eyes-closed EEG of 100 individuals (Pahor and Jaušovec, 2014). The results confirmed the hypothesis, showing a negative correlation between theta–gamma cross-frequency coupling and IQ scores in bilateral frontal and parieto-occipital brain areas. Greater variability of theta–gamma coupling between different EEG channels was also negatively associated with IQ scores. These connectivity patterns suggest that the resting brain of highly intelligent individuals shows lower levels of coupling mainly in frontal and parietal areas, and this also corresponds with the findings reported by Santarnecchi et al. (2015). Altogether the results suggest that lower connectivity of brain areas at rest could be associated with more efficient cognitive functioning. However, this pattern drastically changed when respondents solved the Raven's matrices test. Individuals with low theta–gamma coupling during rest showed an increase in theta–gamma functional coupling (mainly in left frontal areas) as task difficulty increased, which was also associated with better performance on the test. These results imply that good performance on tests of fluid intelligence is associated with an initial low level of theta–gamma coupling during rest followed by a rise in theta–gamma coupling as task difficulty increases. The patterns of theta–gamma functional coupling during task performance presumably reflected temporally correlated activity of neighboring neuronal populations within the frontal lobes. Considering there was no evidence of task-dependent changes in theta–gamma coupling in parieto-occipital areas, the results could be interpreted in terms of the multiple-demand system (Duncan, 2005, 2010, 2013; for more details, see Section 2.3.4), suggesting that individual differences in cognitive abilities depend on the efficiency with which the left frontal lobe organizes and controls structured series of attentional episodes.

The functional role of theta–gamma coupling in frontal and parietal brain areas can be viewed in terms of network theory, which states that there are regions in the brain that are highly connected to other regions (termed hubs), and play a central role in the organization of neural networks. The fronto-parietal network, also called the central-executive network (CEN) has been identified as one of these hubs, particularly in relation to cognitive control (Spreng et al., 2013; Zanto and Gazzaley, 2013). As discussed in the introduction to the chapter, functional and structural connectivity studies suggest that brain hubs are crucial for communication across the brain and for integration of information (van den Heuvel and Sporns, 2013). Furthermore, it has been suggested that oscillatory patterns of neuronal activity play a key role in flexible routing of neural signals (Akam and Kullmann, 2010; Fries, 2005; Salinas and Sejnowski, 2001). In particular, Fries (2009) suggested that structural connectivity interacts with functional connectivity via gamma band synchronization and that this process subserves all cortical computation.

These recent approaches are promising, especially because EEG and MEG are the only direct measures of brain activity with a time resolution on a millisecond scale. Another aspect we must bear in mind is that the action potential is the building block of all our thoughts and consciousness, and probably the only route to follow in the attempt to unravel the brain code.

2.3 BRAIN FUNCTION AND NEUROBIOLOGICAL THEORIES OF INTELLIGENCE

Functional studies examine brain activation patterns during performance on cognitive tasks, ranging from simple to complex. Activity in brain areas and their interconnections can be visualized with two types of techniques: electrophysiological on the one hand and hemodynamic on the other (for a review, see Shibasaki, 2008). The former approach includes EEG, MEG, and also some stimulation techniques like transcranial magnetic stimulation (TMS) and different transcranial electrical stimulation techniques (tDCS, tACS, and tRNS that will be described in more detail in Chapter 4). Although these stimulation techniques are not visualization tools per se, I include them here because they are often used for the stimulation of brain areas related to a cognitive function (e.g., language). The latter group of techniques includes PET or single-photon emission computed tomography (SPECT), fMRI, and near-infrared spectroscopy (NIRS/fNIRS). The advantage of hemodynamic techniques is high spatial resolution, whereas the advantage of electrophysiological techniques is millisecond-level temporal resolution. Sometimes multimodal imaging methods are used to exploit the advantages of each method and in that way enlarge the information potential of the scans (e.g., EEG combined with NIRS or with fMRI). The time restrictions imposed by technical characteristics of scanners and methods have also determined the tasks and research paradigms adopted by researchers. For example, radioisotopes used as tracers of blood flow in SPECT (e.g., technetium 99, iodine 123) have a half-time of several hours; oxygen-15 used in PET has a half-time of 2 min; the blood oxygenation level dependent (BOLD) response in fMRI takes about 5 s to reach its peak after stimulus onset, limiting the time resolution of fMRI to the order of seconds. In contrast, MEG and EEG acquisition have no time limits with some systems allowing for acquisition rates up to 20,000 Hz (Shibasaki, 2008). EEG-based event related potential (ERP) studies have therefore been adopted for mental chronometry approaches. Neurobiological underpinnings of intelligence have been studied by analyzing the characteristics of the first 1000 ms of the EEG response to briefly presented visual or auditory stimuli. More complex tasks are often used in PET and fMRI studies, but also in EEG-based approaches like event-related desynchronization/synchronization (ERD/ERS). It is therefore not surprising that the three most influential neurobiological theories of intelligence: the neural efficiency hypothesis (Haier et al., 1988), the P-FIT (Jung and Haier, 2007), and the multiple-demand theory (Duncan et al., 2000; Duncan, 2010, 2013), were proposed based on data obtained from hemodynamic techniques like PET and fMRI. The contribution of studies based on the electrophysiological approach has been substantial for the neural efficiency theory and to some extent for the P-FIT theory, but was less often used to support the MD theory. The latter is to some extent unexpected as Duncan (2005, 2013) sees intelligent behavior as a structured analysis of attentional episodes on different levels of abstraction in rather restricted time frames.

The following sections will focus on different functional approaches adopted to study the brain–intelligence relation, and their role in theory formation.

2.3.1 ERP Studies

An ERP is difficult to detect in the ongoing EEG because it is hidden in the background noise of the signal that is unrelated to the cognitive process supposedly evoked by the stimulus. Therefore, the stimulus is presented repeatedly ($n > 100$) and the EEG response is averaged. In that way, the background noise is canceled out, making the characteristic features of the resulting waveform—a peak or trough—more visible (Rugg, 2002). Some of the most commonly described ERP components in relation to cognitive processes like intelligence are the P1–N1 complex, the P3 or P300 that can be divided into P3a and P3b components (see Fig. 2.3A–B). Another approach for analyzing the event-related EEG signal is event-related desynchronization/synchronization (ERD/ERS).

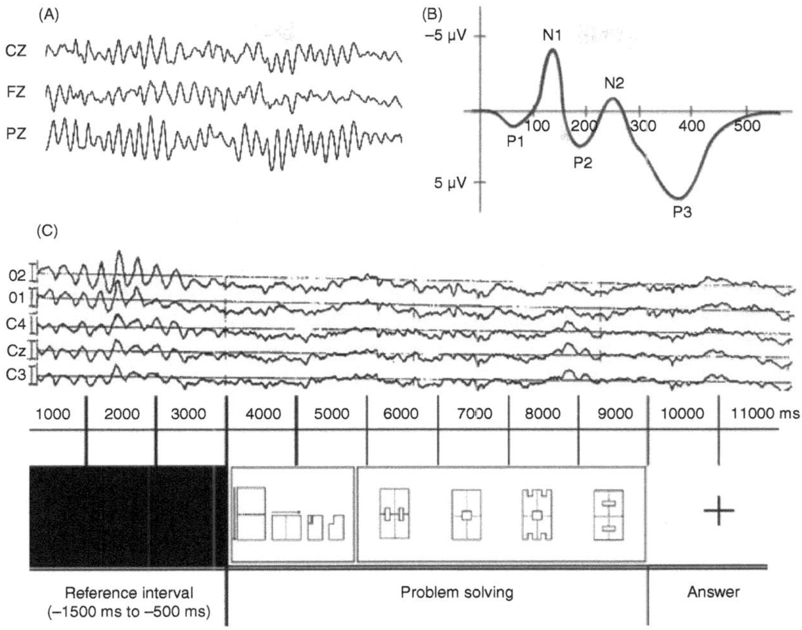

FIGURE 2.4 (A) An ongoing EEG recorded at vertex electrode positions (Fz, Cz, Pz). (B) Averaged EEG signal with characteristic ERP components like the P1–N1 complex and the P3. (C) A schematic display of an ERD/ERS paradigm with a resting (reference interval) preceding the stimulus onset (a task from the paper folding and cutting test PF&C). The *cross* indicates the time point when the stimulus disappears and the respondent must provide an answer.

ERD/ERS is determined by computing the power of induced or evoked event-related EEG activity in a centered narrow frequency band (2 Hz) as a function of time (for more detail, see Pfurtscheller, 1999). Most often a decrease/increase in power (desynchronization/synchronization) is calculated in relation to a resting interval immediately preceding stimulus onset (see Fig. 2.4C).

Numerous studies indicate that the P1–N1 complex reflects sensory and early attention processes (Luck and Girelli, 1999). There is some evidence that the P1 amplitude is associated with the suppression of irrelevant, and the N1 with the processing of the attended, relevant information. Thus, it was suggested that the P1 reflects inhibitory, and the N1 excitatory processes (Hillyard et al., 1994; Klimesch et al., 2004 for review).

The P3 component is usually obtained using *oddball* paradigms. The respondent is instructed to detect a rare stimulus (e.g., a square) in a series of more frequently presented stimuli (e.g., circles). Its functional meaning has been related to the closure of a perceptual process (Verleger, 1988), and to an update of information in working memory (Donchin and Coles, 1988). P3 represents the last phase in the identification of a relevant stimulus caused by its significance or by the requirements of the task (Hillyard and Münte, 1984). As stressed by Kok (2001), the P3 amplitude is a reflection of the degree to which an external stimulus matches the internal representation of the stimulus relevant for the task. In addition, P3 amplitude is sensitive to the amount of attentional resources engaged during task performance. On the other hand, P3 latency is thought to index classification speed, which is proportional to the time required to detect and evaluate a target stimulus. Besides the P3 (or P3b) that is most prominent over parietal electrode sites, a second peak dubbed the P3a observed over frontal electrodes and elicited by an infrequent distracter stimulus inserted randomly into the standard oddball sequence, has been identified. It is further assumed that the P3a originates from attentional frontal mechanisms, while the P3b is assumed to originate in the parietal areas and is related to memory processes (see Polich, 2007 for a review).

P3 latency and amplitude show the most consistent association with intelligence (Vernon et al., 2000). Research relating intelligence and ERPs has shown that peak latencies correlate with IQ (Amin et al., 2015; Barrett and Eysenck, 1994; Caryl, 1994; Chalke and Ertl, 1965; De Pascalis et al., 2008; Ertl, 1971; Jaušovec and Jaušovec, 2000; Shucard and Horn, 1972; Troche et al., 2009). The average correlation found across these studies was about −0.30. P3 latency has been reported to be inversely related to the level of a person's intellectual ability inferring a positive relationship between intelligence and the speed of information processing. Individuals with a higher IQ tend to evaluate and classify information more rapidly compared to those with lower intelligence (Vernon et al., 2000). Research relating P3 latencies with brain pathology provided congruous results: clinical studies on demented and retarded patients showed increased P3 latencies compared to controls (Johnson, 1995).

While P3 latency showed rather stable negative correlations with different measures of intelligence, the relation for P3 amplitude is less clear. Positive correlations between IQ and P3 amplitude were often reported when tasks of sufficient difficulty were used (e.g., Amin et al., 2015; De Pascalis et al., 2008; Jaušovec and Jaušovec, 2000; Troche et al., 2009; Wronka et al., 2013). Another aspect that might have influenced the inconsistency in the relation between P3 amplitude and IQ is that the P3 can overlap with the slow wave. The slow wave is a negative deflection that has a similar time window and is elicited by tasks requiring memory rehearsal (Kok, 2001; Wronka et al., 2013).

Another measure that attracted ERP research in the 1980s was introduced by Hendrickson (Blinkhorn and Hendrickson, 1982; Hendrickson and Hendrickson, 1980), and coined *string length*—a crude measure of the complexity of the ERP curve. The remarkably high correlation of 0.7–0.8 reported between string length and IQ generated a fair amount of interest in the research community, excited at the prospect of discovering a potential biomarker of intelligence. Hendrickson's theoretical explanation was that differences in intelligence reflect the rate at which errors are produced during the train of neural transmissions underlying problem solving. High IQ individuals make fewer errors than low IQ individuals, thus the individual epochs are less variable from trial to trial, preserving more detail of the single evoked response, therefore producing a more complex ERP. Similar high correlations were also reported by Stough et al. (1990). Others reported modest correlations (Gilbert et al., 1991; Haier et al., 1983), correlations near zero (Barrett and Eysenck, 1992, 1994; Batt et al., 1999; Burns et al., 1996; Sandman and Barron, 1986; Shagass et al., 1981; Vogel et al., 1987; Widman et al., 1993), and even negative correlations (Bates and Eysenck, 1993; Bates et al., 1995). Negative correlations between string measure and verbal IQ, and positive correlations with performance IQ were observed by Batt et al. (1999).

The problem with the string measure was that it was unclear what the length actually measures. Researchers have associated the string measure with ERP complexity (e.g., Barrett and Eysenck, 1994), others have criticized it on the basis of its nonspecificity (Robinson, 1993) and dependence on the ERP amplitude (Haier et al., 1983). Based on experimental evidence, Burns et al. (1997) concluded that the string measure is not a valid measure of the ERP. In a study conducted in our lab (Jaušovec and Jaušovec, 2000), a different parameter of ERP complexity with a more firm theoretical and mathematical background was employed. At that time, different measures indicating deterministic chaos (e.g., Lyapunov exponent L1, correlation dimension D2, or Kolmogorov entropy K2) were used to describe differences in EEG patterns. It was found that less chaotic patterns were associated with deep sleep or pathology, intermediate values were found during the awake (eyes closed) state, and the highest values were associated with mental activity (Lutzenberger et al., 1992a, b; Stam et al., 1994; Varghese et al., 1987). Less chaotic EEG patterns have been also observed in more intelligent individuals (Anokhin et al., 1999; Jaušovec, 1998), and in more creative ones (Mölle et al., 1999). The main problem related to the nonlinear analysis of EEG data is a mathematical one. Most of the measures used to determine low deterministic chaos were not designed to be used with relatively short and noisy data sets, such as EEG. Hence, approximate entropy (ApEn) was introduced as a quantification of regularity in time-series data. It has been shown that ApEn can distinguish a wide variety of systems ranging from multiperiodic, to stochastic and mixed systems; furthermore, it is applicable to noisy, medium-sized data sets (Pincus, 1994, 1995). Mathematically, ApEn measures the likelihood that runs of patterns that are close in one observation remain close on the following incremental comparisons.

In our study, several experiments using more and less complex visual and auditory oddball tasks were used to clarify the relation between ERP complexity and intelligence. The correlations indicated that in the attended conditions, more intelligent individuals showed more regular ERP waveforms than less intelligent individuals. These trends were observed both in auditory and visual oddball paradigms. The differences were most pronounced in the first 500 ms after stimulus onset. The negative correlations between ApEn and different intelligence measures obtained in the study suggest that high intelligent individuals have more spatially and temporarily coordinated electrocortical activity when engaged in cognitive tasks. This explanation gained further support from the P3 latencies and peak amplitude analysis. Less intelligent individuals displayed reduced P3 amplitude and increased P3 latencies in the oddball tasks. Given the fact that P3 amplitudes indicate the level of cognitive resources devoted to processing the stimulus, and P3 latencies are a measure of the duration of the stimulus evaluation process, the results of the study lend support to the hypothesis that superior intelligence is related to specific and simultaneously activated neural networks, namely, to more efficient neural processing (Jaušovec and Jaušovec, 2000).

Altogether the ERP analyses of the P3 in relation to IQ suggest that in more intelligent individuals, information processing is faster, complementing behavioral data (shorter reaction time measures), and that allocation of mental resources is more efficient. One could say that more intelligent individuals are less chaotic thinkers (Jaušovec, 1998).

2.3.2 The Neural Efficiency Hypothesis

Haier et al. (1988) compared PET brain scans of 30 individuals while solving different tasks: 8 individuals solved the Raven's advanced progressive matrices, 13 performed a visual vigilance task, and 9 simply watched flashing visual stimuli. The PET scans showed that individuals of the former two groups activated the right hemisphere to a greater extent, however, the most striking finding was that a negative relation between brain activity and performance on Raven's IQ test was observed. This led Haier et al. (1988; p. 199) to conclude:

> *[...] that some individual differences in cognitive ability may be related to efficiency or density of neural circuits.*

This finding was later named the efficiency hypothesis and still serves as an explanation of results obtained from several PET (Andreasen et al., 1995), SPECT (Charlot et al., 1992), fMRI (Ruff et al., 2003), ERP (Jaušovec and Jaušovec, 2000), and ERD/ERS (Neubauer et al., 2005) studies. The gist of the theory summarized by Haier et al. (1992, p. 415) is:

> *Intelligence is not a function of how hard the brain works but rather how efficiently it works. This efficiency may derive from the disuse of many brain areas irrelevant for good task performance as well as the more focused use of specific task relevant areas.*

Based on a review of research pertinent to the relationship between psychometrically determined intelligence and functional characteristics of brain activation observed during cognitive task performance, it can be concluded that most of the reviewed studies have demonstrated a negative correlation between brain activity under cognitive load and intelligence (Brancucci, 2012; Neubauer and Fink, 2009). Some studies have shown a specific topographic pattern of differences related to the level of intelligence. High-ability subjects made relatively greater use of parietal regions, whereas low-ability subjects relied more exclusively on frontal regions (Gevins and Smith, 2000; Jaušovec and Jaušovec, 2004).

The same activation pattern—more able individuals displayed less prefrontal but more parietal cortical activation—was also reported by two fMRI studies (Lee et al., 2006; Rypma et al., 2006). It was further reported that highly intelligent subjects displayed more brain activity in the early stages of task performance, while individuals of average intelligence showed a reverse pattern. This temporal distribution of brain activity suggested that cognitive processes in highly intelligent individuals are faster than in average intelligent individuals (Jaušovec and Jaušovec, 2004). The finding is also in line with the reported shorter P3 peak latencies in more intelligent individuals discussed in the previous section.

Moreover, neural efficiency seems to be corroborated mainly when participants work on tasks of low-to-medium difficulty or complexity, and is particularly prominent in frontal brain regions (Neubauer and Fink, 2009; Nussbaumer et al., 2015). In a study by Doppelmayr et al. (2005), the expected findings of a negative relation between brain activation and intelligence emerged solely for the easier items of the Raven test, whereas a tendency in the opposite direction was observed for the difficult ones. The authors further showed that less intelligent individuals displayed a decrease in activation from easy to difficult tasks, whereas the opposite was true for the brighter participants. It is likely that individuals with low IQs did not even attempt to solve harder problems, and this could explain their lower activation levels compared to those with high IQs. It could be further assumed that individuals with low IQs have to work harder on easy problems than individuals with high IQs.

Another possible moderating variable influencing the neural efficiency hypothesis put forward by Neubauer and Fink (2009) is task content, differentiating between tasks requiring the involvement of long-term memory and tasks with less memory involvement. Klimesch and coworkers conducted a series of ERD/ERS studies, most of which tapped into long-term memory, that showed that more capable individuals displayed brain activation patterns opposite to the neural efficiency hypothesis (Klimesch et al., 1996, 1997). Doppelmayr et al. (2005) suggested that a possible reason might be that more able individuals are better at inhibiting brain activity from task-irrelevant input that would display, for nonsemantic tasks, the same pattern of an inverse relation between brain activity and cognitive competence as the one described by the neural efficiency hypothesis.

The results of two ERP studies conducted in our lab also point in this direction. In one study, men displayed longer P1 latencies and lower (less negative) N1 amplitudes during performance on a mental rotation task in comparison with women. As already mentioned in the previous section, P1 is associated with suppression of irrelevant information reflecting inhibitory processes, and N1 is associated with processing of the relevant information reflecting excitatory processes, suggesting that men devoted more time to the analysis of irrelevant information presented in the rotation tasks that resulted in more efficient (less activity) processing of relevant information (Jaušovec, 2012). In a second study, a similar but reverse pattern of sex differences in the P1–N1 complex was reported (Jaušovec and Jaušovec, 2009). Women displayed longer P1 latencies than men during performance on a simple verbal task, resulting in shorter reaction times. This would further suggest that even in the early stages of information processing, sex differences in brain activity exist, and that these differences are related to the type of task. It seems that men and women process information in a similar way in the domain in which they usually show superior performance—men in mental rotation, and women in verbal processing. I will come back to this issue in the final section of the chapter to address the question of sex differences in intelligence with respect to functional and structural brain characteristics.

The earlier discussion opens up the question: what is neural efficiency? Not more than cheap talk, as dismissed by Deary (2000). In my opinion, it is a bit more than that.

In a review of literature explaining the neural efficiency hypothesis, Haier (1993) proposed that a failure of neural pruning may lead to an overabundance of synaptic connections or redundant brain circuitry that could result in more brain activity displayed in less intelligent individuals. This hypothesis is a more elaborate version of his explanation of the inverse relation between glucose metabolic rates and IQ published in 1988, suggesting that they might be related to a change in interneuronal connections, similar to the changes observed in the frontal cortex during the second decade of development, when a decrease in the density of synapses occurs. A variant of Haier's explanation was the neural plasticity model suggested by Garlick (2002). The argumentation was that a better environmentally adapted brain with stronger and more appropriate network connections would on the one hand perform faster, and on the other more efficiently. In that way, two characteristics of more intelligent individuals could be explained: First, faster and less variable reaction times when solving simple tasks, and second better performance on more complex tasks requiring less brain activity. Miller (1994) put forward the brain myelination hypothesis, suggesting that better myelinated axons would allow for faster communication between brain areas and for less cross-talk between neurons. This would explain faster processing of intelligent individuals, as well as less brain activity in comparison with low intelligent individuals during problem solving. The explanation is tempting because it gains a lot of support by recent diffusion tensor imaging research indicating a rather robust relation between intelligence and axonal organization. Specifically, higher white matter structural integrity has been observed in more intelligent individuals (see Section 2.1.4).

2.3.3 The P-FIT Theory

In an influential paper by Jung and Haier (2007) it was suggested that variations in a distributed network, mainly involving frontal and parietal brain areas, are responsible for individual differences in intelligence. The theory was termed the parieto-frontal integration theory (P-FIT). The P-FIT, as well as the MD theory that will be discussed in the next section, both address the question: where in the brain is intelligence? Two steps are usually involved in an attempt to answer this question. In the first step, a minimal number of converging brain areas involved in intelligent thinking is determined with neuroimaging techniques like PET, MRI, or fMRI. In the second step, knowledge about the function of identified areas derived from different domains (e.g., lesion studies, cognitive theory, and research) is used to provide the explanatory content of the theory and its relation to the brain.

Jung and Haier (2007) included 37 studies in their analysis, 11 of which were structural, and 26 were functional. In a recent metaanalysis, Basten et al. (2015) further divided the functional studies into those that employed a task approach and those that used an individual differences approach. In the task approach, participants are placed in an fMRI scanner and asked to solve cognitive tasks that are supposedly related to intelligence. The analysis is based on brain activation patterns observed for the whole group of individuals, not taking into account the individual intelligence level of the participants. Therefore, some brain areas, although showing increased activity due to cognitive challenge, are not necessarily linked to individual differences in intelligence. In contrast, in the individual differences approach the brain scans of more and less intelligent individuals are compared while they solve a task assumed to be highly saturated with *g*. In Jung and Haier's review, only 10 such studies were included. Their description of the P-FIT model suggested that not only frontal brain areas, but also temporal and occipital areas play a crucial role in early processing of sensory information. This information is then fed forward to the parietal cortex, wherein abstraction and elaboration emerge. All these processes are dependent on the fidelity of the underlying white matter necessary to facilitate rapid and error-free transmission of data from posterior to frontal brain regions (see Fig. 2.5).

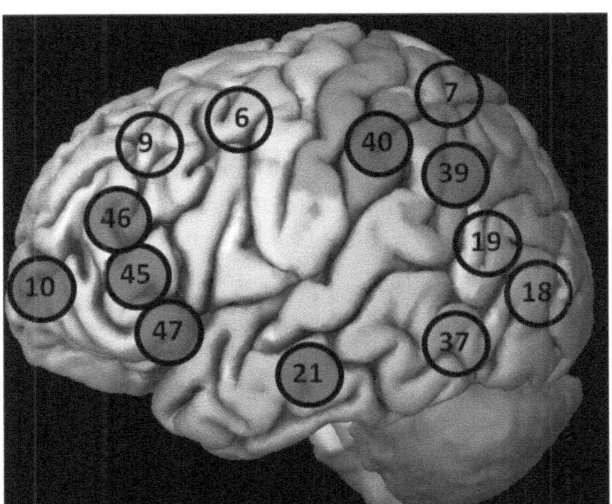

FIGURE 2.5 Brodmann areas *(BA)* proposed by Jung and Haier (2007) defining the P-FIT. Not shown in the figure is the anterior cingulate (BA 32).

The P-FIT describes rather large parts of the association cortices in the brain as relevant for individual differences in intelligence. A critical comment by Basten et al. (2015) was that these areas are probably activated for any kind of higher cognitive demand—even for tasks that have marginal loadings on *g*. The P-FIT somewhat resembles Luria's hierarchical structure model of the brain that involves three cortical zones and the concept that human behaviors arise from the input of all parts and levels of the brain (Luria, 1973). Luria's model was further elaborated by Felleman and Van Essen (1991), although just for the visual cortex of the macaque. However, the main organizational principles, such as distributed connectivity among areas, reciprocity of connections, hierarchical organization, and distinct/intertwined processing, may also apply to humans. It is therefore surprising that attempts to validate the P-FIT failed. In 2009, a special issue of *Intelligence* was devoted to the question: is there a *neuro-g*? Two studies that applied similar methodology, one conducted in New York, United States, and the other in Madrid, Spain, did not correspond very well with respect to brain areas related to *g*. The main problem was that a very small number of discrete brain areas approached 50% of convergence across both studies, even though they employed the same neuroimaging technique (Colom et al., 2009; Haier et al., 2009). When different test batteries were used to derive *g*, brain areas related to *g* changed correspondingly. The only agreement was that the areas related to intelligence are distributed throughout the brain and not localized only in the frontal lobes.

The P-FIT was recently put to an empirical test by a voxel-based quantitative metaanalysis of 12 structural and 16 functional human brain imaging studies (Basten et al., 2015). As already described in Section 2.1.2, the analysis showed overlap with P-FIT over frontal, temporal, and occipital brain areas, whereas no correlation with local amounts of gray matter in parietal areas was observed. The analysis of functional brain imaging studies found an overlap with P-FIT only for frontal and parietal brain regions, but far less overlap in temporal and occipital sensory regions (see Fig. 2.6). The authors explained these differences in terms of study selection; only studies that adopted the individual differences approach were included in their study. Another surprising finding was that no overlap between the results of the functional and structural metaanalyses was established. As put forward by the authors (Basten et al., 2015; p. 21):

> *[...] the dissociation of functional versus structural brain imaging correlates of intelligence is at odds with the principle assumption of the P-FIT that functional and structural studies on neural correlates of intelligence converge to imply the same set of brain regions.*

Several ad-hoc explanations come to one's mind. It could well be that what the scans show is more related to task characteristics than to *g*. There is no one neuro-*g* as commented by Haier et al. (2009)—to our knowledge, this statement has been empirically proven by only one study (Hampshire et al., 2012). Hence, we can conclude: the search for a *neuro-g* continues.

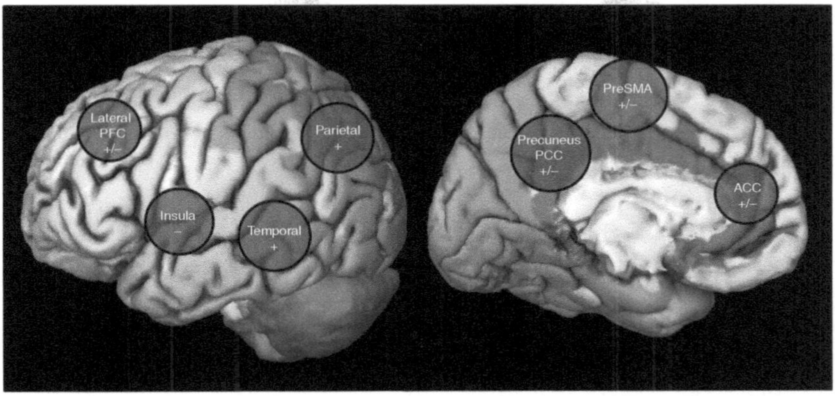

FIGURE 2.6 Functional metaanalysis of voxel-based morphometric studies (Basten et al., 2015). Shown are significant correlations of intelligence test data with local amounts of gray matter. *ACC*, Anterior cingulate cortex; *PCC*, posterior cingulate cortex; *SMA*, supplementary motor area; *PFC*, prefrontal cortex. +, Positive association; –, negative association. The brain images were created with Allen Human Brain Atlas, available from: http://human.brain-map.org.

2.3.4 The Multiple-Demand System of the Brain and *g*

John Duncan's idea of a multiple-demand (MD) system that can explain Spearman's *g* is based on functional neuroimaging data showing that many different tasks activate the same brain regions in the frontal cortex, and to some extent, in the parietal area (Duncan, 2000, 2003, 2005, 2010, 2013; Duncan and Owen, 2000; Duncan et al., 2000). Even though the theory has lately expanded to include parietal areas (e.g., intraparietal sulcus), the focus and level of precision in explanation is mainly on the prefrontal cortex (e.g., Duncan, 2005, 2010, 2013). In my opinion the power of Duncan's explanation of *g* is in the attempt to provide a link between constructs derived from cognitive psychology and machine learning (e.g., hierarchical reinforcement learning, Botvinick et al., 2009; SOAR, Newell, 1990) on the one hand, and the knowledge concerning the function of specific brain areas involved in cognitive processes that coordinate human behavior, on the other. Human behavior is divided into discrete units consisting of subtask sequences which in turn are built of simple actions (Botvinick et al., 2009). Going to the dentist, for example, includes subgoals like: making an appointment, going there—transportation, waiting to be called, and the more unpleasant subgoals that follow and make up the general goal of dental care. These actions can be utilized in different settings—one can also wait in a lawyer's office, or at the airport. The assumption is that the lateral prefrontal cortex produces a temporal representation of these sequences, an "action plan" termed task modeling. In Duncan's opinion *g* is the reflection of this task modeling function.

Duncan et al. (2005) reviewed peak cortical activation from 20 studies that examined a variety of demands: response conflict, task novelty, number of elements in working memory, working memory delay, and perceptual difficulty. Even though the demands differed across studies, similar clustering of brain activation was observed. Three clusters were observed on the surface area. The first in the posterior part of the inferior frontal sulcus, the second in the right hemisphere just anterior to the lateral fissure spreading inward along the frontal operculum to the anterior insula, and the third in the intraparietal sulcus. On the medial surface, the most prominent cluster was in the dorsal part of the anterior cingulate cortex and the supplementary motor area. These clusters are almost identical to those described in Basten's metaanalysis (Basten et al., 2015; see Fig. 2.6). For Duncan, the lack of cognitive specificity—even when no task whatsoever was given to respondents, the same areas were activated—is evidence that *g* is a reflection of frontal lobe function (Duncan et al., 2005; p. 254). The frontal cortex achieves this generality in activation by what Duncan dubbed *adaptive coding*. According to this principle, neurons of the prefrontal cortex have no specific function and, therefore, can respond to multiple demands by emphasizing relevant information and producing a representation of the outer world—a structured series of attentional episodes—allowing the organism to break complex problems into manageable subgoals (Duncan, 2013). Subgoaling is maintained in WM, which further supports the role of the prefrontal cortex in intelligent behavior. There is a large body of evidence suggesting that *g* is related to working memory (for a review, see Buehner et al., 2005), and that WM processes activate the prefrontal cortex (e.g., Petrides, 2000). Variability in the efficiency to build such models could explain individual differences in intelligence. Another aspect of this process, besides dividing the task into subgoals, is maintaining the plan via a sequence of attentional episodes. One has to keep the model and its parts in mind to understand the whole. To give an example from reading: in order to enhance students' reading comprehension, a teacher occasionally asks them to guess what the story is about. Poor readers give answers that are not related to previous guesses. Each guess relates solely to the text segment just read, with no connection to other parts of the text. According to Duncan, *goal neglect*—a mismatch between what should be done and what is actually done (occasionally reported in frontal lobe patients), can under specific circumstances be observed in normal individuals. Indeed, people with below average IQ (−1 SD) tend to neglect certain instructions as complexity increases: they do not perform the actions requested, even though they are able to accurately repeat what they were told to do. This suggests that in low intelligent individuals, an increase in the number of elements (complexity) gives rise to competition in task mode, which in turn leads to a loss of some elements.

In comparison with the P-FIT, the MD system is a much more explanatory model with a broader focus. Intuitively, one is inclined to expect that *g*, because it loads on every mental task, should be related to a limited brain area, such as the lateral prefrontal cortex. Last but not least, factor analysis is a data reduction method. However, common sense has often proven misleading in science. As stressed by Gray and Thompson (2004), identifying a single active area may be of interest for a better understanding of the brain–intelligence relationship, but identifying an active network would be even more informative. Only recently research employing voxel-mirrored homotopic connectivity (Santarnecchi et al., 2015), structural efficiency in predestinated brain networks (Pineda-Pardo et al., 2016), and cross-frequency coupling (Pahor and Jaušovec, 2014) has attempted to uncover these neural-nets and the brain code that makes the connections possible. Thinking occurs in tiny time units, therefore the next step will be a temporarily fine-grained analysis of activated brain networks and their relation to intelligence.

2.4 PROCESSING SPEED AND WORKING MEMORY

The cognitive functions that predict psychometric intelligence have been most often identified as WM and processing speed. As stressed by Deary (2012) they represent an interim level of reductionism between *g* and biological intelligence, providing explanations for the relation between *g* and activated brain areas derived from PET, MRI, fMRI scans and neuroelectric brain activation patterns. A good example is Duncan's elucidation of *g* by the MD system (Duncan, 2005). The following two sections will provide a brief review of the assumed neurobiological backgrounds of processing speed and WM.

2.4.1 Processing Speed

The idea behind the relation between processing speed and *g* is rather straightforward. Processing speed is viewed as a form of cognitive or neural limitation of processing a simple stimulus. The most often studied parameters are reaction time and reaction time variability. Jensen (2006) conceptualized reaction time variability as noise in the information processing system, or as negative neural efficiency/information processing. Furthermore, a distinction has been made between the decision and movement components of overall response times. In order to measure choice reaction time, the "Jensen box" was developed, and later replaced with the so-called finger on buttons response box. The main idea was to minimize time of motor movement in response to a target stimulus. Another aspect in reaction time research is Hick's law (Hick, 1952), which states that reaction time increases with increased response uncertainty: the number of optional answers provided. For example, it will take you less time to place your order in a restaurant that has 6–8 food choices on the menu than in a restaurant that has dozens of options. Not to mention wine lists of encyclopedic lengths.

It has been long assumed that measures of processing speed are inversely related to IQ. As stressed by Deary (2012), it is surprising that it took decades before this was confirmed on a representative data set. In 2001, Deary et al. (2001) published the results of a large ($n = 900$) population representative study that showed that IQ (as measured by the Alice Heim intelligence test) was related to reaction time ($r = -0.31$) and to reaction time intraindividual variability

(r = −0.26). For a four choice reaction time task (Hick-type paradigm), intelligence scores correlated r = −0.49 with reaction time and r = −0.26 with reaction time intraindividual variability. In a metaanalysis of 172 studies that involved a total of 53,542 participants, Sheppard and Vernon (2008) reported similar correlations between different measures of processing speed and intelligence. Besides the standard reaction time measures, visual and auditory inspection times[f] were also examined. The correlations between mean reaction time measures and *g* ranged from −0.22 (1 bit of information-processing) to −0.40 (8 bits of information-processing). Even more pronounced were the correlations for short-term memory tasks (−0.25 to −0.45). The mean correlation between general intelligence and visual inspection time was −0.36 and for auditory inspection time it was −0.31. The main conclusion that can be drawn based on these studies is that diverse measures of mental speed are significantly correlated with measured intelligence. Another trend observed was that the correlations tend to be slightly higher for (1) measures of fluid intelligence than for crystallized intelligence and (2) more complex measures (more response alternatives) as compared to simple ones.

The question with respect to neurocognitive processes is: what do these measures stand for? As already mentioned in Section 2.3.1, the average correlation reported across studies relating P3 latencies with *g* was −0.30. These findings suggest that individuals with a higher IQ tend to evaluate and classify information more rapidly compared to those with lower intelligence (Vernon et al., 2000). A similar conclusion can be also drawn based on studies reporting positive correlations between processing speed and the level of white matter integrity (see Section 2.1.4). Ritchie et al. (2015b) reported that these correlations showed the most pronounced decline with age. A similar relation was also reported by Haász et al. (2013). A finding worth mentioning was that the cognitive trait component of reasoning ability was also related to white matter integrity, albeit to a lesser extent, and shared similar white matter tracts as processing speed, a fact that implies that information processing speed is closely related to reasoning ability. Further support for the neurocognitive relation between mental speed and IQ was provided by a recent study by Schubert et al. (2015). Based on EEG (ERP) data obtained while respondents were solving elementary cognitive tasks, the authors concluded that no unitary construct of mental speed could be determined. However, factor analysis revealed that a mental speed factor (including two more specific speed factors: reaction time and ERP latencies) was substantially associated with *g*.

An even more ambitious relation between mental speed, intelligence, and the central nervous system, far beyond the *brain perspective* proposed in the

f. A psychophysical task not involving reaction speed per se, but recording correct discriminations based on simple stimuli. Typically, pi-shaped stimuli are presented and respondents have to indicate which end of the pi-shape is longer/shorter. In the auditory condition a tone (stereo) is shifted to the right/left and the respondent has to determine its laterality.

introduction to the chapter, was suggested by Jensen (2011). The definition of intelligence in his new theory was:

> *Intelligence is the periodicity of neural oscillation in the action potentials of the brain and central nervous system (Jensen, 2011; p. 173).*

Action potentials are the building blocks of all our thoughts, and also of intelligence. Their periodical fluctuation is a response to incoming stimuli. A simplified example could be the change of alpha waves (8–12 Hz) in an eyes-closed condition to fast oscillations like beta waves (20 Hz <) when eyes are opened. As already explained in previous sections on IAF (Section 2.2), people have different peak alpha frequencies; however, the correlation with intelligence was small in our analysis and indicated a rather complex sex-by-spatial ability pattern. According to Jensen, oscillation rates determine the reaction times of individuals. Those with higher oscillation rates can react faster to stimuli with increasing response uncertainty (Hick's law) in comparison to those with average or low oscillation rates. In that way, reaction time is a pure measure of oscillation rates, and therefore also of *g*. Jensen makes the analogy with a movie where motion is provided by the specific rate of projecting static pictures. The same is true for our perception of the outer world that is composed of a series of stimuli entering the focus of attention in working memory. The explanation comes close to Duncan's (2005) structured series of attentional episodes, describing individual differences in intelligence with goal neglect: increased complexity of instruction makes less intelligent individuals perform a task by abandoning parts of the instruction. The idea is tempting, however only weak empirical support has been presented to date, as will be shown in the next section on WM.

2.4.2 Working Memory

As stressed by Corkin (2002), a lot of what we know about the cognitive and neural organization of memory has to do with Henry Molaison known as patient H.M. Because of his severe epilepsy, a neurosurgical procedure was carried out to reduce it. Although his epileptic seizures were reduced, the surgery also made him globally amnesic. From the time of his operation when he was 27 years old until his death at 82 years of age (2008), he was unable to consciously learn or remember any new information (anterograde amnesia). In addition, he had also a memory loss of 11 years preceding his surgery (retrograde amnesia): his time "stopped" when he was 16 (for a detailed description, see Ogden, 2005). According to Scoville's[g] drawings, H.M. underwent bilateral surgical ablation of medial temporal lobe structures, including the hippocampus, parahippocampal gyrus, and amygdala. When MRI scans became available, it was revealed that the surgery was not as excessive as assumed (Corkin, 2002), and only recent

g. Scowille was the surgeon who performed the ablation.

postmortem imaging provided a clear anatomical verification of the lesion and the pathologic state of surrounding areas (Annese et al., 2014). Henry's verbal and intellectual abilities, as well as personality appeared to be unaffected. For example, he could carry out a conversation, however minutes afterward he was completely unaware of having had the conversation or of the person he was talking with. He also struggled to show anger, to pretend to be angry, or recognize its expression in others (Ogden, 2005). Although his memory loss was global, he could learn sensory motoric skills, and showed priming effects, while being unaware that he had repeatedly performed the task.[h] At times, he was even surprised how well he was doing, as his first impression was that the task was rather difficult. Another unexpected finding was that Henry could draw an exact floor plan of the house where he had been living 5 years after the operation, and could also provide the correct address which is in sharp contrast with his impaired performance on all other spatial and memory recall tasks. Yet another atypical memory performance observed was that Henry could recognize magazine pictures he studied, and he was able to do that even after a period of 6 months, comparable to a control group (Corkin, 2002; Ogden, 2005).

At the other extreme are people with extraordinary memory performance abilities called savants. Probably the best description was provided by Alexander Luria (1968) in his book *The Mind of a Mnemonist.* Most will also remember the popular movie *Rain Man* in which Dustin Hoffman played an autistic savant—Raymond Babbitt. The character was inspired by Kim Peek, known as a "megasavant," who was actually not autistic but had a rare disease causing damage to his brain. Solomon Shereshevsky, known as "S." in Luria's book, was a newspaper reporter who never took notes at briefings, because he could verbatim remember everything he heard. On the other hand, Luria characterized S. as an unambitious person with little ability to reason or understand metaphors. These portrayals demonstrate the complexity of human memory, and the difficulty of converting them into cognitive models with some relation to brain structure and function.

Working memory (WM) relates to storage and information processing components, and it is used very widely (Baddeley, 2012; Baddeley et al., 2011; Cowan, 2000). The concept developed from an earlier one called short-term memory; both terms are still used interchangeably. As put forward by Baddeley (2012), short-term memory is most often used in relation to its literal meaning: keeping a limited amount of information in mind for a short period of time. For example, this process is required when someone tells us a telephone number we are required to memorize. In contrast, WM not only refers to storage of information, but also to manipulation of this information. Not all would agree with this approximate distinction. Engle et al. (1999), for example, attribute the function of short-term memory primarily to processes, such as grouping, chunking, and rehearsal. On the other hand, WM relates to executive attention that controls

h. He was asked to perform a mirror-tracing task: to draw a line between two adjacent outlines of a star that he could see only in a mirror, with left and right reversed.

the content of the active portion of memory (e.g., keeps the telephone number in a temporarily available state). Nevertheless, the main problem in WM research is to distinguish between storage and executive control. Even the classic digit-span test for determining WM storage capacity is not immune to "chunking" of digits—the use of intelligent grouping of elements. This ambiguity has prompted several suggestions what the real WM capacity might be—ranging from Miller's (1956) 7 ± 2 chunks; to the concept that WM is limited by the amount of time that has elapsed rather than by the number of elements that can be held simultaneously (Baddeley, 1986); to today's most popular view that storage limits exist, but they are completely task-specific, with no possibility of extracting a general estimate (Cowan, 2000).

Because WM plays a key role in problem solving and intelligence the indistinctness of capacity and WM component processing is also part of the research on the structure of intelligence. As stressed by Colom et al. (2008), WM and intelligence are highly related constructs, but we still do not know why. In several correlation studies, it was shown that short-term storage largely accounts for the relationship between WM and intelligence (Colom et al., 2008; Martínez et al., 2011). In a recent confirmatory analysis, it was shown that WM shared 83.4% of variance with fluid intelligence measured under time pressure, whereas without time pressure only 58.2% of variance was shared by both constructs (Chuderski, 2015). Still other studies using structural equation modeling suggested that attention and WM predict overlapping parts of intelligence (Schweizer and Moosbrugger, 2004). In yet another structural equation modeling study, it was shown that relational integration has the same predictive value for intelligence as WM storage and processing. Relational integration was defined as the ability to build new relations between elements, thereby creating structural representations (Oberauer et al., 2008). It is surprising that new constructs are being introduced to explain the WM construct, despite the fact that we have immense problems distinguishing between the existing ones.

Similarly, adding new components has characterized the development of WM models. Probably the most popular model of WM is the one proposed by Baddeley and Hitch (1974): the multicomponent model of WM. In its original form, it consisted of three components: the central executive, and two slave systems, the visuospatial sketch pad and the phonological loop. At that time, the three components sufficed to explain most research findings related to WM. For example, it explained why we encounter more problems recalling longer words than shorter ones, or words that sound similar compared to those that sound different. The phonological loop is probably the most explored component of the model. It has two functions, a phonological store, which holds memory traces in acoustic or phonological form, and an articulatory rehearsal function similar to subvocal speech. The function of the latter is to retrieve and rearticulate the contents held in this phonological store, and in that way refresh the memory trace. The function of the visuospatial sketchpad is in maintaining and manipulating visual and spatial information. The visual and spatial subcomponents have independent storage and maintenance/manipulation mechanisms. The visual component is related to perception and visual imagery, while the spatial component shows closer connection to attention and action. The central executive is the most complex component in the model. In the initial model, it was described as a homunculus that does everything that the two other systems could not. Research has, however, demonstrated that the central executive is involved in a number of separable executive functions involved in the storage function of WM, as well as in more general cognitive processes. The central executive operates whenever information within the stores needs to be manipulated. In complex cognitive processes, when solving an intelligence test for instance, its function is attentional control, enabling the focusing of attention to concurrent tasks and attentional switching (Baddeley, 1986, 2012). The central executive lacks storage capacity, and the storage functions of the other two systems cannot explain certain memory functions, such as recall of prose, thus a third storage system was introduced: the episodic buffer. The episodic buffer is assumed as a subsystem between short- and long-term memory. Its function is in binding different sources of information into coherent structures or episodes. In yet another updated version, the episodic buffer received a central position: it remained a passive system but with the crucial function of integrating information from different sources and modalities into chunks or episodes (Baddeley et al., 2011). Here again a parallel can be drawn to Duncan's (2005) MD system, and the structured series of attentional episodes that lie at the core of its function.

More recently, state-based models of WM have gained prominence (D'Esposito and Postle, 2015). These models assume that allocation of attention to different representations in long-term memory (either semantic, sensory, or motor) governs temporary retention in WM. The most well-known among these models is Cowan's embedded-processes model in which WM is defined as a cognitive state that retains information in an accessible state. Activation occurs in long-term memory, is temporary, and fades unless maintained by verbal rehearsal or continued attention. In the core of this new theoretical framework are two constructs: focus of attention and its capacity, that is, scope of attention. A successful measure of the scope of attention must prevent rehearsal and grouping processes, allowing a clearer estimate of how many separate chunks of information the focus of attention circumscribes at once (Cowan et al., 2005).

From a neuropsychological perspective the frontoparietal network has been associated with performance on WM tasks (e.g., Chein and Fiez, 2010; Jonides et al., 2008; Palva et al., 2010; Posner and Petersen, 1990; Schumacher et al., 1996). It has been further suggested that the central executive function of WM is linked to frontal lobes, whereas the WM storage component is associated with parietal areas (Champod and Petrides, 2010; Collette and Van der Linden, 2002; Olson and Berryhill, 2009; Sauseng et al., 2010). Based on evidence from several brain imaging studies, the left intraparietal sulcus (LIS) has been identified as a unique area responsible for amodal or multimodal storage of information (Cowan et al., 2011; Majerus et al., 2006, 2010; Xu and Chun, 2006).

Support for a frontoparietal distinction related to the WM functions of processing and storing of information comes also from research employing neuroelectric brain imaging methods (Klimesch, 1999; Klimesch et al., 2008; Sauseng et al., 2010). Research indicates that theta oscillations are related to WM processes. Furthermore, there is evidence to suggest that theta synchronizes during WM processes and serves as a gating mechanism, providing optimal neural conditions for specific processing (Sauseng et al., 2010). Klimesch et al. (2008) further suggested that upper alpha is related to long-term memory processes. Alpha band activity desynchronizes in relation to task performance (Pfurtscheller and Aranibar, 1977). Until recently alpha synchronization was considered a cortical idling phenomenon (Pfurtscheller, 1999). However, recent evidence suggests that synchronization of alpha activity does not merely indicate cortical deactivation, but reflects internal top-down processes and selective inhibition of brain regions—this idea resembles the concept of WM of Baddeley (2012), and the central executive processes (Fink et al., 2009; Jensen and Mazaheri, 2010; Klimesch et al., 2007). According to Klimesch et al. (2008), an attentional control function maintains focus on highly selective elements of task performance by inhibiting distracting interferences from task-irrelevant brain areas. Thus, working and long-term memory systems have their own types of top-down processes that control access and manipulation of stored information, which are based on the theta and upper alpha frequency bands.

It can be concluded that contemporary cognitive research of WM has favored state-based models because they accommodate well to neuroscience data. The focus of attention account can explain the two main functions of WM: temporary storage of information and maintenance of stored information. The analogy of a searchlight comes to mind: spotting bits of internal or external information, where the frequency of brain oscillations determines the timing at which the focus of attention is shifted. High-frequency components increase the number of bits of information processed, which may explain the relationship between *g* and speed of information processing, as well as the memory capacity, namely, the scope of attention relation. From a neural perspective the lateral prefrontal cortex is important for attentional control: it represents top-down influence on posterior sensory regions reactivating cortical memory traces—the memories themselves (Sreenivasan et al., 2014). The similarity with neural models of intelligence is obvious, but not unexpected given that WM has sometimes been proposed as virtually synonymous with intelligence (e.g., Martínez et al., 2011).

2.5 SEX DIFFERENCES IN INTELLIGENCE, BRAIN STRUCTURE, AND FUNCTION

As stressed by de Vries and Forger (2015), sex differences in the brain are not separable from sex differences in other body parts. Women may see things differently (Jaušovec and Jaušovec, 2009), but they are also perceived and treated by the world differently, creating divergent knowledge and experiences that might have profound consequences for the brain.

2.5.1 Psychometric Evidence

It is unlikely that psychometric evidence could ever provide a definitive answer to the question if these differences exist. Ackerman (2006) noted that whether males or females differ in general intelligence depends on the test used to assess cognitive ability. Test designers had two options, either to use separate norms for males and females, as it is done in personality questionnaires for the personality factors of agreeableness and neuroticism (Caprara et al., 2002), or to remove items that show sex bias. When Terman (1916) provided test norms for the Binet–Simon intelligence test, the sex differences were, in his opinion, negligible and therefore ignored, thus the same IQ scale for boys and girls was adopted. In later years, when the scale was refined, items that showed greater sex differences were eliminated (Terman and Merrill, 1937). The same holds true for WAIS-III (*Wechsler Adult Intelligence Scales*, third edition; 1997, Harcourt Assessment Inc.). As put forward by Irwing (2012), the WAIS-III manual documents state that extensive procedures, such as differential item functioning and expert opinion, were used in the construction of the test in order to eliminate gender bias. The test does not include 3-D mental rotation, for which a large male advantage was found (for a review, see Miller and Halpern, 2014). Even more numerous are methodological problems related to sample selection and the methods used to establish sex differences. In particular, the method of correlated vectors has been criticized suggesting that multigroup confirmatory factor analysis is a better option (Irwing, 2012).

The study of sex differences in intelligence has divided scientists for decades; one academic camp led primarily by Richard Lynn claims a male advantage in general intelligence, whereas the other states that no sex differences in general intelligence exist (Flores-Mendoza et al., 2013). The latter group recognizes sex differences in subtest-specific abilities, but does not attribute them to differences in first-order factors, or to the second-order factor *g* (e.g., van der Sluis et al., 2008). A detailed review of the literature is beyond the scope of this section, therefore I will provide just two examples of more recent findings related to sex differences in intelligence exemplifying the aforementioned problems.

Lynn (1999) proposed a developmental theory in order to explain sex differences in mean performance on measures of general intelligence. Briefly, the theory states that girls mature more rapidly in brain size and neurological development than boys up to the age of 15 years. Faster maturation of girls up to this age compensates for their smaller brain size resulting in little or no sex differences in intelligence. From the age of 16 years onward, the growth rate of girls decelerates, and this may explain a male advantage of about 4 IQ points that is consistent with larger average male brain size. To test this hypothesis, Savage-McGlynn (2012) used multigroup confirmatory factor analysis to assess mean differences in younger (7–14 years) and older (15–18 years) groups of individuals. The groups were taken from a nationally representative sample of children from 85 schools and colleges across the UK with a sample of 663 younger (323 male) and 263 (114 male) older participants. The participants were invited to participate in the standardization of Raven's Standard Progressive Matrices Plus (SPM+). The analysis led Savage-McGlynn (2012, p. 139) to conclude:

> *[...] the current investigation failed to find significant mean sex differences between groups of male and female participants younger than 15 or older than 15 years of age.*

It is worth mentioning that the maximal score obtained on the SPM+ for younger and older males exceeded that of younger and older females for 4 and 3 points, respectively (average *SD* for the four groups = 6.21). A similar trend was also observed for minimal scores (1 point higher for younger males and 3 points higher for older males).

Another study that investigated sex differences in *g* was performed by Irwing in 2012. Multigroup factor analyses were used to analyze the American standardization sample of the WAIS-III, consisting of 2450 individuals aged from 16 to 89 years. Individuals with sensory or motor deficits, drug- or alcohol-dependent individuals, medicated individuals, and individuals with neuropsychological disorders were excluded from the sample. Sample sizes differed for full-scale IQ scores (603 males; 696 females), and for specific factors of verbal comprehension, perceptual organization, WM, and processing speed (1174 males; 1303 females). The analyses performed by Irwing (2012) showed a sex difference favoring men in *g* (0.19–0.22 d), information (0.40 d), arithmetic (0.37–0.39 d), symbol search (0.40–0.30 d), and a sex difference favoring women in processing speed (0.72–1.30 d). These differences were only significant for the main effects of sex but not for age nor for interaction effects. Irwing's (2012) conclusion was that the findings supported Lynn's developmental theory of sex differences.

Although both studies used similar methodology, they provided contradictory findings. At first glance, the most obvious reason might be the different tests used to determine *g*. Irwin's (2012) study used WAIS-III, whereas Savage-McGlynn's (2012) study used SPM+. One might expect different outcomes particularly because WAIS-III lacks 3-D mental rotation tasks, and because items that showed excessive sex differences were eliminated in the construction process. Yet another explanation was put forward by Irwing (2012). Because sex differences were most attenuated between the ages 28 and 60 years (even showing a reverse pattern at age 40; female IQ > male IQ), and because this age span corresponds with the time period that is crucial for professional success, it could be speculated that successful males (being more engaged with their professional career than females) might have been less inclined to participate in standardization studies than intelligent women. All these factors could have led to an underestimation of sex differences in *g*. One could also argue that research on intelligence-related sex differences is limited by its exclusive reliance on IQ tests.

Similar methodological issues have been reported for studies focusing on sex differences in spatial and verbal abilities. A metaanalysis of studies published before 1973 found an average difference of about half a standard deviation (SD) in favor of males on tests of visuospatial ability (Hyde, 1981). Most pronounced gender differences of nearly 1 SD have been reported for mental rotation tasks (Mackintosh and Bennett, 2005; Masters and Sanders, 1993; Voyer et al., 1995). These findings were confirmed in a recent large-scale study including 90,000 females and 111,000 males from 53 nations (Lippa et al., 2010). The results showed significant male advantages in mental rotation ($d = 0.47$) and line angle judgment ($d = 0.49$). A metaanalysis of mental rotation tasks showed that sex differences increased when the tasks were presented under time constraints (Voyer, 2011). The duration of time limits did not influence the results; the difference was observed between test conditions with a time limit of any duration and between those without a time limit (Voyer, 2011).

Perhaps one of the first written accounts of female superiority in verbal ability is found in an ancient Sanskrit book, suggesting that nine shares of speech were given to women and one to men (Nyborg, 1994). Systematic analyses have shown that females surpass males in some, but not necessarily all areas of verbal ability (Garai and Scheinfeld, 1968; Halpern, 2004; Jensen, 1998). Specifically, females seem to have an advantage in episodic memory tasks where verbal processing is required or can be used, as well as in verbal fluency (Herlitz et al., 1997; Maitland et al., 2004). A recent study analyzed data collected by PISA (10 years of data collection)—that included mathematics and reading performance of nearly 1.5 million 15 year olds in 75 countries—showed that the average sex difference in reading was three times larger than the sex difference in mathematics. A second finding was that the overall average difference increased from 32.0 points in 2000 to 38.8 points in 2009 (Stoet and Geary, 2013). Ten points are approximately 1/10th of a standard deviation.

Therefore, the assumptions based on the psychometric-trait approach represent a more or less successful balance between scientific rigor and political correctness. Alternatively, data based on ecologically validated measures of intellectual success in arts and sciences may provide some insight into sex differences in cognitive ability. The most prestigious award in science, medicine and literature is the Nobel Prize, followed by the Wolf Prize granted in Israel for Physics, Chemistry, Medicine, Mathematics, Agriculture, and Art (music, architecture, painting, and sculpture). Table 2.3 summarizes female laureates for both prizes in physics, chemistry, medicine, and literature, and economy (only for the Nobel Prize) from the first time it was awarded till 2015. Two assumptions can be deduced. First, there is a clear difference in the number of female Nobel laureates in literature compared to other domains, especially physics and chemistry, roughly corresponding with the reported male advantage in mathematical reasoning and the female advantage in verbal ability. Second,

TABLE 2.3 Female Nobel Laureates in Physics, Chemistry, Physiology or Medicine, Economy, and Literature From 1901 to 2015; and Female Wolf Prize Foundation Laureates in Physics, Chemistry and Medicine From 1978 to 2015

	Year					
Domain	**1901–40**	**1941–2000**	**2001–15**	**Female Σ**	**Male Σ**	**%**
Nobel Prize laureates						
Physics	1	1	0	2	199	1.0
Chemistry	2	1	1	4	168	2.3
Physiology or medicine	0	6	6	12	198	5.7
Economy	0	0	1	1	76	1.3
Literature	4	5	5	14	94	13.0
Wolf Prize laureates						
Physics		0	0	0	56	0
Chemistry		0	0	0	46	0
Medicine		4	1	5	49	10.2

the data do not indicate that this difference decreased or has changed over the past 100 years, since the first Nobel Prize was awarded. A similar trend can also be observed for the Wolf Prize. Can all of these be attributed to sociocultural influence and the glass ceiling effect? Probably not. The first female Nobel Prize in physics was awarded to Marie Curie Sklodowska in 1903, and the second to Maria Goeppert Mayer in 1963. One can speculate that in 1903 and even in 1963, there was much more gender inequality and male chauvinism than at the present time. Thus it seems that scientific excellence was always recognized as being male or female.

2.5.2 Neurobiological Evidence

The focus of this section will be on structural and functional sex differences in the central nervous system that might relate to the observed differences in cognitive performance—general intelligence, verbal ability, and mental rotation.

On average, males have larger total brain volumes than females. This difference is rather robust as demonstrated in several studies and metaanalyses (Leonard et al., 2008; Ruigrok et al., 2014; Yan et al., 2011). The difference remains unchanged also after correcting for body size as demonstrated by Ankney (1992). The average difference is 100 g when corrected for body size, and 140 g when uncorrected (Rushton and Ankney, 2009), or 4 billion neurons as estimated by

Pakkenberg and Gundersen (1997), and not related to body size. Several recent studies suggested that in predicting species' cognitive capabilities absolute brain size outperforms any other relative measure (Deaner et al., 2007; Pietschnig et al., 2015). In a recent metaanalysis of 126 studies including newborns to individuals over 80 years old (best represented for individuals 18–59 years old), Ruigrok et al. (2014) found that on average, males have larger absolute intracranial volumes (12%), total brain volume (11%), cerebrum (10%), gray matter (9%), white matter (13%), cerebrospinal fluid (11.5%), and cerebellum (9%) than females.

A second rather stable observation, as reported in Section 2.1, is that brain size positively correlates with intelligence (Pietschnig et al., 2015; Ritchie et al., 2015a; Rushton and Ankney, 2009; Vernon et al., 2000). Combining these two propositions: (1) males have larger brains than females, and (2) brain size correlates positively with intelligence; led Lynn (1999) and Nyborg (2005) to conclude that males are slightly more intelligent than females. Lynn (1999) predicted that males should outperform females for about 3.8 IQ points. This prediction comes rather close to the one obtained by Nyborg (2005), who analyzed the results of two studies and showed a male advantage of 3.6 points. In a recent study based on the dataset from the NLSY97 survey, representing more than 15 million 12–17 year old adolescents living in the United States in 1997, Nyborg (2015) estimated an IQ male advantage of 6.83 points for blacks, 7.03 for Hispanics, and 3.6 points for whites (mean $SD = 1.7$)

At first glance, the data and the explanation seem straightforward. However, as admonished by Darwin (1871), intellect cannot be accurately determined by just looking at the cubic content of a skull. Size matters but so does organization; moreover, some studies suggest that it is the interaction between size and organization that is important (Gong et al., 2009, 2011; Yan et al., 2011). A MRI tractography study found that female brains are much more efficient than male brains (Gong et al., 2009). For females, the difference was especially pronounced in six brain regions (five of which were in the left hemisphere) and for males just in two, both in the right hemisphere. This finding is in line with the reported male advantage in spatial processing and female superiority in verbal ability. In another study that replicated the finding that females have higher local efficiency in cortical anatomical networks than males, it was further shown that individuals with smaller brains had higher local efficiency than those with larger brains (Yan et al., 2011). The correlation was found only in females suggesting an interaction between sex and brain size. Moreover, a large-scale diffusion tensor imaging (DTI) study by Ingalhalikar et al. (2014) that included 949 individuals (8–22 years, 428 males and 521 females), found that on the cerebral level, male brains displayed more connectivity within the hemispheres (intrahemispheric), whereas female brains displayed more connectivity between hemispheres (interhemispheric), while in the cerebellum a reverse pattern was observed, with males showing stronger connections between the left cerebellar hemisphere and the contralateral cortex. The differences were most pronounced in young adults (17–22 years). The findings led the uthors to conclude that male brains are structured to facilitate connectivity between perception and coordinated action, whereas female brains are designed to facilitate communication between analytical and intuitive processing (Ingalhalikar et al., 2014).

The most often reported sex differences in brain substructures are gray matter proportion (women have a higher proportion of gray matter), the relative size of the corpus callosum (larger in women), and the sizes of Brocca and Wernicke speech areas (proportionally larger in women) (e.g., Allen et al., 1991, 2003; Knaus et al., 2006). When these differences were adjusted for brain volume the following picture emerged: both women and men with lower cerebral volumes had higher proportions of gray matter (this relationship is slightly stronger for women), the same holds true for callosal volume—individuals with smaller brains had relatively larger callosa (Leonard et al., 2008). The analysis of speech areas yielded similar results, with the exception of planum temporale in the left hemisphere. In women, planar size did not depend on cerebral volume, whereas in men it did (Leonard et al., 2008). The planum temporale lies posterior to the primary auditory cortex within the lateral fissure that forms the heart of Wernicke's area. A very simplified model of the neural basis of language assumes that the meaning of words is represented in Wernicke's area (Kolb and Whishaw, 2009). These findings provide further evidence for female superiority in verbal ability.

Taken together, brain structure and in particular, brain size can to some extent explain the reported sex differences in *g*, as well as female superiority in verbal processing.

Sex-related differences have also been observed in studies analyzing brain function. During performance on numerical and figural tasks, males are more likely to produce cortical activation patterns that are in line with the neural efficiency hypothesis, whereas no significant differences have been reported for females (Lipp et al., 2012; Neubauer and Fink, 2003; Neubauer et al., 2002, 2005). Perhaps the most important finding of these studies is that the inverse intelligence–activation relationship (i.e., neural efficiency) appears to be moderated by task content and gender. Males and females display the expected inverse IQ–activation relationship in the domain in which they usually perform better: females in verbal domains, and males in the visuospatial domain. Not only ability factors, but also situational factors, such as implicitly activated stereotypes, can impair performance (Steele, 1997), hence, the observed differences could also have emerged due to male/female negative stereotypes. However, a recent study demonstrated that sex differences in neural efficiency during performance on visuospatial tasks do not result from the stereotype threat effect (Dunst et al., 2013).

A possible explanation for the observed sex-related efficiency patterns in spatial performance could be hippocampal shape as reported by Colom et al. (2013). The hippocampus is a structure in the limbic lobe of the cortex that is thought to mediate spatial memory and spatial navigation (Kolb and Whishaw, 2009). Colom reported a positive association between hippocampal shape and performance on spatial tasks in males (larger radial distance: higher score), whereas for females this relation was negative (shorter radial distance: higher score).

Sex differences in neural efficiency have also been reported for emotional intelligence (Jaušovec and Jaušovec, 2008), which refers to the ability to recognize emotion, reason with emotion, and emotion-related information, and process emotional information as part of general problem solving (Mayer et al., 1999). Females tend to surpass males on tests of emotional intelligence (Mayer et al., 2008). In our study (Jaušovec and Jaušovec, 2008), the most robust sex-related difference in brain activity was observed in lower-2 alpha band (8–10 Hz). Males and females displayed an inverse IQ activation relationship in the domain in which they usually perform better: females in the emotional intelligence domain, and males in the visuospatial ability domain. Because activity in the lower-2 alpha band is related to attentional processes (Klimesch, 1999), it was suggested that high ability representatives of both sexes compensate for their inferior problem solving skills (males in emotional tasks, and females in spatial rotation tasks) by increasing their level of attention.

Men demonstrate distinct advantages on a broad range of visual tasks. The most pronounced sex differences of nearly 1SD have been reported for mental rotation, the ability to imagine visual shapes rotated to an orientation other than the one in which they are presented (Mackintosh and Bennett, 2005; Masters and Sanders, 1993; Voyer et al., 1995). Mental rotation tasks can therefore serve as a useful probe for investigating the neural underpinnings of sex differences in cognition. Neuroimaging studies investigating this phenomenon have produced diverse results.

Several fMRI studies have reported significant differences in brain activity between males and females during performance on mental rotation tasks, even though no differences on the behavioral level were observed (Butler et al., 2006; Hugdahl et al., 2006; Jordan et al., 2002). On the other hand, some studies reported no significant sex differences in brain activity, yet reported that men outperformed women on the spatial rotational tasks used during brain imaging (Bell et al., 2006; Halari et al., 2005). Furthermore, there are still other studies that employed neuroelectric or hemodynamic brain imaging techniques, and reported sex differences both in performance and in brain activation patterns (Gootjes et al., 2008a,b; Gur et al., 2000; Hahn et al., 2010; Jaušovec, 2012; Lipp et al., 2012). Moreover, research suggests that males show predominantly parietal activation during performance on mental rotation tasks, whereas females also show frontal brain activity (Butler et al., 2006; Hugdahl et al., 2006). A metaanalysis of early studies on sex differences in brain lateralization in relation to mental rotation indicated a strong preference for the right hemisphere in males, while females showed no hemispheric advantage (Vogel et al., 2003). On the other hand, certain studies revealed that during spatial processing, men show greater and more bilateral activation than women (Gur et al., 2000), and that such differences can even be observed in preschool children (Hahn et al., 2010).

Despite these inconsistent findings, it has often been suggested that men and women apply different strategies while solving mental rotation tasks (Butler et al., 2006; Gootjes et al., 2008a, b; Hugdahl et al., 2006; Jordan et al., 2002). It was hypothesized that women performed mental rotation tasks in an effortful, "top-down" way, whereas men used an automatic, "bottom-up" strategy. This corresponds to the reported negative relation of time limits when solving rotation tasks observed in females (Voyer, 2011). Top-down processing has been linked to activity in the prefrontal cortex that plays a role in decision making and spatial working memory. In contrast, bottom-up processing tends to be associated with less prefrontal involvement, less frontal control of task performance, and a greater involvement of primary sensory brain regions (Butler et al., 2006). A similar distinction was proposed by Hugdahl et al. (2006), who associated the frontal lobe activity of females with a serial reasoning strategy and activity in the parietal lobe with a "gestalt" perceptual strategy, as displayed by men. In contrast, Gootjes et al. (2008a, b) related male right hemispheric activity with a holistic strategy, and a more intense left hemispheric activity observed in females with an analytic strategy.

In a recent study conducted in our lab (Jaušovec and Jaušovec, 2012) we were able to show that the variable sex, when controlled for differences in ability levels and hormonal changes, had no influence on performance on a mental rotation task. In contrast, significant differences in brain activity patterns between males and females were observed for participants who displayed high mental rotation ability. In comparison to males, females displayed more theta synchronization. This difference was especially pronounced in frontal brain areas. Moreover, females displayed more intense interhemispheric coupling between left fronto-central and right parieto-occipital areas. Theta activity is relatively specific for control in WM (Klimesch et al., 2008; Sauseng et al., 2010), thus the obtained differences in brain activity patterns suggest that females solved the rotation tasks via intense WM processing (theta synchronization). The frontal pattern of brain activity in females further points to more controlled processing that involves the central executive.

It can be concluded that research has uncovered a considerable amount of data pointing toward sex differences in brain structure, organization, and function in relation to general intelligence. Thus, it is difficult to believe that they are not reflected in test measures of general intelligence. Interestingly, researchers that investigate sex differences in brain structure either enounce that these differences do not exist on a behavioral level (e.g., de Vries and Forger, 2015; Dunst et al., 2014; Pietschnig et al., 2015), or make no explicit statements about the matter. Hence we are faced with a paradox: differences in the origin of intelligent behavior, the brain, are not visible in the psychometric model that represents it—*g*.

REFERENCES

Aboitiz, F., Scheibel, A., Fisher, R., Zaidel E, 1992. Fiber composition of the human corpus callosum. Brain Res. 598, 143–153.

Ackerman, P.L., 2006. Cognitive sex differences and mathematics and science achievement. Am. Psychol. 61 (7), 722–723.

Ackerman, P.L., Beier, M.E , Boyle, M.O., 2005. Working memory and intelligence: the same or different constructs? Psychol. Bull. 131 (1), 30–60.

Akam, T., Kullmann, D.M., 2010. Oscillations and filtering networks support flexible routing of information. Neuron 67 (2), 308–320.

Allen, J.S., Damasio, H., Grabowski, T.J., Bruss, J., Zhang, W., 2003. Sexual dimorphism and asymmetries in the gray–white composition of the human cerebrum. NeuroImage 18 (4), 880–894.

Allen, L.S., Richey, M.F., Chai, Y.M., Gorski, R.A., 1991. Sex differences in the corpus callosum of the living human being. J. Neurosci. 11, 933–942.

Amin, H.U., Malik, A.S., Kamel, N., Chooi, W.-T., Hussain, M., 2015. P300 correlates with learning memory abilities and fluid intelligence. J. NeuroEng. Rehabil. 12 (1), 87.

Anderson, B., Harvey, T., 1996. Alterations in cortical thickness and neuronal density in the frontal cortex of Albert Einstein. Neurosci. Lett. 210, 161–164.

Andreasen, N.C., O'Leary, D.S., Arndt, S., Cizadlo, T., Rezai, K., Watkins, G.L., Boles Ponto, L.L., Hichwa, R.D., 1995. PET studies of memory: novel and practiced free recall of complex narratives. NeuroImage 2, 284–295.

Angelakis, E., Lubar, J.F., Stathopoulou, S., 2004a. Electroencephalographic peak alpha frequency correlates of cognitive traits. Neurosci. Lett. 371 (1), 60–63.

Angelakis, E., Lubar, J.F., Stathopoulou, S., Kounios, J., 2004b. Peak alpha frequency: an electroencephalographic measure of cognitive preparedness. Clin. Neurophysiol. 115 (4), 887–897.

Ankney, C.D., 1992. Sex differences in relative brain size: the mismeasure of woman, too? Intelligence 16, 329–336.

Annese, J., Schenker-Ahmed, N.M., Bartsch, H., Maechler, P., Sheh, C., Thomas, N., et al., 2014. Postmortem examination of patient H.M. 's brain based on histological sectioning and digital 3D reconstruction. Nat. Commun. 5, 3122.

Anokhin, A.P., Lutzenberger, W., Birbaumer, N., 1999. Spatiotemporal organization of brain dynamics and intelligence: an EEG study in adolescents. Int. J. Psychophysiol. 33, 259–273.

Anokhin, A.P., Müller, V., Lindenberger, U., Heath, A.C., Myers, E., 2006. Genetic influences on dynamic complexity of brain oscillations. Neurosci. Lett. 397, 93–98.

Anokhin, A.P., Vogel, F., 1996. EEG alpha rhythm frequency and intelligence in normal adults. Intelligence 23, 1–14.

Aurlien, H., Gjerde, I.O., Aarseth, J.H., Eldoen, G., Karlsen, B., Skeidsvoll, H., Gilhus, N.E., 2004. EEG background activity described by a large computerized database. Clin. Neurophysiol. 115, 665–673.

Axmacher, N., Henseler, M.M., Jensen, O., Weinreich, I., Elger, C.E., Fell, J., 2010. Cross-frequency coupling supports multi-item working memory in the human hippocampus. Proc. Natl. Acad. Sci. USA 107, 3228–3233.

Baddeley, A.D., 1986. Working Memory. Clarendon Press; Oxford University Press, Oxford; New York.

Baddeley, A., 2012. Working memory: theories, models, and controversies. Annu. Rev. Psychol. 63 (1), 1–29.

Baddeley, A.D., Allen, R.J., Hitch, G.J., 2011. Binding in visual working memory: the role of the episodic buffer. Neuropsychologia 49 (6), 1393–1400.

Baddeley, A.D., Hitch, G.J., 1974. Working memory. In: Bower, G.A. (Ed.), The Psychology of Learning and Motivation: Advances in Research and Theory. Academic Press, New York, NY, pp. 47–89.

Barrett, P.T., Eysenck, H.J., 1992. Brain evoked potentials and intelligence: the Hendrikson paradigm. Intelligence 16, 361–381.

Barrett, P.T., Eysenck, H.J., 1994. The relationship between evoked potential component amplitude, latency, contour length, variability, zero-crossings and psychometric intelligence. Pers. Individ. Diff. 16, 3–32.

Basten, U., Hilger, K., Fiebach, C.J., 2015. Where smart brains are different: a quantitative meta-analysis of functional and structural brain imaging studies on intelligence. Intelligence 51, 10–27, https://doi.org/10.1016/j.intell.2015.04.009.

Basten, U., Stelzel, C., Fiebach, C.J., 2013. Intelligence is differentially related to neural effort in the task-positive and the task-negative brain network. Intelligence 41 (5), 517–528.

Bates, T., Eysenck, H.J., 1993. String length attention and intelligence: focussed attention reverses the string length-IQ relationship. Pers. Individ. Diff. 15, 363–371.

Bates, T., Stough, C., Mangan, G., Pellett, O., 1995. Intelligence and complexity of the averaged evoked potential: an attentional theory. Intelligence 20, 27–39.

Batt, R., Nettelbeck, T., Cooper, C.J., 1999. Event related potential correlates of intelligence. Pers. Individ. Diff. 27, 639–658.

Bell, E.C., Willson, M.C., Wilman, A.H., Dave, S., Silverstone, P.H., 2006. Males and females differ in brain activation during cognitive tasks. NeuroImage 30 (2), 529–538.

Belluscio, M.A., Mizuseki, K., Schmidt, R., Kempter, R., Buzsaki, G., 2012. Cross-frequency phase-phase coupling between theta and gamma oscillations in the hippocampus. J. Neurosci. 32 (2), 423–435.

Benninger, C., Matthis, P., Scheffner, D., 1984. EEG development of healthy boys and girls. Results of a longitudinal study. Electroencephalogr. Clin. Neurophysiol. 57, 1–12.

Blinkhorn, S.F., Hendrickson, D.E., 1982. Averaged evoked responses and psychometric intelligence. Nature 295, 596–597.

Bloom, J.S., Hynd, G.W., 2005. The role of the corpus callosum in interhemispheric transfer of information: excitation or inhibition? Neuropsychol. Rev. 15 (2), 59–71.

Boring, E.G., 1923. Intelligence as the tests test it. New Rep. 35, 35–37.

Botvinick, M.M., Niv, Y., Barto, A.C., 2009. Hierarchically organized behavior and its neural foundations: a reinforcement learning perspective. Cognition 113 (3), 262–280.

Boutros, N.N., Arfken, C., Galderisi, S., Warrick, J., Pratt, G., Iacono, W., 2008. The status of spectral EEG abnormality as a diagnostic test for schizophrenia. Schizophr. Res. 99, 225–237.

Brancucci, A., 2012. Neural correlates of cognitive ability. J. Neurosci. Res. 90 (7), 1299–1309, https://doi.org/10.1002/jnr.23045.

Bressler, S.L., Menon, V., 2010. Large-scale brain networks in cognition: emerging methods and principles. Trends Cogn. Sci. 14 (6), 277–290.

Brouwer, R.M., Hedman, A.M., van Haren, N.E.M., Schnack, H.G., Brans, R.G.H., Smit, D.J.A., et al., 2014. Heritability of brain volume change and its relation to intelligence. NeuroImage 100, 676–683.

Buehner, M., Krumm, S., Pick, M., 2005. Reasoning = working memory≠attention. Intelligence 33 (3), 251–272.

Burgaleta, M., Johnson, W., Waber, D.P., Colom, R., Karama, S., 2014a. Cognitive ability changes and dynamics of cortical thickness development in healthy children and adolescents. NeuroImage 84, 810–819.

Burgaleta, M., MacDonald, P.A., Martínez, K., Román, F.J., Álvarez-Linera, J., González, A.R., et al., 2014b. Subcortical regional morphology correlates with fluid and spatial intelligence: basal ganglia and cognitive abilities. Hum. Brain Map. 35 (5), 1957–1968.

Burns, N.R., Nettelbeck, T., Cooper, C.J., 1996. The string measure of the event-related potential IQ and inspection time. Pers. Individ. Diff. 21, 563–572.

Burns, N.R., Nettelbeck, T., Cooper, C.J., 1997. The string measure of the ERP: what does it measure? Int. J. Psychophysiol. 27, 43–53.

Butler, T., Imperato-McGinley, J., Pan, H., Voyer, D., Cordero, J., Zhu, Y.-S., et al., 2006. Sex differences in mental rotation: top–down versus bottom–up processing. NeuroImage 32 (1), 445–456.

Caprara, V., Barbaranello, C., Borgogni, L., Bucik, V., Boben, D., 2002. Model velikih pet, Cenetr za psihodiagnostična sredstva. Ljubljana.

Caryl, P.G., 1994. Early event-related potentials correlate with inspection time and intelligence. Intelligence 18, 15–46.

Chalke, F., Ertl, J., 1965. Evoked potentials and intelligence. Life Sci. 4, 1319–1322.

Champod, A.S., Petrides, M., 2010. Dissociation within the frontoparietal network in verbal working memory: a parametric functional magnetic resonance imaging study. J. Neurosci. 30 (10), 3849–3856.

Charlot, V., Tzourio, N., Zilbovicius, M., Mazoyer, B., Denis, M., 1992. Different mental imagery abilities result in different regional cerebral blood flow activation patterns during cognitive tasks. Neuropsychologia 30, 565–580.

Chein, J.M., Fiez, J.A., 2010. Evaluating models of working memory through the effects of concurrent irrelevant information. J. Exp. Psychol. Gen. 139 (1), 117–137.

Chiang, M.-C., Barysheva, M., Shattuck, D.W., Lee, A.D., Madsen, S.K., Avedissian, C., et al., 2009. Genetics of brain fiber architecture and intellectual performance. J. Neurosci. 29 (7), 2212–2224.

Chiang, A.K.I., Rennie, C.J., Robinson, P.A., van Albada, S.J., Kerr, C.C., 2011. Age trends and sex differences of alpha rhythms including split alpha peaks. Clin. Neurophysiol. 122 (8), 1505–1517.

Christensen, H., Anstey, K.J., Parslow, R.A., Maller, J., Mackinnon, A., Sachdev, P., 2007. The brain reserve hypothesis, brain atrophy and aging. Gerontology 53 (2), 82–95.

Chuderski, A., 2015. The broad factor of working memory is virtually isomorphic to fluid intelligence tested under time pressure. Pers. Individ. Dif. 85, 98–104.

Clark, C.R., Veltmeyer, M.D., Hamilton, R.J., Simms, E., Paul, R., Hermens, D., et al., 2004. Spontaneous alpha peak frequency predicts working memory performance across the age span. Int. J. Psychophysiol. 53, 1–9.

Clarke, A.R., Barry, R.J., McCarthy, R., Selikowitz, M., 2001. Age and sex effects in the EEG: development of the normal child. Clin. Neurophysiol. 112, 806–814.

Clayden, J.D., Jentschke, S., Munoz, M., Cooper, J.M., Chadwick, M.J., Banks, T., et al., 2012. Normative development of white matter tracts: similarities and differences in relation to age, gender, and intelligence. Cereb. Cortex 22 (8), 1738–1747.

Cohn, N.B., Kircher, J., Emmerson, R.Y., Dustman, R.E., 1985. Pattern reversal evoked potentials: age, sex and hemispheric asymmetry. Electroencephalogr. Clin. Neurophysiol. 62 (6), 399–405.

Colgin, L.L., Denninger, T., Fyhn, M., Hafting, T., Bonnevie, T., Jensen, O., et al., 2009. Frequency of gamma oscillations routes flow of information in the hippocampus. Nature 462 (7271), 353–357.

Collette, F., Van der Linden, M., 2002. Brain imaging of the central executive component of working memory. Neurosci. Biobehav. Rev. 26 (2), 105–125.

Colom, R., Abad, F.J., Quiroga, M.Á., Shih, P.C., Flores-Mendoza, C., 2008. Working memory and intelligence are highly related constructs, but why? Intelligence 36 (6), 584–606.

Colom, R., Jung, R.E., Haier, R.J., 2006. Distributed brain sites for the g-factor of intelligence. NeuroImage 31 (3), 1359–1365, https://doi.org/10.1016/j.neuroimage.2006.01.006.

Colom, R., Haier, R.J., Head, K., Álvarez-Linera, J., Quiroga, M.Á., Shih, P.C., Jung, R.E., 2009. Gray matter correlates of fluid, crystallized, and spatial intelligence: testing the P-FIT model. Intelligence 37 (2), 124–135, https://doi.org/10.1016/j.intell.2008.07.007.

Colom, R., Román, F.J., Abad, F.J., Shih, P.C., Privado, J., Froufe, M., et al., 2013. Adaptive n-back training does not improve fluid intelligence at the construct level: gains on individual tests suggest that training may enhance visuospatial processing. Intelligence 41 (5), 712–727.

Colombo, J.A., Reisin, H.D., Miguel-Hidalgo, J.J., Rajkowska, G., 2006. Cerebral cortex astroglia and the brain of a genius: a propos of A. Einstein's. Brain Res. Rev. 52 (2), 257–263.

Corkin, S., 2002. What's new with the amnesic patient HM? Nat. Rev. Neurosci. 3 (2), 153–160.

Cowan, N., 2000. The magical number 4 in short-term memory: a reconsideration of mental storage capacity. Behav. Brain Sci. 24 (1), 87–114.

Cowan, N., Elliott, E.M., Scott Saults, J., Morey, C.C., Mattox, S., Hismjatullina, A., Conway, A.R.A., 2005. On the capacity of attention: Its estimation and its role in working memory and cognitive aptitudes. Cogn. Psychol. 51 (1), 42–100.

Cowan, N., Li, D., Moffitt, A., Becker, T.M., Martin, E.A., Saults, J.S., Christ, S.E., 2011. A neural region of abstract working memory. J. Cogn. Neurosci. 23 (10), 2852–2863.

D'Esposito, M., Postle, B.R., 2015. The cognitive neuroscience of working memory. Annu. Rev. Psychol. 66 (1), 115–142.

Darwin, C., 1871. The Descent of Man, and Selection in Relation to SexJohn Murray, London, UK.

De Pascalis, V., Varriale, V., Matteoli, A., 2008. Intelligence and P3 components of the event-related potential elicited during an auditory discrimination task with masking. Intelligence 36 (1), 35–47.

de Vries, G.J., Forger, N.G., 2015. Sex differences in the brain: a whole body perspective. Biol. Sex Dif. 6 (1), 15.

Deaner, R.O., Isler, K., Burkart, J., van Schaik, C., 2007. Overall brain size, and not encephalization quotient, best predicts cognitive ability across non-human primates. Brain Behav. Evol. 70 (2), 115–124.

Deary, I.J., 2000. Looking Down on Human Intelligence: From Psychometrics to the Brain. Oxford University Press, Oxford; New York.

Deary, I.J., 2012. Intelligence. Annu. Rev. Psychol. 63 (1), 453–482.

Deary, I.J., Der, G., Ford, G., 2001. Reaction times and intelligence differences: a population-based cohort study. Intelligence 29 (5), 389–399.

Deary, I.J., Whalley, L.J., Lemmon, H., Crawford, J.R., Starr, J.M., 2000. The stability of individual differences in mental ability from childhood to old age: follow-up of the 1932 Scottish Mental Survey. Intelligence 28, 49–55.

del Río, D., Cuesta, P., Bajo, R., García-Pacios, J., López-Higes, R., del-Pozo, F., Maestú, F., 2012. Efficiency at rest: magnetoencephalographic resting-state connectivity and individual differences in verbal working memory. Int. J. Psychophysiol. 86 (2), 160–167.

Diamond, M.C., Scheibel, A.B., Murphy, G.M., Harvey, T., 1985. On the brain of a scientist: Albert Einstein. Exp. Neurol. 88, 198–204.

Donchin, E., Coles, M.G.H., 1988. Is the P300 component a manifestation of context updating? Behav. Brain Sci. 11 (03), 357–374.

Doppelmayr, M., Klimesch, W., Hödlmoser, K., Sauseng, P., Gruber, W., 2005. Intelligence related upper alpha desynchronization in a semantic memory task. Brain Res. Bull. 66 (2), 171–177.

Duncan, J., et al., 2000. A neural basis for general intelligence. Science 289 (5478), 457–460.

Duncan, J., 2003. Intelligence tests predict brain response to demanding task events. Nat. Neurosci. 6, 207–208.

Duncan, J., 2005. Prefrontal cortex and Sperman's g. In: Duncan, J., McLeod, P., Phillips, L. (Eds.), Measuring the Mind: Speed, Control, and Age,. first ed. Oxford University Press, Oxford, New York, pp. 249–271.

Duncan, J., 2010. The multiple-demand (MD) system of the primate brain: mental programs for intelligent behaviour. Trends Cogn. Sci. 14 (4), 172–179.

Duncan, J., 2013. The structure of cognition: attentional episodes in mind and brain. Neuron 80 (1), 35–50.

Duncan, J., McLeod, P., Phillips, L. (Eds.), 2005. Measuring the Mind: Speed, Control, and Age. first ed. Oxford University Press, Oxford; New York.

Duncan, J., Owen, A.M., 2000. Common regions of the human frontal lobe recruited by diverse cognitive demands. Trends Neurosci. 23 (10), 475–483.

Dunst, B., Benedek, M., Bergner, S., Athenstaedt, U., Neubauer, A.C., 2013. Sex differences in neural efficiency: are they due to the stereotype threat effect? Pers. Individ. Dif. 55 (7), 744–749.

Dunst, B., Benedek, M., Koschutnig, K., Jauk, E., Neubauer, A.C., 2014. Sex differences in the IQ-white matter microstructure relationship: a DTI study. Brain Cogn. 91, 71–78.

Elliott, H.C., 1969. Textbook of Neuroanatomy. Lippincott, Philadelphia, PA.

Engle, R.W., Tuholski, S.W., Laughlin, J.E., Conway, A.R.A., 1999. Working memory, shortterm memory and general fluid intelligence: a latent variable approach. J. Exp. Psychol. Gen. 128, 309–331.

Ertl, J.P., 1971. IQ, evoked responses and Fourier analysis. Nature 241, 209–210.

Falk, D., Lepore, F.E., Noe, A., 2013. The cerebral cortex of Albert Einstein: a description and preliminary analysis of unpublished photographs. Brain 136 (4), 1304–1327.

Felleman, D.J., Van Essen, D.C., 1991. Distributed hierarchical processing in the primate cerebral cortex. Cereb. Cortex 1 (1), 1–47.

Fernández, A., Hornero, R., Mayo, A., Poza, J., Gil-Gregorio, P., Ortiz, T., 2006. EEG spectral profile in Alzheimer's disease and mild cognitive impairment. Clin. Neurophysiol. 117, 306–314.

Fink, A., Grabner, R.H., Benedek, M., Reishofer, G., Hauswirth, V., Fally, M., et al., 2009. The creative brain: investigation of brain activity during creative problem solving by means of EEG and FMRI. Hum. Brain Map. 30 (3), 734–748.

Flores-Mendoza, C., Widaman, K.F., Rindermann, H., Primi, R., Mansur-Alves, M., Pena, C.C., 2013. Cognitive sex differences in reasoning tasks: evidence from Brazilian samples of educational settings. Intelligence 41 (1), 70–84.

Fries, P., 2005. A mechanism for cognitive dynamics: neuronal communication through neuronal coherence. Trends Cogn. Sci. 9 (10), 474–480.

Fries, P., 2009. Neuronal gamma-band synchronization as a fundamental process in cortical computation. Annu. Rev. Neurosci. 32 (1), 209–224.

Friese, U., Köster, M., Hassler, U., Martens, U., Trujillo-Barreto, N., Gruber, T., 2013. Successful memory encoding is associated with increased cross-frequency coupling between frontal theta and posterior gamma oscillations in human scalp-recorded EEG. NeuroImage 66, 642–647.

Galderisi, S., Mucci, A., Volpe, U., Boutros, N., 2009. Evidence-based medicine and electrophysiology in schizophrenia. Clin. EEG Neurosci. 40, 62–77.

Ganjavi, H., Lewis, J.D., Bellec, P., MacDonald, P.A., Waber, D.P., Evans, A.C., et al., 2011. Negative associations between corpus callosum midsagittal area and IQ in a representative sample of healthy children and adolescents. PLoS One 6 (5), e19698.

Garai, J.E., Scheinfeld, A., 1968. Sex differences in mental and behavioral traits. Genet. Psychol. Monogr. 77, 169–299.

Garcés, P., Vicente, R., Wibral, M., Pineda-Pardo, J.Á., López, M.E., Aurtenetxe, S., et al., 2013. Brain-wide slowing of spontaneous alpha rhythms in mild cognitive impairment. Front. Aging Neurosci. 5, 100.

Garlick, D., 2002. Understanding the nature of the general factor of intelligence: the role of individual differences in neural plasticity as an explanatory mechanism. Psychol. Rev. 109 (1), 116–136.

Gevins, A., Smith, M.E., 2000. Neurophysiological measures of working memory and individual differences in cognitive ability and cognitive style. Cereb. Cortex 10 (9), 829–839.

Gilbert, D.G., Johnson, S., Gilbert, B., McColloch, M.A., 1991. Event-related potential correlates of IQ. Pers. Individ. Dif. 12, 1183–1184.

Gmehlin, D., Thomas, C., Weisbrod, M., Walther, S., Pfüller, U., Resch, F., Oelkers-Ax, R., 2011. Individual analysis of EEG background-activity within school age: impact of age and sex within a longitudinal data set. Int. J. Dev. Neurosci. 29 (2), 163–170.

Gong, G., He, Y., Evans, A.C., 2011. Brain connectivity: gender makes a difference. Neuroscientist 17 (5), 575–591.

Gong, G., Rosa-Neto, P., Carbonell, F., Chen, Z.J., He, Y., Evans, A.C., 2009. Age- and gender-related differences in the cortical anatomical network. J. Neurosci. 29 (50), 15684–15693.

Gootjes, L., Bruggeling, E.C., Magnee, T., Van Strien, J.W., 2008a. Sex differences in functional connectivity during mental rotation: an EEG study. Int. J. Psychophysiol. 69 (3), 228.

Gootjes, L., Bruggeling, E.C., Magnée, T., Van Strien, J.W., 2008b. Sex differences in the latency of the late event-related potential mental rotation effect. NeuroReport 19 (3), 349–353.

Grandy, T.H., Werkle-Bergner, M., Chicherio, C., Lövdén, M., Schmiedek, F., Lindenberger, U., 2013. Individual alpha peak frequency is related to latent factors of general cognitive abilities. NeuroImage 79, 10–18.

Gray, J.R., Thompson, P.M., 2004. Neurobiology of intelligence: science and ethics. Nat. Rev. Neurosci. 5 (6), 471–482.

Grazioplene, R.G., G. Ryman, S., Gray, J.R., Rustichini, A., Jung, R.E., DeYoung, C.G., 2015. Subcortical intelligence: caudate volume predicts IQ in healthy adults: caudate volume and intelligence. Hum. Brain Map. 36 (4), 1407–1416.

Gur, R.C., Alsop, D., Glahn, D., Petty, R., Swanson, C.L., Maldjian, J.A., et al., 2000. An fMRI study of sex differences in regional activation to a verbal and a spatial task. Brain Lang. 74 (2), 157–170.

Haász, J., Westlye, E.T., Fjær, S., Espeseth, T., Lundervold, A., Lundervold, A.J., 2013. General fluid-type intelligence is related to indices of white matter structure in middle-aged and old adults. NeuroImage 83, 372–383.

Hahn, N., Jansen, P., Heil, M., 2010. Preschoolers' mental rotation: sex differences in hemispheric asymmetry. J. Cogn. Neurosci. 22 (6), 1244–1250.

Haier, R.J., 1993. Cerebral glucose metabolism and intelligence. In: Vernon, P.A. (Ed.), Biological Approaches to the Study of Human Intelligence. Ablex, Norwood, NJ.

Haier, R.J., Karama, S., Colom, R., Jung, R., Johnson, W., 2014. A comment on "Fractionating Intelligence" and the peer review process. Intelligence 46, 323–332.

Haier, R.J., Karama, S., Leyba, L., Jung, R.E., 2009. MRI assessment of cortical thickness and functional activity changes in adolescent girls following three months of practice on a visual-spatial task. BMC Res. Notes 2 (1), 174.

Haier, R.J., Neuchterlein, K.H., Hazlett, E., Wu, J.C., Paek, J., Browning, H.L., Buchsbaum, M.S., 1988. Cortical glucose metabolic rate correlates of abstract reasoning and attention studied with positron emission tomography. Intelligence 12, 199–217.

Haier, R.J., Robinson, D.L., Braden, B., Williams, D., 1983. Electrical potentials of the cerebral cortex and psychometric intelligence. Pers. Individ. Dif. 4, 591–599.

Haier, R.J., Siegel, B., Tang, C., Abel, L., Buchsbaum, M.S., 1992. Intelligence and changes in regional cerebral glucose mezabolic rate following learning. Intelligence 16, 415–426.

Halari, R., Sharma, T., Hines, M., Andrew, C., Simmons, A., Kumari, V., 2005. Comparable fMRI activity with differential behavioural performance on mental rotation and overt verbal fluency tasks in healthy men and women. Exp. Brain Res. 169 (1), 1–14.

Halpern, D.F., 2004. A Cognitive-process taxonomy for sex differences in cognitive abilities. Curr. Dir. Psychol. Sci. 13 (4), 135–139.

Hampshire, A., Highfield, R.R., Parkin, B.L., Owen, A.M., 2012. Fractionating human intelligence. Neuron 76 (6), 1225–1237.

Hampshire, A., Owen, A.M., 2014. RE: Comment about "Fractionating Human Intelligence". Non-existent flaws in the original article and their relation to limitations of the P-FIT model. Intelligence 46, 333–340.

Hampshire, A., Parkin, B., Highfield, R., Owen, A.M., 2014. Brief response to Ashton and colleagues regarding Fractionating Human Intelligence. Pers. Individ. Dif. 60, 16–17.

Harmony, T., Marosi, E., Diaz de Leon, A.E., Becker, J., Fernandez, T., 1990. Effect of sex, psychosocial disadvantages and biological risk factors on EEG maturation. Electroencephalogr. Clin. Neurophysiol. 75, 482–491.

Hedman, A.M., van Haren, N.E.M., Schnack, H.G., Kahn, R.S., Hulshoff Pol, H.E., 2012. Human brain changes across the life span: a review of 56 longitudinal magnetic resonance imaging studies. Hum. Brain Map. 33 (8), 1987–2002.

Hendrickson, D.E., Hendrickson, A.E., 1980. The biological basis for individual differences in intelligence. Pers. Individ. Dif. 1, 3–33.

Herlitz, A., Nilsson, L.G., Bäckman, L., 1997. Gender differences in episodic memory. Mem. Cogn. 25 (6), 801–811.

Hick, W.E., 1952. On the rate of gain of information. Quart. J. Exp. Psychol. 4, 11–26.

Hillyard, S., Münte, T., 1984. Selective attention to color and location: an analysis with event-related brain potentials. Percept. Psychophys. 36, 185–198.

Hillyard, S.A., Luck, S.J., Mangun, G.R., 1994. The cuing of attention to visual field locations: analysis with ERP recordings. In: Heinze, H.J., Münte, T.F., Mangun, G.R. (Eds.), Cognitive Electrophysiology. Birkhäuser, Boston, MA, pp. 1–25.

Hindriks, R., van Putten, M.J.A.M., 2013. Thalamo-cortical mechanisms underlying changes in amplitude and frequency of human alpha oscillations. NeuroImage 70, 150–163.

Hines, T., 2014. Neuromythology of Einstein's brain. Brain Cogn. 88, 21–25.

Hooper, G.S., 2005. Comparison of the distributions of classical and adaptivelyaligned EEG power spectra. Int. J. Psychophysiol. 55, 179–189.

Huang, Y.-Z., Sommer, M., Thickbroom, G., Hamada, M., Pascual-Leonne, A., Paulus, W., et al., 2009. Consensus: new methodologies for brain stimulation. Brain Stimul. 2 (1), 2–13.

Hugdahl, K., Thomsen, T., Ersland, L., 2006. Sex differences in visuo-spatial processing: an fMRI study of mental rotation. Neuropsychologia 44 (9), 1575–1583.

Hutchinson, A.D., Mathias, J.L., Jacobson, B.L., Ruzic, L., Bond, A.N., Banich, M.T., 2009. Relationship between intelligence and the size and composition of the corpus callosum. Exp. Brain Res. 192 (3), 455–464.

Hyde, J.S., 1981. How large are cognitive gender differences? A meta-analysis using ω^2 and D. Am. Psychol. 36 (8), 892–901.

Ingalhalikar, M., Smith, A., Parker, D., Satterthwaite, T.D., Elliott, M.A., Ruparel, K., et al., 2014. Sex differences in the structural connectome of the human brain. Proc. Natl. Acad. Sci. USA 111 (2), 823–828.

Irwing, P., 2012. Sex differences in g: an analysis of the US standardization sample of the WAIS-III. Perso. Individ. Dif. 53 (2), 126–131.

Isaacson, W., 2007. Einstein: His Life and Universe. Simon & Schuster, New York, NY.

Jann, K., Federspiel, A., Giezendanner, S., Andreotti, J., Kottlow, M., Dierks, T., Koenig, T., 2012. Linking brain connectivity across different time scales with electroencephalogram, functional magnetic resonance imaging, and diffusion tensor imaging. Brain Connect. 2 (1), 11–20.

Jann, K., Koenig, T., Dierks, T., Boesch, C., Federspiel, A., 2010. Association of individual resting state EEG alpha frequency and cerebral blood flow. NeuroImage 51 (1), 365–372.

Jaušovec, N., 1996. Differences in EEG alpha activity related to giftedness. Intelligence 3, 159–173.

Jaušovec, N., 1997. Differences in EEG alpha activity between gifted and non-identified individuals: Insights into problem solving. Gifted Child Quart. 1, 26–32.

Jaušovec, N., 1998. Are gifted individuals less chaotic thinkers? Pers. Individ. Dif. 25, 253–267.

Jaušovec, N., 2000. Differences in cognitive processes between gifted, intelligent, creative and average individuals while solving complex problems: an EEG study. Intelligence 28, 213–237.

Jaušovec, N., 2012. Sex differences in event-related potential components during the solution of complex mental rotation tasks. NeuroReport 23 (6), 360–363.

Jaušovec, N., Jaušovec, K., 2000. Correlations between ERP parameters and intelligence: a reconsideration. Biol. Psychol. 55 (2), 137–154.

Jaušovec, N., Jaušovec, K., 2004. Differences in induced brain activity during the performance of learning and working-memory tasks related to intelligence. Brain Cogn. 54 (1), 65–74.

Jaušovec, N., Jaušovec, K., 2008. Spatial rotation and recognizing emotions: gender related differences in brain activity. Intelligence 36 (5), 383–393.

Jaušovec, N., Jaušovec, K., 2009. Gender related differences in visual and auditory processing of verbal and figural tasks. Brain Res. 1300, 135–145.

Jaušovec, N., Jaušovec, K., 2010. Resting brain activity: differences between genders. Neuropsychologia 48 (13), 3918–3925.

Jaušovec, N., Jaušovec, K., 2012. Sex differences in mental rotation and cortical activation patterns: Can training change them? Intelligence 40 (2), 151–162, https://doi.org/10.1016/j.intell.2012.01.005.

Jensen, A.R., 1998. The *g* Factor. The Science of Mental Ability. Praeger, Westport, CT.

Jensen, A.R., 2006. Clocking the Mind: Mental Chronometry and Individual Differences, first ed. Elsevier, Amsterdam; Boston; London.

Jensen, A.R., 2011. The theory of intelligence and its measurement. Intelligence 39 (4), 171–177.

Jensen, O., Mazaheri, A., 2010. Shaping functional architecture by oscillatory alpha activity: gating by inhibition. Front. Hum. Neurosci. 4, 186.

Johnson, Jr., R., 1995. Event-related potentials insights into the neurobiology of memory systems. In: Johnson, Jr., R., Baron, J.C. (Eds.), Handbook of Neuropsychology. Elsevier, Amsterdam, The Netherlands, pp. 135–164.

Johnson, W., Bouchard, Jr., T., 2005. The structure of human intelligence: It is verbal, perceptual, and image rotation (VPR), not fluid and crystallized. Intelligence 33 (4), 393–416.

Johnson, W., Bouchard, T.J., 2007a. Sex differences in mental abilities: *g* masks the dimensions on which they lie. Intelligence 35 (1), 23–39.

Johnson, W., Bouchard, T.J., 2007b. Sex differences in mental ability: a proposed means to link them to brain structure and function. Intelligence 35 (3), 197–209.

Johnson, W., Jung, R.E., Colom, R., Haier, R.J., 2008. Cognitive abilities independent of IQ correlate with regional brain structure. Intelligence 36 (1), 18–28.

Jones, D.K., Leemans, A., 2011. Diffusion tensor imaging. In: Modo, M., Bulte, J.W.M. (Eds.), Magnetic Resonance Neuroimaging Methods and Protocols. Springer, London, UK, pp. 127–144.

Jonides, J., Lewis, R.L., Nee, D.E., Lustig, C.A., Berman, M.G., Moore, K.S., 2008. The mind and brain of short-term memory. Annu. Rev. Psychol. 59, 193–224.

Jordan, K., Wüstenberg, T., Heinze, H.J., Peters, M., Jäncke, L., 2002. Women and men exhibit different cortical activation patterns during mental rotation tasks. Neuropsychologia 40 (13), 2397–2408.

Jung, R.E., Haier, R.J., 2007. The Parieto-Frontal Integration Theory (P-FIT) of intelligence: converging neuroimaging evidence. Behav. Brain Sci. 30 (02), 135.

Kamiński, J., Brzezicka, A., Wróbel, A., 2011. Short-term memory capacity (7 ± 2) predicted by theta to gamma cycle length ratio. Neurobiol. Learn. Mem. 95 (1), 19–23.

Karama, S., Ad-Dab'bagh, Y., Haier, R.J., Deary, I.J., Lyttelton, O.C., Lepage, C., Evans, A.C., 2009. Positive association between cognitive ability and cortical thickness in a representative US sample of healthy 6 to 18 year-olds. Intelligence 37 (2), 145–155.

Klimesch, W., 1999. EEG alpha and theta oscillations reflect cognitive and memory performance: a review and analysis. Brain Res. Brain Res. Rev. 29 (2–3), 169–195.

Klimesch, W., 2012. Alpha-band oscillations, attention, and controlled access to stored information. Trends Cogn. Sci. 16 (12), 606–617.

Klimesch, W., Doppelmayr, M., Pachinger, T., Russegger, H., 1997. Event-related desynchronization in the alpha band and the processing of semantic information. Cogn. Brain Res. 6, 83–94.

Klimesch, W., Freunberger, R., Sauseng, P., Gruber, W., 2008. A short review of slow phase synchronization and memory: evidence for control processes in different memory systems? Brain Res. 1235, 31–44.

Klimesch, W., Schimke, H., Doppelmayr, M., Ripper, B., Schwaiger, J., Pfurtscheller, G., 1996. Event-related desynchronization (ERD) and the Dm-effect: does alpha desynchronization during encoding predict later recall performance? Int. J. Psychophysiol. 24, 47–60.

Klimesch, W., Sauseng, P., Hanslmayr, S., 2007. EEG alpha oscillations: the inhibition–timing hypothesis. Brain Res. Rev. 53 (1), 63–88.

Klimesch, W., Schimke, H., Pfurtscheller, G., 1993. Alpha frequency, cognitive load and memory performance. Brain Topogr. 5 (3), 241–251.

Klimesch, W., Schack, B., Schabus, M., Doppelmayr, M., Gruber, W., Sauseng, P., 2004. Phase-locked alpha and theta oscillations generate the P1-N1 complex and are related to memory performance. Cogn. Brain Res. 19 (3), 302–316.

Knaus, T.A., Bollich, A.M., Corey, D.M., Lemen, L.C., Foundas, A.L., 2006. Variability in perisylvian brain anatomy in healthy adults. Brain Lang. 9, 219–232.

Kok, A., 2001. On the utility of P3 amplitude as a measure of processing capacity. Psychophysiology 38 (3), 557–577.

Kolb, B., Whishaw, I.Q., 2009. Fundamentals of Human Neuropsychology, sixth ed. Worth Publishers, New York, NY.

Kondacs, A., Szabó, M., 1999. Long-term intra-individual variability of the background EEG in normals. Clin. Neurophysiol. 110, 1708–1716.

Langer, N., Pedroni, A., Gianotti, L.R.R., Hänggi, J., Knoch, D., Jäncke, L., 2012. Functional brain network efficiency predicts intelligence. Hum. Brain Map. 33 (6), 1393–1406.

Lansbergen, M.M., Arns, M., van Dongen-Boomsma, M., Spronk, D., Buitelaar, J.K., 2011. The increase in theta/beta ratio on resting-state EEG in boys with attention-deficit/hyperactivity disorder is mediated by slow alpha peak frequency. Progr. NeuroPsychopharmacol. Biol. Psychiatry 35 (1), 47–52.

Lebedev, A.N., 1994. The neurophysiological parameters of human memory. Neurosci. Behav. Physiol. 24 (3), 254–259.

Lee, K.H., Choi, Y.Y., Gray, J.R., Cho, S.H., Chae, J.-H., Lee, S., Kim, K., 2006. Neural correlates of superior intelligence: Stronger recruitment of posterior parietal cortex. NeuroImage 29, 578–586.

Leonard, C.M., Towler, S., Welcome, S., Halderman, L.K., Otto, R., Eckert, M.A., Chiarello, C., 2008. Size matters: cerebral volume influences sex differences in neuroanatomy. Cereb. Cortex 18 (12), 2920–2931.

Li, J., Yu, C., Li, Y., Liu, B., Liu, Y., Shu, N., et al., 2009. COMT val158met modulates association between brain white matter architecture and IQ. Am. J. Med. Genet. B Neuropsych. Genet. 150B (3), 375–380.

Lipp, I., Benedek, M., Fink, A., Koschutnig, K., Reishofer, G., Bergner, S., et al., 2012. Investigating neural efficiency in the visuo-spatial domain: an fMRI study. PLoS One 7 (12), e51316.

Lippa, R.A., Collaer, M.L., Peters, M., 2010. Sex differences in mental rotation and line angle judgments are positively associated with gender equality and economic development across 53 nations. Arch. Sex. Behav. 39 (4), 990–997.

Lisman, J.E., Jensen, O., 2013. The theta–gamma neural code. Neuron 77 (6), 1002–1016.

Lopes da Silva, F.H., Vos, J.E., Mooibroek, J., Van Rotterdam, A., 1980. Relative contributions of intracortical and thalamo-cortical processes in the generation of alpha rhythms, revealed by partial coherence analysis. Electroencephalogr. Clin. Neurophysiol. 50, 449–456.

Luck, S.J., Girelli, M., 1999. Electrophysiological approaches to the study of selective attention in the human brain. In: Parasuraman, R. (Ed.), The Attentive Brain. MIT-Press, Cambridge, MA, pp. 71–94.

Luders, E., Gaser, C., Narr, K.L., Toga, A.W., 2009. Why Sex Matters: brain size independent differences in gray matter distributions between men and women. J. Neurosci. 29 (45), 14265–14270.

Luders, E., Narr, K.L., Bilder, R.M., Szeszko, P.R., Gurbani, M.N., Hamilton, L., et al., 2008. Mapping the relationship between cortical convolution and intelligence: effects of gender. Cereb. Cortex 18 (9), 2019–2026.

Luders, E., Narr, K.L., Bilder, R.M., Thompson, P.M., Szeszko, P.R., Hamilton, L., Toga, A.W., 2007. Positive correlations between corpus callosum thickness and intelligence. NeuroImage 37 (4), 1457–1464.

Luders, E., Thompson, P.M., Narr, K.L., Zamanyan, A., Chou, Y.-Y., Gutman, B., et al., 2011. The link between callosal thickness and intelligence in healthy children and adolescents. NeuroImage 54 (3), 1823–1830.

Luria, A.R., 1968. The Mind of a Mnemonist. Basic Books, New York, NY.

Luria, A.R., 1973. The Working Brain. Penguin, Harmondsworth, England.

Lutzenberger, W., Birbaumer, N., Flor, H., Rockstroh, B., Elbert, T., 1992a. Dimensional analysis of the human EEG and intelligence. Neurosci. Lett. 143, 10–14.

Lutzenberger, W., Elbert, T., Birbaumer, N., Ray, W.J., 1992b. The scalp distribution of the fractal dimension of the EEG and its variation with mental tasks. Brain Topologr. 5, 27–34.

Lynn, R., 1999. Sex differences in intelligence and brain size: a developmental theory. Intelligence 27 (1), 1–12.

Mackintosh, N.J., Bennett, E.S., 2005. What do Raven's Matrices measure? An analysis in terms of sex differences. Intelligence 33 (6), 663–674.

Maitland, S.B., Herlitz, A., Nyberg, L., Bäckman, L., Nilsson, L.-G., 2004. Selective sex differences in declarative memory. Mem. Cogn. 32 (7), 1160–1169.

Majerus, S., D'Argembeau, A., Martinez Perez, T., Belayachi, S., Van der Linden, M., Collette, F., et al., 2010. The commonality of neural networks for verbal and visual short-term memory. J. Cogn. Neurosci. 22 (11), 2570–2593.

Majerus, S., Poncelet, M., Van der Linden, M., Albouy, G., Salmon, E., Sterpenich, V., et al., 2006. The left intraparietal sulcus and verbal short-term memory: focus of attention or serial order? NeuroImage 32 (2), 880–891.

Marshall, P.J., Bar-Haim, Y., Fox, N.A., 2002. Development of the EEG from 5 months to 4 years of age. Clin. Neurophysiol. 113, 1199–1208.

Martín-Loeches, M., Bruner, E., de la Cuétara, J.M., Colom, R., 2013. Correlation between corpus callosum shape and cognitive performance in healthy young adults. Brain Struct. Funct. 218 (3), 721–731.

Martínez, K., Burgaleta, M., Román, F.J., Escorial, S., Shih, P.C., Quiroga, M.Á., Colom, R., 2011. Can fluid intelligence be reduced to "simple" short-term storage? Intelligence 39 (6), 473–480.

Masters, M.S., Sanders, B., 1993. Is the gender difference in mental rotation disappearing? Behav. Genet. 23 (4), 337–341

Mayer, J.D., Caruso, D.R., Salovey, P., 1999. Emotional intelligence meets traditional standards for an intelligence. Intelligence 27 (4), 267–298.

Mayer, J.D., Roberts, R.D., Barsade, S.G., 2008. Human abilities: emotional intelligence. Annu. Rev. Psychol. 59 (1), 507–536.

Mayer, J.D., Salovey, P., Caruso, D.R., 2002. Mayer-Salovey-Caruso emotional intelligence test (MSCEIT). MHS, Toronto.

Mazaheri, A., Jensen, O., 2010. Rhythmic pulsing: linking ongoing brain activity with evoked responses. Front. Hum. Neurosci. 4, 177.

Men, W., Falk, D., Sun, T., Chen, W., Li, J., Yin, D., et al., 2014. The corpus callosum of Albert Einstein's brain: another clue to his high intelligence? Brain 137 (4), e268.

Miller, D.I., Halpern, D.F., 2014. The new science of cognitive sex differences. Trends Cogn. Sci. 18 (1), 37–45.

Miller, E.M., 1994. Intelligence and brain myelination—a hypothesis. Pers. Individ. Dif. 17, 803–832.

Miller, G.A., 1956. The magical number seven, plus or minus two: some limits on our capacity for processing information. Psychol. Rev. 63 (no. 2), 81–97.

Mölle, M., Marshall, L., Wolf, B., Fehm, H., Born, J., 1999. EEG complexity and performance measures of creative thinking. Psychophysiology 36, 95–104.

Moreno, M.B., Concha, L., González-Santos, L., Ortiz, J.J., Barrios, F.A., 2014. Correlation between corpus callosum sub-segmental area and cognitive processes in school-age children. PLoS One 9 (8), e104549.

Moretti, D., 2004. Individual analysis of EEG frequency and band power in mild Alzheimer's disease. Clin. Neurophysiol. 115 (2), 299–308.

Mori, S., Zhang, J., 2006. Principles of diffusion tensor imaging and its applications to basic neuroscience research. Neuron 51 (5), 527–539.

Mundy-Castle, A.C., Nelson, G.K., 1960. Intelligence, personality and brain rhythms in a socially isolated community. Nature 185, 484–485.

Narr, K.L., Woods, R.P., Thompson, P.M., Szeszko, P., Robinson, D., Dimtcheva, T., et al., 2007. Relationships between IQ and regional cortical gray matter thickness in healthy adults. Cereb. Cortex 17 (9), 2163–2171.

Navas-Sánchez, F.J., Alemán-Gómez, Y., Sánchez-Gonzalez, J., Guzmán-De-Villoria, J.A., Franco, C., Robles, O., et al., 2014. White matter microstructure correlates of mathematical giftedness and intelligence quotient: white matter microstructure. Hum. Brain Map. 35 (6), 2619–2631.

Neubauer, A.C., Fink, A., 2003. Fluid intelligence and neural efficiency: effects of task complexity and sex. Pers. Individ. Dif. 35 (4), 811–827.

Neubauer, A.C., Fink, A., 2009. Intelligence and neural efficiency. Neurosci. Biobehav. Rev. 33 (7), 1004–1023.

Neubauer, A.C., Fink, A., Schrausser, D.G., 2002. Intelligence and neural efficiency: the influence of task content and sex on the brain–IQ relationship. Intelligence 30, 515–536.

Neubauer, A.C., Grabner, R.H., Fink, A., Neuper, C., 2005. Intelligence and neural efficiency: further evidence of the influence of task content and sex on the brain–IQ relationship. Cogn. Brain Res. 25 (1), 217–225.

Newell, A., 1990. Unified Theories of Cognition. Harvard University Press, Cambridge, MA.

Nikulin, V.V., Brismar, T., 2005. Long-range temporal correlations in electroencephalographic oscillations: relation to topography, frequency band, age and gender. Neuroscience 130, 549–558.

Nunez, P.L., 1995. Mind, brain, and electroencephalography. In: Nunez, P.L. (Ed.), Neocortical Dynamics and Human EEG Rhythms. Oxford University Press, New York, NY, pp. 133–194.

Nunez, P.L., Wingeier, B.M., Silberstein, R.B., 2001. Spatial–temporal structures of human alpha rhythms: theory, microcurrent sources, multiscale measurements, and global binding of local networks. Hum. Brain Map. 13, 125–164.

Nussbaumer, D., Grabner, R.H., Stern, E., 2015. Neural efficiency in working memory tasks: the impact of task demand. Intelligence 50, 196–208.

Nyborg, H., 1994. The neuropsychology of sex-related differences in brain and specific abilities: hormones, developmental dynamics and new paradigm. In: Vernon, P.A. (Ed.), The Neuropsychology of Individual Differences. Academic Press Inc, London, UK, pp. 59–113.

Nyborg, H., 2005. Sex-related differences in general intelligence *g*, brain size, and social status. Person. Individ. Dif. 39 (3), 497–509.

Nyborg, H., 2015. Sex differences across different racial ability levels: theories of origin and societal consequences. Intelligence 52, 44–62.

Oberauer, K., Süβ, H.-M., Wilhelm, O., Wittmann, W.W., 2008. Which working memory functions predict intelligence? Intelligence 36 (6), 641–652.

Ogden, J.A., 2005. Fractured Minds: A Case-Study Approach to Clinical Neuropsychology, second ed. Oxford University Press, New York, NY.

Olson, I.R., Berryhill, M., 2009. Some surprising findings on the involvement of the parietal lobe in human memory. Neurobiol. Learn. Mem. 91 (2), 155–165.

Osipova, D., Takashima, A., Oostenveld, R., Fernández, G., Maris, E., Jensen, O., 2006. Theta and gamma oscillations predict encoding and retrieval of declarative memory. J. Neurosci. 26 (28), 7523–7531.

Pahor, A., Jaušovec, N., 2014. Theta–gamma cross-frequency coupling relates to the level of human intelligence. Intelligence 46, 283–290.

Pahor, A., Jaušovec, N., 2016. Making Brains run Faster: are they Becoming Smarter? The Spanish journal of psychology, 19.

Pakkenberg, B., Gundersen, H.J., 1997. Neocortical neuron number in humans: effect of sex and age. J. Comp. Neurol. 384 (2), 312–320.

Palva, J.M., Monto, S., Kulashekhar, S., Palva, S., 2010. Neuronal synchrony reveals working memory networks and predicts individual memory capacity. Proc. Natl. Acad. Sci. 107 (16), 7580–7585.

Petrides, M., 2000. Mapping prefrontal cortical systems for the control of cognition. In: Toga, A.W., Mazziotta, J.C. (Eds.), Brain Mapping: The Systems. Academic Press, San Diego, CA, pp. 159–176.

Petsche, H., 1996. Approaches to verbal, visual and musical creativity by EEG coherence analysis. Int. J. Psychophysiol. 24, 145–159.

Pfurtscheller, G., 1999. Quantification of ERD and ERS in the time domain. Handbook of Electroencephalography and Clinical Neuropsychology, vol. 6, Elsevier, New York; Amsterdam, pp. 89–105.

Pfurtscheller, G., Aranibar, A., 1977. Event-related cortical desynchronization detected by power measurements of scalp EEG. Electroencephalogr. Clin. Neurophysiol. 42, 817–826.

Pietschnig, J., Penke, L., Wicherts, J.M., Zeiler, M., Voracek, M., 2015. Meta-analysis of associations between human brain volume and intelligence differences: how strong are they and what do they mean? Neurosci. Biobehav. Rev. 57, 411–432.

Pincus, S., 1994. Approximate entropy (ApEn) as a complexity measure. Chaos 5, 110–117.

Pincus, M.S., 1995. Quantifying complexity and regularity of neurobiological systems. Meth. Neurosci. 28, 336–363.

Pineda-Pardo, J.A., Martínez, K., Román, F.J., Colom, R., 2016. Structural efficiency within a parieto-frontal network and cognitive differences. Intelligence 54, 105–116.

Polich, J., 2007. Updating P300: an integrative theory of P3a and P3b. Clin. Neurophysiol. 118 (10), 2128–2148.

Posner, M.I., Petersen, S.E., 1990. The attention system of the human brain. Annu. Rev. Neurosci. 13, 25–42.

Posthuma, D., Neale, M.C., Boomsma, D.I., 2001. Are smarter brains running faster? Heritability of alpha peak frequency, IQ, and their interrelation. Behav. Genet. 31 (6), 567–579.

Raichle, M.E., 2006. The brain's dark energy. Science 314 (5803), 1249.

Raichle, M.E., MacLeod, A.M., et al., 2001. A default mode of brain function. Proc. Natl. Acad. Sci. USA 98, 676–682.

Rhein, C., Mühle, C., Richter-Schmidinger, T., Alexopoulos, P., Doerfler, A., Kornhuber, J., 2014. Neuroanatomical correlates of intelligence in healthy young adults: the role of basal ganglia volume. PLoS One 9 (4), e93623.

Ritchie, S.J., Bastin, M.E., Tucker-Drob, E.M., Maniega, S.M., Engelhardt, L.E., Cox, S.R., et al., 2015b. Coupled changes in brain white matter microstructure and fluid intelligence in later life. J. Neurosci. 35 (22), 8672–8682.

Ritchie, S.J., Booth, T., Valdés Hernández, M., del, C., Corley, J., Maniega, S.M., Gow, A.J., et al., 2015a. Beyond a bigger brain: multivariable structural brain imaging and intelligence. Intelligence 51, 47–56.

Robinson, D.L., 1993. The EEG and intelligence: an appraisal of methods and theories. Pers. Individ. Dif. 15, 695–716.

Román, F.J., Abad, F.J., Escorial, S., Burgaleta, M., Martínez, K., Álvarez-Linera, J., et al., 2014. Reversed hierarchy in the brain for general and specific cognitive abilities: a morphometric analysis: reversed hierarchy in the brain. Hum. Brain Map. 35 (8), 3805–3818.

Roux, F., Uhlhaas, P.J., 2014. Working memory and neural oscillations: alpha–gamma versus theta–gamma codes for distinct WM information? Trends Cogn. Sci. 18 (1), 16–25.

Ruff, C.C., Knauff, M., Fangmeier, T., Spreer, J., 2003. Reasoning and working memory: common and distinct neural processes. Neuropsychologia 41, 1241–1253.

Rugg, M.D. (Ed.), 2002. Electrophysiology of Mind: Event-Related Brain Potentials and Cognition (Reprinted). Oxford University Press, Oxford.

Ruigrok, A.N.V., Salimi-Khorshidi, G., Lai, M.-C., Baron-Cohen, S., Lombardo, M.V., Tait, R.J., Suckling, J., 2014. A meta-analysis of sex differences in human brain structure. Neurosci. Biobehav. Rev. 39, 34–50.

Rushton, J.P., Ankney, C.D., 2009. Whole brain size and general mental ability: a review. Int. J. Neurosci. 119 (5), 692–732.

Rypma, B., Berger, J.S., Prabhakaran, V., Bly, B.M., Kimberg, D.Y., Biswal, B.B., D'Esposito, M., 2006. Neural correlates of cognitive efficiency. NeuroImage 33, 969–979.

Salinas, E., Sejnowski, T.J., 2001. Correlated neuronal activity and the flow of neural information. Nat. Rev. Neurosci. 2 (8), 539–550.

Salinsky, M.C., Oken, B.S., Morehead, L., 1991. Test-retest reliability in EEG frequency analysis. Electroencephalogr. Clin. Neurophysiol. 79, 382–392.

Samson-Dollfus, D., Delapierre, G., Do Marcolino, C., Blondeau, C., 1997. Normal and pathological changes in alpha rhythms. Int. J. Psychophysiol. 26 (1–3), 395–409.

Sandman, C.A., Barron, J.L., 1986. Parameters of the event-related potential are related to functioning in the mentally retarded. Int. J. Neurosci. 29, 37–44.

Sandman, C.A., Head, K., Muftuler, L.T., Su, L., Buss, C., Davis, E.P., 2014. Shape of the basal ganglia in preadolescent children is associated with cognitive performance. NeuroImage 99, 93–102.

Santarnecchi, E., Tatti, E., Rossi, S., Serino, V., Rossi, A., 2015. Intelligence-related differences in the asymmetry of spontaneous cerebral activity: intelligence and brain functional asymmetry. Hum. Brain Map. 36 (9), 3586–3602.

Sarnthein, J., 2005. Increased EEG power and slowed dominant frequency in patients with neurogenic pain. Brain 129 (1), 55–64.

Sauseng, P., Griesmayr, B., Freunberger, R., Klimesch, W., 2010. Control mechanisms in working memory: a possible function of EEG theta oscillations. Neurosci. Biobehav. Revi. 34 (7), 1015–1022.

Savage-McGlynn, E., 2012. Sex differences in intelligence in younger and older participants of the Raven's Standard Progressive Matrices Plus. Pers. Individ. Dif. 53 (2), 137–141.

Schnack, H.G., van Haren, N.E.M., Brouwer, R.M., Evans, A., Durston, S., Boomsma, D.I., et al., 2015. Changes in thickness and surface area of the human cortex and their relationship with intelligence. Cereb. Cortex 25 (6), 1608–1617.

Schubert, A.-L., Hagemann, D., Voss, A., Schankin, A., Bergmann, K., 2015. Decomposing the relationship between mental speed and mental abilities. Intelligence 51, 28–46.

Schumacher, E.H., Lauber, E., Awh, E., Jonides, J., Smith, E.E., Koeppe, R.A., 1996. PET evidence for an amodal verbal working memory system. NeuroImage 3 (2), 79–88.

Schweizer, K., Moosbrugger, H., 2004. Attention and working memory as predictors of intelligence. Intelligence 32 (4), 329–347.

Sederberg, P.B., Kahana, M.J., Howard, M.W., Donner, E.J., Madsen, J.R., 2003. Theta and gamma oscillations during encoding predict subsequent recall. J. Neurosci. 23 (34), 10809–10814.

Seeley, W.W., Menon, V., Schatzberg, A.F., Keller, J., Glover, G.H., Kenna, H., et al., 2007. Dissociable intrinsic connectivity networks for salience processing and executive control. J. Neurosci. 27 (9), 2349–2356.

Shagass, C., Roemer, R.A., Straumanis, J., Josiassen, R.C., 1981. Intelligence as a factor in evoked potential studies of psychopathology: comparison of low and high IQ subjects. Biol. Psychiatry 16, 1007–1030.

Shaw, J.C., 2004. The Brain's Alpha Rhythms and the Mind. Elsevier Science, Amsterdam, The Netherlands.

Shaw, P., Greenstein, D., Lerch, J., Clasen, L., Lenroot, R., Gogtay, N., et al., 2006. Intellectual ability and cortical development in children and adolescents. Nature 440 (7084), 676–679.

Sheppard, L.D., Vernon, P.A., 2008. Intelligence and speed of information-processing: a review of 50 years of research. Pers. Individ. Dif. 44 (3), 535–551.

Shibasaki, H., 2008. Human brain mapping: hemodynamic response and electrophysiology. Clin. Neurophysiol. 119 (4), 731–743.

Shucard, D.W., Horn, J.L., 1972. Evoked cortical potentials and measurement of human abilities. J. Comp. Physiol. Psychol. 78, 59–68.

Singer, W., 2013. Cortical dynamics revisited. Trends Cogn. Sci. 17 (12), 616–626.

Smit, C.M., Wright, M.J., Hansell, N.K., Geffen, G.M., Martin, N.G., 2006. Genetic variation of individual alpha frequency (IAF) and alpha power in a large adolescent twin sample. Int. J. Psychophysiol. 61 (2), 235–243.

Spearman, C., 1927. The Abilities of Man. Macmillan, London, UK.

Spitzka, E.A., 1907. A study of the brains of six eminent scientists and scholars belonging to the American Anthropometric Society, together with a description of the skull of Professor E. D. Cope. Trans. Am. Philos. Soc. 21 (4), 175.

Spreng, R.N., Sepulcre, J., Turner, G.R., Stevens, W.D., Schacter, D.L., 2013. Intrinsic architecture underlying the relations among the default, dorsal attention, and frontoparietal control networks of the human brain. J. Cogn. Neurosci. 25 (1), 74–86.

Sreenivasan, K.K., Curtis, C E., D'Esposito, M., 2014. Revisiting the role of persistent neural activity during working memory. Trends Cogn. Sci. 18 (2), 82–89.

Sridharan, D., Levitin, D.J., Menon, V., 2008. A critical role for the right fronto-insular cortex in switching between central-executive and default-mode networks. Proc. Natl. Acad. Sci. USA 105 (34), 12569–12574.

Stam, K.J., Tavy, D.L.J., Jelles, B., Achtereekte, H.A.M., Slaets, J.P.J., Keunen, R.W.M., 1994. Non-linear dynamical analysis of multichannel EEG: clinical applications in dementia and Parkinson's disease. Brain Topogr. 7, 141–150.

Steele, C.M., 1997. A threat in the air: how stereotypes shape intellectual identity and performance. Am. Psychol. 52 (6), 613–629.

Stoet, G., Geary, D.C., 2013. Sex differences in mathematics and reading achievement are inversely related: within- and across-nation assessment of 10 years of PISA data. PLoS One 8 (3), e57988.

Stough, C.K.K., Nettelbeck, T., Cooper, C.J., 1990. Evoked brain potentials, string length and intelligence. Pers. Individ. Dif. 11, 401–406.

Tamnes, C.K., Østby, Y., Walhovd, K.B., Westlye, L.T., Due-Tønnessen, P., Fjell, A.M., 2010. Intellectual abilities and white matter microstructure in development: a diffusion tensor imaging study. Hum. Brain Map. 31 (10), 1609–1625.

Terman, L.M., 1916. The Measurement of Intelligence. Houghton Mifflin Company, Boston, MA.

Terman, L.M., Merrill, M.A., 1937. Measuring Intelligence. Houghton Mifflin, Boston, MA.

Troche, S.J., Houlihan, M.E., Stelmack, R.M., Rammsayer, T.H., 2009. Mental ability, P300, and mismatch negativity: analysis of frequency and duration discrimination. Intelligence 37 (4), 365–373.

Valdés-Hernández, P.A., Ojeda-González, A., Martínez-Montes, E., Lage-Castellanos, A., Virués-Alba, T., Valdés-Urrutia, L., Valdes-Sosa, P.A., 2010. White matter architecture rather than cortical surface area correlates with the EEG alpha rhythm. NeuroImage 49 (3), 2328–2339.

van den Heuvel, M.P., Sporns, O., 2013. Network hubs in the human brain. Trends Cogn. Sci. 17 (12), 683–696.

van der Sluis, S., Derom, C., Thiery, E., Bartels, M., Polderman, T.J., Verhulst, F.C., et al., 2008. Sex differences on the WISC-R in Belgium and the Netherlands. Intelligence 36 (1), 48–67.

Varghese, L., Nampoori, V.P.N., Pratap, R., 1987. Nonlinear analysis of an electroencephalogram (EEG) during epileptic seizure. Curr. Sci. 20, 1039–1041.

Verleger, R., 1988. Event-related potentials and cognition: a critique of the context updating hypothesis and an alternative interpretation of P3. Behav. Brain Sci. 11, 343–356.

Vernon, P.A., Wickett, J.C., Bazana, P.C., Stelmack, R.M., 2000. The neuropsychology and psychophysiology of human intelligence. In: Sternberg, R.J. (Ed.), Handbook of Intelligence. Cambridge University Press, Cambridge, UK, pp. 245–264.

Vogel, F., Kruger, J., Schalt, E., Schnobel, R., Hassling, L., 1987. No consistent relationship between oscillations and latencies of visual and auditory evoked EEG potentials and measures of mental performance. Hum. Neurobiol. 6, 173–182.

Vogel, J.J., Bowers, C.A., Vogel, D.S., 2003. Cerebral lateralization of spatial abilities: a meta-analysis. Brain Cogn. 52 (2), 197–204.

Vogel, W., Broverman, D.M., 1964. Relationship between EEG and test intelligence: a critical review. Psychol. Bull. 62, 132–144.

Voyer, D., 2011. Time limits and gender differences on paper-and-pencil tests of mental rotation: a meta-analysis. Psychonom Bull. Rev. 18 (2), 267–277.

Voyer, D., Voyer, S., Bryden, M.P., 1995. Magnitude of sex differences in spatial abilities: a meta-analysis and consideration of critical variables. Psychol. Bull. 117 (2), 250–270.

Waber, D.P., Forbes, P.W., Almli, C.R., Blood, E.A., The Brain Development Cooperative Group, 2012. Four-year longitudinal performance of a population-based sample of healthy children on a neuropsychological battery: the NIH MRI study of normal brain development. J. Int. Neuropsychol. Soc. 18 (02), 179–190.

Widman, K.F., Carlson, J.S., Saetermoe, C.L., Galbraith, G.C., 1993. The relationship of auditory evoked potentials to fluid and crystallized intelligence. Pers. Individ. Dif. 15, 205–217.

Witelson, S.F., Kigar, D.L., Harvey, T., 1999. The exceptional brain of Albert Einstein. Lancet 353 (9170), 2149–2153.

Wolf, D., Fischer, F.U., Fesenbeckh, J., Yakushev, I., Lelieveld, I.M., Scheurich, A., et al., 2014. Structural integrity of the corpus callosum predicts long-term transfer of fluid intelligence-related training gains in normal aging: callosal integrity predicts training transfer. Hum. Brain Map. 35 (1), 309–318.

Wronka, E., Kaiser, J., Coenen, A.M.L., 2013. Psychometric intelligence and P3 of the event-related potentials studied with a 3-stimulus auditory oddball task. Neurosci. Lett. 535, 110–115.

Xu, Y., Chun, M.M., 2006. Dissociable neural mechanisms supporting visual short-term memory for objects. Nature 440 (7080), 91–95.

Yan, C., Gong, G., Wang, J., Wang, D., Liu, D., Zhu, C., et al., 2011. Sex- and brain size-related small-world structural cortical networks in young adults: a DTI tractography study. Cereb. Cortex 21 (2), 449–458.

Zanto, T.P., Gazzaley, A., 2013. Fronto-parietal network: flexible hub of cognitive control. Trends Cogn. Sci. 17 (12), 602–603.

Zappasodi, F., Pasqualetti, P., Tombini, M., Ercolani, M., Pizzella, V., Rossini, P.M., Tecchio, F., 2006. Hand cortical representation at rest and during activation: gender and age effects in the two hemispheres. Clin. Neurophysiol. 117 (7), 1518–1528.

Chapter 3

Raising Intelligence by Means of Behavioral Training

The first wave of research that attempted to train intelligence was triggered by the studies of Klingberg et al. (2002) and Jaeggi et al. (2008). This might be disputed, but as revealed in Chapter 1, most of the investigations that could be classified as interventions to enhance cognitive performance in the 20th century and earlier were designed for educational school settings, and aimed at a socially and intellectually deprived population of children being at risk to be identified as cultural-familial mentally retarded (e.g., the Abecedarian project, the Milwaukee Project, or the Head Start program, to name just the ones conducted in the United States). The training program developed by Klingberg et al. (2002),[a] although targeting working memory capacity and, therefore, classified as a core training or domain general approach (e.g., Morrison and Chein, 2011), was designed for children with special needs diagnosed with attention deficit hyperactivity disorder (ADHD). Since this chapter will mainly discuss approaches aiming to raise IQ in normal young adults, the first study that can be classified as such is the one by Jaeggi et al. (2008), which involved 70 healthy, young individuals with a mean age of 25.6 years. A second important feature is that the intervention had a far transfer effect on fluid intelligence. Or, to satisfy the critiques, provided provisional evidence of increased performance on tests of fluid intelligence. The Jaeggi study deserves the glory of being the first, not just because of the year of publication, but also due to several coincidental circumstances. First, the strong claim made in the title of the paper declaring that working memory (WM) training can increase fluid intelligence was accompanied by a positive opinion by one of the most known experts in the field of intelligence research, Robert Sternberg (2008) suggesting that after all, it is possible to increase fluid intelligence. Second, the rather weak methodology used in the experiment offered a lot of possibilities for attack from skeptics (e.g., Moody, 2009), and by that initiating a bulk of studies that tried to replicate the findings.

---

a. A training protocol that combines numerous tasks with diverse stimulus types targeting multiple components of the WM system, which became part of the Cogmed computerized WM training http://www.cogmed.com/ (e.g., Hulme and Melby-Lervåg, 2012; Klingberg et al., 2005; Shipstead et al., 2012a).

Increasing Intelligence. http://dx.doi.org/10.1016/B978-0-12-809407-5.00003-8

This reminds me of the highly controversial Mozart effect described by Rauscher et al. (1993), which at the end of the last century provoked a Mozart research mania. The effect describes the phenomenon that after 10 min of listening to Mozart's sonata (K. 448), students scored 8–9 points higher on a Stanford–Binet spatial subtest than students who had listened to a relaxation tape or listened to nothing. The IQ effects did not persist beyond the 10–15 min testing session (for a review see Pietschnig et al., 2010). The similarity between Rauscher's and Jaeggi's study is not just on the surface, that is, the immense media and scientific interest the findings and claims provoked, but also on a theoretical level. Both approaches explained the observed influence on ability with an overlap of the intervention (e.g., WM processes, or the specific structure of the Mozart's sonata) with components of the constructs they tried to influence—fluid intelligence or mathematical and visuospatial ability.

A second reason for a more detailed discussion of the Mozart effect are the suggestions put forward by critiques of Jaeggi's study, namely that the intervention used just one training task (Sternberg, 2008), and the opinion of some researchers that future training approaches should incorporate complexity, novelty, and diversity within training designs (e.g., Moreau and Conway, 2014). This could be achieved by combining cognitive demands with motor function, such as playing music (Moreno et al., 2011) or sports (Moreau and Conway, 2014). Even more challenging is the idea put forward by Martin et al. (2014): using transcranial direct current stimulation (tDCS) to enhance cognitive training, a proposal I will discuss in more detail in the next chapter.

3.1 THE MOZART EFFECT

To use music as a cognitive enhancer is not a new idea. For centuries, music has been used for healing and stimulating emotions. The Greeks at Asclepius placed an ill person in the center of the amphitheater, and used specific voices to heal that individual. Much later the beneficial influence of music on epileptiform activity in patients with seizures (Hughes et al., 1998; Hughes, 2002), as well as in individuals with hearing loss (Tomatis, 1996) and early childhood autism (Rimland and Edelson, 1995) was reported. It was further suggested that music can accelerate learning of foreign languages, reading and mathematics (Lozanov, 1978), retention of terminology (Panksepp, 1998), and creative ability (Adaman and Blaney, 1995; Jaušovec, 1994). Furthermore, it was suggested that neural changes accompanying musical training during childhood fundamentally alter the nervous system persisting into adulthood after auditory training has ceased (Skoe and Kraus, 2012; George and Coch, 2011). Moreover, Fedorenko et al. (2012) in a functional magnetic resonance imaging (fMRI) study showed that areas sensitive to musical structure across the brain were located bilaterally in the anterior and posterior superior temporal gyrus, and bilaterally in the premotor cortex, and the supplementary motor area. All of these areas have also been associated with intelligent behavior (Chapter 2).

The effects of specific musical pieces on human behavior were explained as a consequence of their impact on positive mood and arousal (Husain et al., 2002; Nantanis and Schellenberg, 1999; Panksepp and Bernatzky, 2002; Thompson et al., 2001). Listening to music affects arousal (degree of physiological activation), mood (long lasting emotions), and listeners' enjoyment, which in turn influence performance on a variety of cognitive tasks.

Rauscher et al. (1993, 1995) provided a different explanation for the beneficial influence of music on human performance. In their view, the link between listening to music and reasoning is subserved by similarities in neural activation, as specified by the Trion model of cortical organization (McGrann et al., 1994; Leng and Shaw, 1991; Shaw et al., 1985; Shenoy et al., 1993). Music acts as an exercise for exciting and priming the common repertoire and sequential flow of the cortical firing patterns responsible for higher brain functions. Leng and Shaw (1991) proposed that music is a "prelanguage" available at an early age, which can access these inherent firing patterns and enhance the cortex's ability to accomplish pattern development, thus improving other higher brain functions. Rauscher's explanation differs markedly from the neuroaffective explanations and was criticized because no evidence exists of cross-modal priming between unrelated stimuli (Husain et al., 2002). Studies have shown that visual events are not readily primed by preexposure to auditory events, even if both stimuli are related (Green et al., 2001). In a study conducted in our lab (Jaušovec and Habe, 2004), it was found that auditory background stimulation can influence visual brain activity, even if both stimuli are unrelated. Students who solved a simple visual task while listening to Mozart's music displayed more coherent brain activity (mainly in the gamma band), whereas a dissociation of brain areas in the gamma band was observed while respondents solved the same task in silence. Oscillations in the gamma band play a crucial role in music perception (Bhattacharya et al., 2001a,b; Bhattacharya and Petsche, 2001), and are associated with the binding hypothesis—the linking of separate nodes of activity in the cortex, thereby allowing the identification of the object as a whole (Singer and Gray, 1995; Tallon-Baudry and Bertrand, 1999). It can be assumed that listening to a certain type of music (e.g., Mozart) increases the coupling of specific brain areas and in that way facilitates the selection and "binding" of pertinent aspects of a sensory stimulus into a perceived whole. It can be further assumed that if such a pattern of activated brain areas coincides with the pattern needed for task completion, an increase in task performance could be the result. This hypothesis was also supported by a second study we carried out in our lab (Jaušovec and Habe, 2005). We showed that Mozart's music had a beneficial influence on respondents' performance of spatial rotation tasks, and a slightly negative influence on performance of numerical tasks. On the physiological level, a general effect of Mozart's music on brain activity in the induced gamma band was observed, accompanied by a more specific effect in the induced lower-2 alpha band, which was only present while respondents solved the numerical tasks.

Further support for Rauscher's priming theory is provided by several fMRI and positron emission tomorgraphy (PET) studies (for a review see Janata and Grafton, 2003; Patel, 2003) showing a considerable overlap of regions implicated in the perception–action cycle of music and areas involved in the perception, memorization, and production of abstract sequences as well as language and syntax. In addition, significant differences in activation by the Mozart sonata (in comparison to Beethoven's *Für Elise*) in the dorsolateral prefrontal cortex, occipital cortex, and cerebellum (all expected to be important for spatiotemporal reasoning) were reported in an fMRI study by Bodner et al. (2001).

One aspect, which in our opinion has been neglected in studying the Mozart effect, is the interaction between different learning stages and music. Changes in synapses resulting from the simultaneous (or near-simultaneous) activation of neurons are thought to underlie all changes in behavior due to experiences, including those that involve learning. Hebb's (1949) notion of the cell assembly was based on evidence suggesting that learning is a time-dependent process, thus it can be influenced at different stages. Priming refers to the facilitative effect of performing one task on the subsequent performance of the same or similar tasks; whereas consolidation refers to the posttraining period during which the hypothesized process of synaptic change occurs and transforms from a labile state into a more permanent one. Most of the studies investigating the Mozart effect (e.g., McKelvie and Low, 2002; Jaušovec and Habe, 2005; Rideout and Laubach, 1996; Rideout and Taylor, 1997; Rideout et al., 1998) used a procedure in which the same participants were exposed to the control and experimental conditions. These conditions were counterbalanced, therefore music—beside its priming effect—could also have had an influence on the consolidation of memory. Hence the main objective of a third study we conducted (Jaušovec et al., 2006) was to investigate the influence of Mozart's music on different phases of learning—represented by Hebb's recurrent activation phase (priming) and synaptic change (memory consolidation).

In short, the design of the study was as follows. In the first part of the study, 56 students were divided into 4 groups equalized with respect to gender, verbal, and performance intelligence (WAIS-R), emotional intelligence (MSCEIT, Mayer et al., 2002), and the personality factors of extraversion and neuroticism (BFQ, Caprara et al., 2002). In the second part of the experiment, 36 respondents were assigned to 3 groups equalized on the same variables as respondents in the first part of the study. The respondents in the control group relaxed before and after training, whereas the respondents in the experimental groups either listened to music before and after training, listened to music and subsequently relaxed, or vice versa. All groups of individuals were trained first how to solve different spatial rotation tasks (training session), and then asked to solve similar tasks (test session). The spatial rotation tasks were based on a well-established Slovene intelligence test- BTI (Mihelič, 1972), on the Spatial-Temporal Animation Reasoning tasks (STAR) designed by Peterson (2000), and on the PF&C subtest of the Stanford–Binet IQ test (Rideout and Laubach, 1996). The training

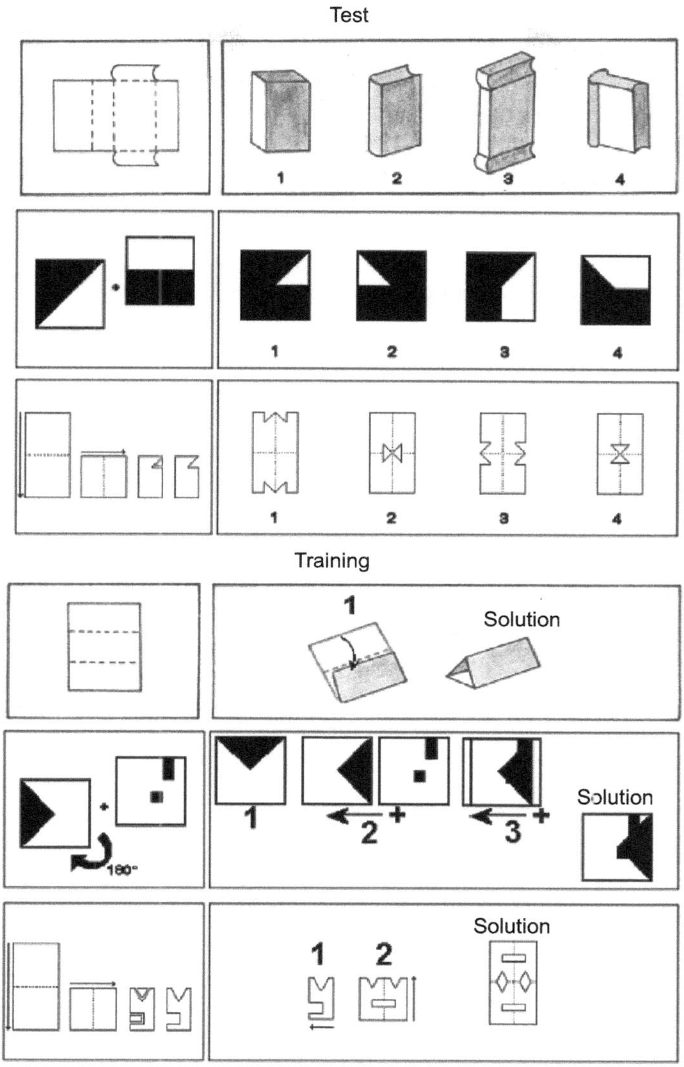

FIGURE 3.1 Examples of the spatial rotation tasks used in the test and training sessions: based on the Slovene intelligence test—BTI (Mihelič, 1972), on the Spatial-Temporal Animation Reasoning tasks (Peterson, 2000), and on the PF&C subtest of the Stanford–Binet IQ measure (Rideout and Laubach, 1996).

session consisted of 45 items which showed the solution principles in a stepwise manner (lower part of Fig. 3.1). The test also consisted of 45 items (upper part of Fig. 3.1). The respondents had to judge which of the four figures on the right corresponded to the figure in the left frame. All items were presented on a computer screen positioned in front of the respondent. The tasks were presented at fixed 11 s interstimulus intervals. They were exposed for 7 s following a 2 s interval, when a cross was presented. During this time the students were instructed to press a button (1–4) which indicated their answer.

Respondents of all groups received the same instruction. They were told that they would first learn how to solve different spatial tasks, which they would then solve after a short break. They were shown the three types of training and test items and provided with a brief explanation of the solution principles. The respondents of the control group were instructed to close their eyes and relax for approximately 8 min prior to and after the training session. The respondents of the other groups, the Mozart group and the Brahms group (second part of the experiment), on both occasions listened to music prior to and after the training session; or were exposed to a combination of either listening to music or relaxing. The music excerpt consisted of the first movement of Mozart's sonata for two pianos in D major (K. 448 for 8 min duration) and Brahms' Hungarian dance no. 5 (played continuously for 8 min). The music was presented through headphones while respondents sat in a reclining chair with their eyes closed. In order to become familiar with the testing procedure prior to the real test a short trial of five items was presented to the subjects.

The obtained results for the behavioral data supported the hypothesized beneficial influence of Mozart music on both learning phases. However, the results showed that all experimental groups outperformed the controls, hence the test of the priming/consolidation dichotomy was not positive. A possible reason for this could be that the tasks used were not sensitive enough to measure the potential differences. The same was true for the physiological measures used—no differences in EEG patterns between the control group and the two experimental groups with the combination of relaxation and listening to music pre- and posttraining were observed. Only respondents who listened to Mozart's sonata showed reliable physiological changes before and after training in relation to the behavioral data. One probable explanation (which was also confirmed in the second part of the study) could be that the prolonged exposure of respondents to music might have had a more marked and permanent influence on brain activity subserving spatiotemporal reasoning, which could be detected by the EEG methodology.

The displayed pattern of brain activity (lower approximate entropy; a quantification of regularity in time-series data, more alpha and gamma band synchronization) of the respondents who listened to Mozart's music before and after training is similar to findings reported in studies investigating neurophysiological differences in brain activity related to verbal and performance components of intelligence (Anokhin et al., 1999; Jaušovec, 1996, 1998; Jaušovec and Jaušovec, 2000; Neubauer and Fink, 2003; Neubauer et al., 1995, 1999, 2002; for more detail see Chapter 2) creativity (Mölle et al., 1999), emotional intelligence (Jaušovec and Jaušovec, 2005), as well as research comparing brain activity between musicians and nonmusicians while performing different tasks (e.g., listening to music or text, and mental rotation). While listening to music and during spatial imagination, musicians displayed higher levels of gamma band synchronicity than nonmusicians (Bhattacharya et al., 2001a,b).

Considering the functional relevance of the frequency bands for various cognitive processes, Doppelmayr et al. (2005) suggested that if the respective frequency band is relevant for task processing, a weaker activation indicates a more efficient processing, thus supporting the neural efficiency hypothesis. If on the other hand, the investigated frequency band is of no relevance for the respective task, a weaker activation indicates better inhibition of processes not necessary for the task. Thus, weaker event-related desynchronization for more intelligent subjects can either be interpreted in terms of more efficient processing (neural efficiency), or in terms of a task-specific inhibition of irrelevant processes (inhibition hypothesis).

In the light of this theoretical background it could be speculated that Mozart's music facilitated the activation of specific brain areas relevant for spatiotemporal reasoning. Based on the specific temporal and spatial distribution of brain activity displayed by individuals of the Mozart group in the narrow alpha and broad gamma bands could be therefore explained in line with the inhibition hypothesis. Respondents who listened to Mozart's music showed less lower-1 alpha band event-related desynchronization, which is mainly related to the level of alertness (Klimesch, 1999). This decreased alertness was accompanied by less upper-alpha desynchronization in the first 3 s of problem solving and an overall gamma synchronization. This activity was mainly located in the parietooccipital brain areas. Given that: (1) upper alpha band activity is modulated by semantic memory processes (Klimesch et al., 1997), (2) the gamma band is associated with the binding hypothesis (Tallon-Baudry and Bertrand, 1999), and (3) parietooccipital brain areas play a central role in spatial perception and imagery (Cabeza and Nyberg, 2000), one could suggest that the solution strategy of the Mozart group of respondents at the beginning of the solution process was more figural and less verbal-semantic, whereas at the end of the solution process, when respondents were prompted to provide an answer, their strategy became more semantic. In relation to the demands of the tasks used in the study, this strategy seems adequate.

With respect to alternative explanations of the influence of music on brain functioning, the results of our study supported Rauscher's et al. (1993) priming explanation. The EEG data provided no evidence pointing to an increased level of arousal (degree of physiological activation) in the Mozart group. By contrast, respondents in this group displayed a decreased level of alertness, nevertheless, they significantly outperformed the control group and the Brahms group. Overall, the data suggest that the increased performance in spatial reasoning was a result of the activation of specific task-relevant brain areas, as well as the inhibition of task irrelevant brain areas provoked by listening to Mozart's sonata K. 448. In music, spatial and temporal sequences of information must be unified thereby activating specific brain areas which might overlap with areas involved in other cognitive processes (Janata and Grafton, 2003; Patel, 2003).

A challenging question for further research would be: are there certain music pieces which are beneficial for specific cognitive processes? Music that would activate frontoparietal brain areas, which are assumed to be related to working memory processes and intelligence, would be of special interest for the study of working memory training-related changes in fluid intelligence.

This brief musical mind wandering has highlighted the strength of mixed training approaches. However, not all would agree that a multifaceted package for IQ training presents an advantage over the use of single task variants. For example, Morra and Borella (2015) anticipated a third urgently needed wave of research on WM training—one that would focus on clarifying the effects that such programs achieve. From a scientific standpoint, Morrison and Chein (2011) also favored the single task variant over what they labeled a "kitchen-sink" approach (e.g., COGITO or Cogmed) with a variety of tasks, stimuli, and engaged processes, that make it difficult to determine which components caused subsequent cognitive improvements. Putting the ecological and motivational aspects aside, it is questionable whether the complex relationship between working memory and intelligence can be clarified by such a mechanistic approach. For example, although the *n*-back task is one of the most often used training tasks, the cognitive mechanisms involved in performance on this task are not well understood, and its relationship to other complex WM tasks is also unclear (Shipstead et al., 2012a). Similarly unclear are the processes involved in the running memory span task (Pollack et al., 1959). Furthermore, the repetition of just one type of training task enables the respondents to develop specific strategies, which may not be related to the cognitive process they are supposed to target. An example of this was the respondent SF in the study by Chase and Ericsson (1982) who, after several months of training, was able to recall sequences of more than 80 digits. However, when the digits were presented at a faster rate, his scores returned to normal levels. The reason for SF's superior performance when the digits were presented at a slower pace was that he developed a mnemonic technique that allowed him to remember long number sequences over the time of training.

This leads to another problem: how to bring order to the many training packages that are currently available on the market or are being developed for specific research purposes? The following sections will provide some examples of behavioral training tasks and approaches that have shown that intelligence can be raised. The focus will be on programs developed for a young and normal population, with no ambition of a detailed metaanalysis. Several comprehensive review papers and metaanalyses have been published that highlight the topic from different viewpoints or focus on different groups of individuals: an educational perspective (Schwaighofer et al., 2015), *n*-back training only (Au et al., 2015; Bogg and Lasecki, 2015), changes in brain activity and structure (Buschkuehl et al., 2012; Taya et al., 2015), individuals with impaired memory performance (Weicker et al., 2016), individuals with intellectual disabilities (Danielsson et al., 2015), older individuals (Karbach and Verhaeghen, 2014; Lustig et al., 2009), children (Bryck and Fisher, 2012), and criticizing studies that claimed that WM training "worked" (Redick, 2015). An overview of these metaanalyses along with their main outcomes with respect to near/far transfer effects, methodological concerns, conclusions, opinions, and suggestions for further research will be presented, thereby providing a balance between generalization and detailed research. Furthermore, a special section will focus on computer games and packages following the motto: with fun to more intelligent brains.

3.2 TYPES OF TRAINING: THEORY, MEMORY, BRAIN, AND PERSONAL BIAS

Training can be defined as a process by which someone is taught the skills that are needed for an art, profession, or job. The synonyms are: drill, practice, routine, exercise, and workout [b]. Similarly broad are the criteria used to classify interventions for increasing intelligence or working memory performance. The classifications suggested by authors of the 20 metaanalyses and review papers included in the next section (Tables 3.1 and 3.4) make it clear that the most precise typology of interventions would be a description of each single task used in the training, which of course misses the main purpose of a typology, namely to form groups and provide order, system, and organization. Therefore, I will use a hierarchical approach, starting with a rough approximation which will become more fine grained and exact, hopefully in a stepwise manner. At first, this might seem like hairsplitting. However, classification criteria to a great extent determine the inclusion/exclusion of studies into a metaanalysis, which can have a major influence on the outcome of the analysis, and the suggestions and guidelines put forward with respect to the central question: is there evidence for a far transfer effect of training on intelligence? Either evidence or insinuation for a biased inclusion/exclusion of studies in metaanalyses—and by that influencing the proof that after all intelligence can be increased—has been reported in the recent papers by Au et al. (2016), Melby-Lervåg and Hulme (2016), and to some extent the minianalysis by Redick (2015).

A first approximation would be to distinguish between training that uses any kind of memory tasks (memory training), from training without memory tasks (other training). With a broad enough definition of memory involving also the sensory register, we would probably end up with an empty set in the other training group. Therefore, some examples for this type of training are needed: (1) the first group of approaches includes pharmacological interventions, brain stimulation using electrical or magnetic transcranial simulation (tDCS, TMS), cardiovascular training (sport, aerobics), music training, neuro-feedback, and meditation. (2) The second group represents approaches like video games, changing life style (participating in classes that teach new skills, such as quilting, photography; engagement in cognitively demanding activities like bridge playing; or participation in volunteer work), and repetition-priming (faster

b. Resource: http://www.merriam-webster.com/dictionary/training

and more accurate responses to stimuli after repeated exposures that does not require conscious memory). Similar classification criteria were suggested by Buschkuehl and Jaeggi (2010), Lampit et al. (2014), Lustig et al. (2009), and Morrison and Chein (2011) (Table 3.1). The first group of training approaches will be considered in more detail in Chapters 4 and 5.

A problem with the classification of the memory-training group, as stressed by Morrison and Chein (2011), is that the definition of WM influences the selection criteria of metaanalyses. For example, Schwaighofer et al. (2015) excluded the studies by Klingberg et al. (2002) and by Schmiedek et al. (2010) because episodic memory and inhibition tasks were used in their training. Nevertheless, most authors agree that memory training can be further divided into strategy and core training (e.g., Danielsson et al., 2015; Lustig et al., 2009; Morrison and Chein, 2011; von Bastian and Oberauer, 2013). On the other hand, von Bastian and Oberauer (2013) based their classification of training approaches on transfer mechanisms, enhancing either memory efficiency or memory capacity. Strategy training is usually defined as targeting task-specific strategies that might help trainees to improve memory performance, most often resembling different mnemonic techniques. In von Bastian and Oberauer's (2013) view, this transfer yields increased memory efficiency providing the possibility for acquiring knowledge and skills, such as strategy usage, chunk learning, and automation of basic processes and in that way decreasing the time needed to move the focus of attention between single items. Examples of such training approaches are elaborative encoding where subjects are instructed about more efficient methods of grouping items into chunks (St Clair-Thompson et al., 2010), or inventing a mental story with items to remember (McNamara and Scott, 2001), and using imagery to make items more salient (Carretti et al., 2007). In contrast, core training involves the repetition of demanding working memory tasks that target domain-general WM mechanisms (Morrison and Chein, 2011). It is assumed that these domains overlap with the same cognitive resources needed in complex problem solving and in that way yield to performance increases on intelligence tests—far transfer effects (von Bastian and Oberauer, 2013). In addition, it is proposed that repetitive cognitive demands that exceed resource limits induce changes in brain regions related to WM performance (Taya et al., 2015). Working memory and intelligence share a common capacity constraint, thus by increasing WM capacity, intelligence should also increase (Buschkuehl and Jaeggi, 2010).

Furthermore, characteristics that define core training and its success were also suggested. Buschkuehl and Jaeggi (2010) summarized the key points for successful core training as: (1) minimizing the possibility for the development of task-specific strategies, (2) the training must be adaptive to individual's varying levels of proficiency minimizing automation, and (3) complex tasks should be used to train several different processes at once. Morrison and Chein (2011) added further characteristics that should be fulfilled by core training tasks, such as maintenance accompanied by interference (e.g., complex memory span tasks), enforcing rapid working memory encoding and retrieval demands (e.g., *n*-back tasks), and including high cognitive work load (e.g., dual-*n*-back tasks).

Core training has been further classified with respect to the number of different tasks used and the processes these tasks are assumed to activate. For the former, von Bastian and Oberauer (2013) suggested the division into: (1) single-paradigm regimens where just one task type is trained (e.g., dual *n*-back); (2) multiparadigm regimens where more task types are used (WM tasks mixed with simple and complex memory span tasks), and (3) multifactorial regimens using heterogeneous tasks targeting multiple cognitive skills, such as WM and executive functions.

Classifications based on the processes involved in WM training are for several reasons far more difficult and complex. First, as already mentioned, they depend on memory theories adopted for the classification, second, for several tasks used in trainings (e.g., single/dual *n*-back, running memory span) the cognitive mechanisms involved are not well understood (Shipstead et al., 2012a), and third, as outlined in the WM section in Chapter 2, it is rather difficult to distinguish between different WM processes. The distinction made by authors was most often between simple and complex memory span tasks and working memory tasks (Morrison and Chein, 2011; Shipstead et al., 2010; Takeuchi et al., 2010; von Bastian and Oberauer, 2013; Weicker et al., 2016). The classifications based on processes markedly differed, and were sometimes inconsistent (e.g., the same training task was associated with different WM processes). The WM processes used for classification were: (1) attention related to task-switching and dual-tasking (e.g., Lampit et al., 2014; Lustig et al., 2009) and further divided into sustained, selective, and divided attention (Rapport et al., 2013); (2) memory updating in single/dual *n*-back tasks (Shipstead et al., 2010; Takeuchi et al., 2010; Weicker et al., 2016); (3) executive functioning (Buschkuehl and Jaeggi, 2010; Karbach and Verhaeghen, 2014; Rapport et al., 2013; von Bastian and Oberauer, 2013). The classification of tasks based on working memory processes also differed noticeably with respect to the level of generality (e.g., executive functioning vs. different aspects of attentional control). Yet another aspect considered in some classifications was the content of WM tasks used, such as verbal or spatial (e.g., Danielsson et al., 2015; Schwaighofer et al., 2015).

3.2.1 A Neurocognitive Perspective

Neuroscience can provide additional insight into the theoretical background of far transfer training effects on fluid intelligence, and can also clarify some ambiguities in relation to classification criteria (D'Esposito and Postle, 2015; Eriksson et al., 2015; Postle, 2015). As mentioned previously, central to transfer between working memory training and intelligence are two assumptions: first, WM and intelligence share a common capacity constraint, and second, repeated performance above this limit can increase capacity. The concept is simple, and was therefore criticized as rather naïve, relying mainly on a "physical–energetic" model, resembling physical training to strengthen muscles (e.g., Melby-Lervåg and Hulme, 2013). On the other hand, several studies analyzing the influence training (repeated performance of activities) has on brain function and structure actually lend support to this simplified view. Throughout a broad spectrum of activities, it was demonstrated that training of specific skills altered brain areas assumed to be related with the trained activity (either cognitive or motor). To give just one well-known example, 3 months of juggling training induced changes in brain areas associated with visual motion (Boyke et al., 2008; Draganski et al., 2004; for a detailed discussion see Section 3.3.1). Therefore, a description of the neurocognitive architecture of WM may be helpful for this theoretical discussion.

Working memory capacity is limited, whereas long-term memory (LTM) does not appear to have any capacity limitations. This limitation is estimated to three or four simple elements (e.g., Cowan, 2000; Luck and Vogel, 1997). The explanation for this capacity limit is provided by two theories of WM, the model of discrete slots, each with all-or-none properties (Luck and Vogel, 2013), and the more recent continuous single resource model (Ma et al., 2014). The slot model suggested that WM capacity is limited by a maximal number of slots (K_{max}). If the sensory input is greater than K_{max}, then K_{max} elements are stored in working memory, while all the residual elements are not. For instance if one's $K_{max} = 3$, and six elements are presented, then three elements are stored, and three are not (Fig. 3.2). When one's memory is probed and the element had not been stored in one of the slots, the person will make a guess about its identity. In contrast, the continuous single resource model assumes that WM capacity is a flexible resource that can be spread among all elements in the sensory input. However, an increased number of elements makes the WM representation less precise, because the resource must be divided among all the elements in the sensory input. In relation to the example presented earlier, the WM representation for the same person and sensory input includes all six elements, although with less precision or resolution (Fig. 3.2). In that case, the person won't guess about the identity of the elements because they are all represented in WM; however, the person's answers will be erroneous and resemble guessing due to imprecise information (Eriksson et al., 2015).

Empirical evidence from cognitive research lends support to both theories (for detailed reviews, see Ma et al., 2014; Luck and Vogel, 2013). Indeed, it is difficult to differentiate between answers that are based on memories that include imprecise information (continuous resource model) and guessing based on no information (discrete slot model). Recent work using a Bayesian

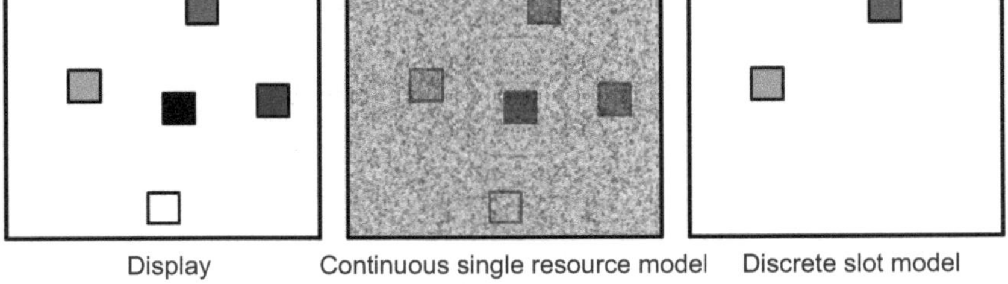

FIGURE 3.2 **Core concept of the continuous resource and discrete slot models of WM (see the main text for a more exhaustive account).**

approach found more supporting evidence for the single slot model (e.g., Donkin et al., 2013; Rouder et al., 2008). However, when the test paradigm changes from requesting recognition to asking subjects to recall the probe, there is more evidence for the continuous resource model (van den Berg et al., 2012).

On the other hand, neural models as well as research based on EEG [event-related potential (ERP)] and fMRI data, at the moment support the single slot model (Ma et al., 2014; Luck and Vogel, 2013). Slot models would predict that an increase in the number of sensory input elements would result in increased brain activity reaching a plateau when K_{max} is attained. In contrast, for continuous resource models the number of elements in sensory input has no influence on the engaged resource. The same amount of resource is engaged in one element or divided among six input elements (Ma et al., 2014). Brain imaging studies have shown that brain activity is sensitive to the number of elements in sensory input reaching an abrupt plateau when memory load exceeds K_{max} (e.g., Anderson et al., 2011; Vogel and Machizawa, 2004).

The explanation provided by neural network models is based on recurrent feedback loops between sets of neurons, which maintain single elements or bits of information in an active state. Oscillatory activity of a set of neurons (referred to as a cell assembly) prevents collapsing of elements into a single representation when there are multiple elements in sensory input. In that way, synchrony within a cell assembly (or slot) keeps the element available in WM representations, whereas asynchrony between cell assemblies enables more elements to be stored in a WM representation. Multiple firing initiates decay hence the elements vanish from the WM representations. This represents the capacity constraint of working memory. On the one hand, repetitive firing of the cell assembly must be fast enough to prevent loss of the element, and on the other, the firing rate must differ from the rates of the other slots to prevent mixing-up of elements in a single representation (e.g., Lisman and Idiart, 1995; Raffone and Wolters, 2001). These models present a challenge for empirical testing due to the number of neurons involved and the neural noise that is present as set size increases. Nevertheless, in a series of experiments conducted in our lab, we showed that the relation between theta and gamma band activity can to some extent predict memory span data (Pahor and Jaušovec, 2016). From a neurocognitive perspective, training aiming to increase WM capacity and consequently IQ would have to increase the number of simultaneously firing cell assemblies. In light of Jensen's (2011) theory that oscillation rates are the pure measure of *g*, just by increasing their rate an increase in WM capacity and IQ could be achieved. Even though this also corresponds with the slot or cell assembly model, it is questionable whether it would work. I will come back to this topic in the next chapter.

Another aspect that has to be addressed in relation to the classification of training approaches are the processes that operate on stored information. From a neurocognitive position, no process or representation is unique to WM, and can also act in any other arrangement. The combination of processes and representations are therefore just a temporary constellation of enhanced accessibility determined by the sensory input and interaction with internal information (e.g., LTM). The same holds true for the involved brain structures, which are also not unique or specific to WM. However, a rough division can be made: processes (e.g., attention) are "located" in anterior brain regions, whereas representations are presumably located more posterior in the same sensory areas where they are processed (e.g., Eriksson et al., 2015; Sreenivasan et al., 2014).

In papers dealing with working memory, there is no shortage of proposed processes (Chapter 2). However, in both most influential cognitive theories of WM, the multicomponent model proposed by Baddeley and Hitch (1974) as well as in state-based models (e.g., Cowan, 2000; Oberauer, 2002), the cornerstone among WM processes is attention (e.g., the central executive or focus of attention). As stressed by Jonides et al. (2008), attention is the process that can explain the main functions of WM: bringing information from perception into the focus of attention (encoding), keeping this information in an active state—resisting interference (maintenance), and bringing it back to attention when needed (retrieval). The key brain regions involved in this process are the prefrontal and superior parietal cortices (e.g., Eriksson et al., 2015; Koenigs et al., 2009; Rigotti et al., 2013; Vogel and Machizawa, 2004), which brings us back to the concept of shared resources between WM and *g*, and the theoretical origin for far transfer effects of WM training.

This concise review of WM characteristics including capacity limits, processes operating on stored information, and their suggested neurocognitive background may be of help for a better understanding of the classification criteria used by authors of the metaanalyses outlined in the next section.

3.3 METAANALYSES AND REVIEW PAPERS: IMPROVING FLUID INTELLIGENCE WITH TRAINING

Table 3.1 provides a synopsis of review papers and metaanalyses on the effectiveness of different training approaches aiming to increase performance on WM and intelligence tests. The studies are rather heterogeneous with respect to their methodology: (1) participating individuals (clinical/nonclinical or mixed; with a broad age range, or focusing just on specific populations—children, younger or older adults). (2) Training tasks used—covering computer games (e.g., Rise of Nations), commercially available interventions (e.g., Cogmed, Jungle Memory), and a wide range of tasks designed for study purposes ranging from interventions involving music, aerobic exercise, repetition-priming, and drugs to more conventional ones like strategic and core training approaches including tasks of simple and complex memory span; running span; single/dual *n*-back, executive function, attention, and shifting. (3) Experimental design, such as the type of control group (active or no contact), or no control group employed, assignment to groups (randomized, quasiexperimental), the number of participating individuals, and the type and amount of remuneration given. All these variables have an essential influence on observed training effects and the conclusions that have been put forward by the authors.

TABLE 3.1 Overview of Metaanalyses on the Effectiveness (Near and Far Transfer Effects on Cognition) of Behaviorally Based Training Approaches. Papers Are Rank Ordered by Year of Publication. Effect Sizes (Cohen's *d* or Hedges's *g*) Were Classified as Small ($d/g \geq 0.2$), Moderate ($d/g \geq 0.5$), or Large ($d/g \geq 0.8$) (Cohen, 1988)

Lustig et al. (2009)
Methodology: Included were 61 studies; study recruitment till 2009; focus on healthy, normally aging adults. Cognitive interventions were classified as: (1) strategy—mnemonic training (11 studies); (2) multimodal—changing lifestyle, including social and cognitive components (14 studies); (3) cardiovascular—aerobic exercise training (9 studies); (4) process—training of specific sensory or cognitive processes, such as discrimination, task-switching, dual-tasking, memory training (22 studies); (5) repetition priming—priming effects for fragmental pictures, semantic judgment, and similar (5 studies).   *Results*: Strategy training showed long-lasting effects on trained tasks with limited transfer. Multimodal training showed moderate-to-large far transfer effects. Cardiovascular training showed most pronounced effects on executive function (moderate to large), simple reaction time tasks, tests of motor learning, and cognitive assessments (moderate). Process training showed moderate far transfer effects. Repetition priming effects are mainly interesting for research and neuroimaging purposes.   *Discussion*: Most promising for the elderly are cardiovascular- and process-based training approaches. Transfer effects could be increased by the use of neuroimaging data identifying core mental processes, which could then be targeted by cognitive interventions.
Shipstead et al. (2010)
Methodology: Included were 17 studies; study recruitment till 2009; focus on healthy (nontherapeutic) and clinical (therapeutic) populations (children with ADHD, stroke patients). Interventions were classified as working memory capacity training (WMC) divided into: (1) simple/complex span training (Cogmed, Jungle Memory), (2) WM updating (dual/single *n*-back, running span).   *Results*: Therapeutic (span training)—little evidence for long-term changes; no evidence for far transfer effects. Mixed findings for alleviation of symptoms. Nontherapeutic (span training)—far transfer to reading comprehension and attention, some evidence for brain-related changes. Nontherapeutic (memory updating)–some evidence for far transfer effects.   *Discussion*: Evidence for near transfer effects, mixed results for far transfer effects. When robust methodology is used little to no far transfer effects were observed. A methodological issue was that no data were provided for the effects of WMC training on changes in WM capacity beyond simple memory span measures. A more appropriate experimental design (active control group) and assessment of targeting constructs (test-batteries instead of single tests) should be used in the future.
Takeuchi et al. (2010)
Methodology: Included were 18 studies; study recruitment till 2010; no division between clinical and non clinical studies was made. Interventions were classified into four groups: (1) basic tasks—simple memory span tasks; (2) *n*-back tasks and other updating tasks like running span tasks; (3) complex span tasks in which respondents have to remember presented stimuli (e.g., letters, numbers) in the course of solving different tasks (e.g., judge if two matrices are symmetrical); (4) other types, such as dual *n*-back task.

(Continued)

TABLE 3.1 *(cont.)*

Results: Near transfer effects were consistently reported, some transfer to other WM tasks. Some far transfer to nonverbal reasoning (fluid intelligence), but only for the *n*-back tasks. Mixed results for performance on inhibition tasks (e.g., Stroop test), for interventions using basic and complex WM tasks. No improvements in measures of processing speed. Reported were also effects of interventions on brain function in the frontoparietal network, white matter integrity and gray matter structure. *Discussion*: Methodological issues considering the use of an active control group. This problem is also related to morphometric data showing changes in brain structure in the absence of behavioral improvements. Further research should investigate training effects on specific clinical populations (e.g., affective disorders), the possibility of enhancing abilities of experts in specific fields, and improving novice learning of specific skills.
Buschkuehl and Jaeggi (2010)
Methodology: Included were 11 studies; study recruitment till 2009; no division between clinical and nonclinical studies was made. Interventions were classified into: (1) WM or executive function training (six studies), (2) computer games and cognitive stimulation (three studies), (3) music training (one study), (4) pharmacological intervention—ceratine (one study). *Results*: All studies showed far transfer effects on intelligence. *Discussion*: Shortcomings are the repeated use of same instruments on pre/post testing, no ecological validation of training effects, and no active control group. An ideal research design was suggested that should include: (1) a significant number of participants, (2) randomized assignment of the participants to experimental and active control groups, (3) use of multiple test indicators to assess intelligence, (4) assessment of long-term effects in all groups.
Morrison and Chein (2011)
Methodology: Included were 37 studies; study recruitment till 2011; no division between clinical and nonclinical samples was made. Interventions were classified into other approaches (attention, speed of processing, neuro-feedback, dual-task, perceptual training) and WM training, further divided into: (1) domain-specific (strategy training—11 studies), (2) domain-general (core training—26 studies), further divided into multifaceted packages and single task approaches (e.g., memory span, memory updating). *Results*: Limited near transfer of strategy training approaches. Mainly beneficial for populations with memory deficits (e.g., aging individuals; Down syndrome, fetal alcohol disorders). Core training shows robust near transfer effects; whereas far transfer effects are less often observed (e.g., Cogmed, COGITO, and dual/single *n*-back). Neuroimaging results of training effects are consistent with the claim that core-training approaches target executive mechanisms. *Discussion*: Obtained training effects could be due to expectancy bias, related to cognitive investment and motivational factors—variables that should be controlled for in future research. Many studies use no control group or just a no-contact control group. Training effects are assessed with just one test interpreting improvements as an indication for increased ability. There is also a lack of consistency in experimental methods and findings, which makes the comparison between study outcomes difficult. Future research should first employ a standardized pre- and posttraining assessment, and second, use an active control group. From a theoretical standpoint, there is need to explicitly clarify the specific mechanisms that beget training gains.

TABLE 3.1 *(cont.)*

Shipstead et al. (2012a)
Methodology: Included were 37 studies; study recruitment till 2011. A division based on age: children (16 studies), young adults (14 studies), and old adults (7 studies) was made. Interventions were classified as WM training (mainly Cogmed training package, single/dual *n*-back, running span, complex span). *Results*: Children—several studies report near transfer effects, and some far transfer effects to intelligence and attention. Young adults—training based on simple span tasks had no transfer effects; *n*-back tasks showed small far transfer effects on intelligence. Older adults—few near transfer effects, no far transfer effects on intelligence. In general, near-transfer effects remained stable over time. No long-term transfer effects on intelligence were observed. *Discussion*: The studies failed to prove that WM training increases WM capacity, which is the main prerequisite for a far transfer effect. Furthermore, inadequate WM capacity measures were employed. Suggestions for improvement are: (1) multiple measures for targeting effects should be used, (2) near transfer should be estimated with valid WM capacity measures that differ from trained tasks, (3) active control groups should be used, and (4) subjective measures of change should be assessed by raters blind to condition assignment.
Rapport et al. (2013)
Methodology: Included were 25 studies; study recruitment till 2012; included were studies obtained on children/adolescents with a diagnosis of ADHD and other attention, hyperactivity, and impulsivity problems. Interventions were classified as: (1) short-term memory training (e.g., short-term memory storage and rehearsal abilities—nine studies), (2) mixed executive functioning training (e.g., short-term memory and behavioral inhibition—nine studies), (3) attention training (e.g., sustained, selective, and divided attention—six studies), (4) set shifting training (minimizing interference while switching between two mental sets—one study). *Results*: Immediately after training, children with ADHD exhibited small improvements on tasks similar to training tasks and medium improvements on follow up. Conclusion: Training is associated with medium magnitude improvements on nontrained memory tasks, which are maintained across 3–6 month follow-up. Individuals with ADHD exhibited small to none far transfer effects—small magnitude improvements for cognitive test performance, but no significant changes in objectively measured academic achievement. The results depended on the type of rating—blinded or not. *Discussion*: Claims that training is beneficial for children with ADHD are not supported by empirical evidence.
Melby-Lervåg and Hulme (2013)
Methodology: Included were 23 studies with 30 independent group comparisons; study recruitment till 2012. A division based on age: younger children (under the age of 10 years), older children (11–18 years), young adults (younger than 50 years), and older adults (51 years or older). Interventions were classified as WM training programs including commercially available programs: (1) Cogmed, Jungle Memory, Cognifit, Memory Booster; (2) study-developed program training; and (3) single/dual *n*-back tasks.

(Continued)

TABLE 3.1 *(cont.)*

Results: Observed were large immediate gains on measures of verbal WM (larger for young children), which were not maintained at follow up (9 months after training). Moderate immediate gains on measures of visuospatial WM (most pronounced for Cogmed training), which were maintained at follow-up. Immediate gains on nonverbal ability were small, and not present for experimental designs with an active control group, whereas there were no significant effects for verbal ability. Small-to-moderate training effects on inhibitory processes in attention were observed.   *Discussion*: Studies suffer from several methodological issues: (1) no control groups, or just no-contact control groups; (2) studies show far-transfer effects on fluid intelligence in the absence of near-transfer effects on measures of WM; (3) no random assignment to training/control groups. Furthermore, a central problem is the lack of theoretical background for the designed training programs, which mainly adopted a "physical-energetic" model: if you train a process, you will gain improvements in that process.
von Bastian and Oberauer (2013)
Methodology: Included were 46 studies; study recruitment till 2013. A division based on age: children; young adults, middle-aged adults, older adults. Interventions were classified as: (1) training employing single paradigms (e.g., single/dual *n*-back; running span), (2) training using multiple paradigms (e.g., short term and WM—Cogmed, Cognifit), and (3) multifactorial training targeting multiple cognitive skills (e.g., WM and executive functions—COGITO).   *Results*: Effects were analyzed with respect to mechanisms of transfer: classified as either aiming to expand WM capacity or increase WM efficiency. The former led to broader far transfer effects. More impressive far transfer effects were observed for multifactorial training regimens. Most interventions used 20 sessions lasting 30 min with adaptive training regimes. Stronger training effects in younger than older adults were reported. It was further suggested that genetic predispositions, such as the availability of dopamine and the brain-derived neurotrophic factor (BDNF) involved in hippocampal plasticity could affect WM training-related benefits. Furthermore, motivation and personality factors could have influenced training outcomes (e.g., positive associations with motivation and conscientiousness; negative associations with neuroticism).   *Discussion*: No final answer can be given whether WM training interventions were successful in increasing WM capacity. In future, the theoretical training background should be elaborated in more detail to answer the question under which circumstances (type of training, duration, and regimen) and for which individuals (initial level of ability, motivation, personality factors, genetic predispositions) WM training is beneficial.
Karbach and Verhaeghen (2014)
Methodology: Included were 49 studies with 61 independent samples; study recruitment till 2013. A division based on age: younger adults (17–31 years); older adults (63–87 years); target group older adults (>60 years). Interventions were classified as: (1) working memory training and (2) executive function training.

TABLE 3.1 *(cont.)*
Results: In the training groups, near- and far-transfer effects were all significantly larger than zero and larger than in the passive or active control groups. Further, near transfer effects were more pronounced than far transfer effects, and were also larger than far transfer effects observed for either control group. Active and passive control treatments yielded statistically indistinguishable effects. Moreover, net treatment effects (near/far transfer) of WM or executive function training were significantly larger than those of either passive or active control groups. The net training effect was not related with the type of control condition (active, passive). *Discussion*: The analysis revealed that for normally aging adults, WM and executive function training is effective, showing small to medium near/far transfer effects.
Lampit et al. (2014)
Methodology: Included were 52 studies with 52 independent comparisons; study recruitment till 2014. Selected were studies with mean participants' age ≥60 years. Interventions were defined as practice on standardized computerized tasks or video games with clear cognitive rationale (e.g., Cogmed, Lumosity, Cognifit, Rise of Nations, Tetris, n-back, and others) and classified into five content types: (1) speed of processing training (nine studies), (2) WM training (nine studies), (3) attention training (six studies), (4) multidomain training (24 studies), and (5) video games for entertainment purposes (four studies). *Results*: The overall effect on cognition was small and significant. Small and significant effects were further observed for: verbal, nonverbal short term memory, and WM as well as visuospatial skills, while moderate effects were observed for processing speed. Not significant were effects for executive function and attention. More effective were training regimens with three or less sessions per week lasting between 30–60 min, and those performed in groups as compared to home-based training. It was further shown that video games were most effective, whereas there was no significant effect when only WM training was used. *Discussion*: The main finding was that computerized cognitive training shows small but positive effects on cognition in healthy older adults.
Danielsson et al. (2015)
Methodology: Included were 10 studies with 28 group comparisons; study recruitment till 2014. Included were individuals with mental disabilities: Down syndrome (six studies), intellectual disabilities (three studies), and borderline intellectual functioning (one study). Training interventions were classified as: (1) visuospatial working memory training (five studies), (2) verbal WM training (two studies), (3) mixed WM training including strategic training (two studies), (4) short-term memory training (one study). *Results*: The overall training effect on cognitive outcome measures was medium (training group) to small (comparison with included control group). Visuospatial training showed only a near transfer effect on tasks measuring visuospatial WM, being close to zero when compared to control group effects. The mixed training approach showed a strong near transfer effect on verbal WM, which remained stable also when compared to control group effects. *Discussion*: The only significant effect was shown for training designs that included a mix of verbal and visuospatial tasks. The effect was greater on verbal and visuospatial short-term memory tasks than on WM tasks. The explanation was that individuals with intellectual disabilities show a deficit on both, verbal and nonverbal WM, and metacognitive training helps them to acquire new strategies. A shortcoming is a lack of analysis for transfer on everyday activities and clinically relevant measures.

(*Continued*)

TABLE 3.1 *(cont.)*

Schwaighofer et al. (2015)
Methodology: Included were 47 studies with 65 group comparisons; study recruitment till 2014. Included were individuals of any cognitive status being less than 75 years old. Interventions were classified as WM training programs including commercially available ones: (1) Cogmed, Jungle Memory, Cognifit, Memory Booster, Brainstim; (2) study-developed program training; and (3) single/dual *n*-back tasks. *Results*: An immediate near-transfer effect was observed on verbal and visuospatial short-term memory as well as on verbal and visuospatial WM which remained present at follow up. There was a small immediate transfer effect on nonverbal and verbal ability, which was not sustained at follow up. No immediate or sustained training effects were observed for mathematical ability and word decoding. Regarding the tested moderators, no effect was observed for trainees' age; frequency of training per week; training interval between single sessions; training modality (verbal, visuospatial or mixed); instructional support; feedback provided. Significant effects were observed for total and single session duration—longer was more beneficial than shorter. Training carried out in the laboratory was better than training at school, and training at home only for transfer effects to nonverbal ability. A positive effect was also observed for supervised training. With respect to training type, Cogmed showed a more pronounced effect on WM, whereas *n*-back training showed a larger effect on nonverbal ability. Type of control group was not a significant moderator. *Discussion*: No effects of training to educationally relevant aspects were observed.
Au et al. (2015)
Methodology: Included were 20 studies, with 24 group comparisons; study recruitment till 2014. Only studies conducted on healthy young adults between the ages of 18 and 50 years, using single/dual *n*-back training tasks were analyzed. *Results*: The treatment/control group difference in fluid intelligence at posttest was significantly greater than the treatment/control group difference in fluid intelligence at baseline. Two moderators were also significant: (1) studies conducted outside the United States showed more pronounced effects on fluid intelligence than those conducted in the United States, and (2) studies that used passive controls demonstrated more net transfer than those with active controls. There was no difference in effects for either single or dual *n*-back tasks used. Further, no effects of location (training carried out in the laboratory or at home) were observed. Higher remuneration for study participation was negatively correlated with effect size. *Discussion*: The metaanalysis confirmed a small but statistically significant net effect of *n*-back training on fluid intelligence. Because of sampling from restricted ranges the observed effect size represents a low-end estimate of the influence of *n*-back training on intelligence. Future studies should explore the characteristics of training improvements in intelligence beyond test scores relying on evidence from practical, real-world settings.
Melby-Lervåg and Hulme (2016)
Methodology: A reanalysis of the metaanalysis performed by Au et al. (2015) and Karbach and Verhaeghen (2014). *Results*: Reanalysis (Au et al., 2015): the pooled (treated and untreated controls) effect size was small and significant, but not significant when analyzed just for studies using a treated control group. Reanalysis (Karbach and Verhaeghen, 2014): after removal of outliers the effect size is small and close to zero.

TABLE 3.1 *(cont.)*

Discussion: Taken together there is no evidence that WM training increases performance on measures of intelligence.
Au et al. (2016)
Methodology: A reply to Melby-Lervåg and Hulme (2016). *Results*: The issue of control groups was clarified indicating that insignificant effect sizes observed in studies with an active control group as compared to studies with no-contact controls are not because of the Hawthorne or placebo effect. In studies with active controls, the net increase (from baseline to posttest) in treated groups is smaller than in studies with no-contact controls; the same trend can be observed for the controls showing higher net gains in no-contact controls than in active controls, which is at odds with the assumption underlying a placebo or Hawthorne effect. *Discussion*: There is convincing evidence that WM training increases performance on IQ tests.
Weicker et al. (2016)
Methodology: Included were 103 studies with 112 independent group comparisons; study recruitment till 2015. Studies were classified either as involving healthy subjects (children, young, and elderly adults) or patients (children and adolescents with WM deficits, patients with acquired brain injuries, and otherwise diagnosed patients). Interventions were classified as: (1) simple and complex memory span tasks, (2) Cogmed, and (3) updating training. *Results*: Most studies showed improvements in trained tasks, moderate near transfer effects on WM function. In general, small but significant transfer effects were found on other cognitive domains—reasoning and intelligence; cognitive control, and executive functioning; attention and processing speed; long-term memory and the quality of everyday life assessed mainly for the patient studies. Reported were also moderate long term near transfer effects and small far transfer effects—reasoning and intelligence; cognitive control and executive function; whereas no long-term effect on attention and processing speed, LTM and the quality of everyday life was observed. The inclusion of passive or active control groups had no influence on the overall WM training effects, the same was true for the moderator training adaptability and type of training, whereas number of training sessions ($n > 20$) significantly influenced training outcomes. *Discussion*: Methodological issues, such as lack of randomization, lack of active control groups, small sample size (patient studies), single, and sometimes inadequate measures of WM function are still present.

Despite these shortcomings some tentative conclusions can be drawn: (1) in general, interventions show a positive near transfer effect on trained tasks, and a positive effect on tasks that are similar to the trained task measuring the targeted ability (e.g., simple memory span task on measures of working memory). (2) Far transfer effects on measures of fluid intelligence are small and observed typically after *n*-back training or training using multifactorial interventions targeting multiple cognitive skills. (3) Training duration often had a positive influence on outcomes, suggesting the use of more than 20 sessions (3 or less per week) with a duration of 30–60 min. (4) Strategy training approaches were mainly beneficial for clinical (intellectual disabilities) and elderly populations, whereas core-training approaches showed more near and far transfer effects. (5) Supervised training was more effective than if a person was just present at the training location. (6) Several moderators, such as training feedback, instructional support, and the type of control group (active vs. passive) had no influence on training outcomes.

In almost all metaanalyses and review papers, a central critique and suggestion for further research addressed the issue of the control group, which was most fiercely advocated by Melby-Lervåg and Hulme (2016), demanding that WM training studies that employed merely an untreated control group should no longer be published (p. 328). An interesting argument with respect to the type of the control group was addressed by von Bastian and Oberauer (2013). Passive and active control groups verify the pre-/posttest increase in the trained target characteristic, whereas the active control group additionally monitors motivational, placebo, and Hawthorne [c] effects, and generic training effects, such as use of a computer and regular attendance of training sessions. The critical question is what type of activity the participants in the active control group should exercise. The activity should be similar to the training intervention as much as possible, but should not enhance the target characteristic that is being trained. One possibility is to let controls practice on low-level tasks used in the training sessions (e.g., in an *n*-back paradigm just on tasks with $n = 1$; whereas the intervention group can go as far as $n = 5$), an approach that can be rather boring for the controls resulting in a lower motivational level on the posttest assessment. Another possibility is to design a more interesting task for the controls that is not related to the training target. In our study, for example, subjects assigned to the active control group were trained in emotional intelligence using communication and social skills tasks (Jaušovec and Jaušovec, 2012b). As stressed by von Bastian and Oberauer (2013), the problem with this type of training is two fold. First, participating in the control activity might be more fun than participating in strategy or core training, which might increase the motivation of controls, and second, these types of task could overlap (i.e., have shared variance) with those used to train target characteristics (e.g., in our study, the control training aimed to increase emotional intelligence). A nonadaptive active control group would probably overestimate training effects, whereas an adaptive one would underestimate the intervention effects, and should be the more conservative choice (von Bastian and Oberauer, 2013).

c. A type of reactivity in which individuals modify or improve an aspect of their behavior in response to their awareness of being observed (McCarney et al., 2007).

Several metaanalyses reported that the type of control group had no influence on training effects (e.g., Au et al. 2015; Karbach and Verhaeghen, 2014; Schwaighofer et al., 2015; Weicker et al., 2016), although studies that used an active control group showed a negligible far transfer effect on intelligence as compared to a medium-to-small effect observed in studies with a no-contact control group. This point has to be explained in more detail. As stressed by Au et al. (2016), if a zero difference between the treatment and control groups at the base test is assumed then the results show that the net increase (pre- to posttest) of the treatment group is lower at posttest when an active control group is used, whereas the net treatment effect in the treatment group is higher when a no-contact control group is used. Further, the net increase in the no-contact control group is at posttest higher than the one observed in the active control group coming close to the one observed in the treatment group with an active control group. This finding is of course at odds with the theoretical assumptions why an active control group should be used as proposed by von Bastian and Oberauer (2013) and several other papers reported in the metaanalyses. Au et al. (2015, 2016) suggested that other factors must have contributed to this difference. One possibility is a regional one. Studies that most often used an active control group were conducted in the United States, and most US studies also showed lower far transfer training effects than those conducted in Europe, although no differences in baseline effect sizes were observed. It was suggested that US students are less compliant, which is a crucial prerequisite for the success of a training program. In this direction point also the analyses of training dropouts, which were much higher for the United States (21%) than observed in European studies (5%). Another aspect is that most US universities (at least the top ones where the studies were conducted) have a much higher budget than European universities (Guisan, 2005) therefore are much better technically equipped than the European ones where some of the studies were conducted. This reminds me of a personal experience several years ago when I was a member of an international commission board evaluating the educational curriculum that should be introduced in the Netherlands. The discussion was about Lozanov's (1978) suggestopedia, and the problems they had when they decided to introduce it in Lichtenstein. Suggestopedia is a technique for teaching students to read. One of the interventions was to display panels with pictures of animals and their written names. Children in Bulgaria [d] were fascinated by the panels and tried to find out what the written signs meant, thereby learning to write without being explicitly instructed. In contrast, their peers from Lichtenstein did not even notice the panels until they were made more attractive with some electronic supplements. Thus, the same training tasks that worked in Europe (e.g., *n*-back) could be less appealing for US students, which could also

d. Georgi Lozanov (1926–2012) was a Bulgarian psychiatrist and educator.

explain the higher dropout rates of these students (see also Section 3.7 on the statistics of video game playing in the United States).

Yet another point worth mentioning is that different metaanalyses, although including roughly the same studies, came to contradictory conclusions about the training effects they have on intelligence. This was evident in a recent critique and discussion between Melby-Lervåg and Hulme (2016) and Au et al. (2015, 2016). Besides the aforementioned difference in the argumentation with respect to the type of control groups that should be used and their effect on study outcomes, the inclusion of studies into metaanalyses also sparked a debate. Melby-Lervåg and Hulme (2016) criticized Au et al. (2015) for not including the study by Colom et al. (2010) in their metaanalysis, who divided an impressive sample of 288 individuals into a training and active control group, and showed that the intervention, consisting of different short-term and WM tasks (experimental condition), and several processing speed and attention tasks (active control group), had no far transfer effect on intelligence. However, this critique is extremely misleading. Au et al. (2015) stated explicitly that only studies using single or dual n-back training tasks were included, hence the Colom study was excluded because no n-back tasks were used in their intervention (Table 3.1). The inclusion of the study by Colom et al. (2010) in a reanalysis performed by Melby-Lervåg and Hulme (2016) had a pronounced negative influence (because of the large sample size) on the calculation of the effect size for the far transfer on fluid intelligence.

The study by Schwaighofer et al. (2015) is yet another example of study selection where the argumentation of inclusion/exclusion criteria is difficult to follow. The authors excluded the studies by Schmiedek et al. (2010) and Klingberg et al. (2002), both of which showed intense far transfer effects on intelligence. The explanation for the exclusion of the former was based on the use of episodic memory tasks, and for the latter a combination of reaction time and inhibition tasks. These exclusion criteria are puzzling as most WM tasks include the process of inhibition, especially the complex WM tasks that were included in the metaanalysis (for a detailed discussion see Eriksson et al., 2015). One could further argue that every WM task requires episodic memory as opposed to semantic memory (e.g., in the WM model by Baddeley, 2012, the central executive coordinates the operations of slave systems including the episodic buffer). Based on these criteria our study[e] should have also been excluded, yet it was not.

Another example of arbitrary study selection is the paper by Redick (2015). In this review, five studies were analyzed reporting significant training effects on posttest measures of fluid intelligence (effect sizes ranged from d= 0.39 to d = 1.41). However, as revealed by Redick, all five studies have several

e. The matrices tasks used in our study (Jaušovec and Jaušovec, 2012b) are similar to the episodic memory tasks used by Schmiedek et al. (2010); see also Sections 3.5 and 3.6.

methodological shortcomings. Beside the rather low number of participants (less than 10 in 2 studies) and few training sessions administered (3 and 9 sessions in 2 studies), in all five studies the reported significant gains in IQ scores compared to controls could be obtained because IQ scores of control individuals were on posttest significantly lower than on pretest. Multiple factors could have affected the IQ drops, some of which were random but some were also systematic. For example, the groups differed in their baseline scores. The critique put forward by Redick (2015) is therefore adequate. The problem I have with this paper is that it does not provide an example of a study with an appropriate experimental design that shows far transfer effects. Is there none? Redick (2015, p. 18) states:

> Certainly, the five studies highlighted here are not representative of the entire literature of WM training studies and intelligence transfer (for review, see Haier, 2014; Melby-Lervåg and Hulme, 2013; Shipstead et al., 2012b).

I agree, but the provided references are another example of a tendentious selection. The Haier (2014) paper is not a review but a two-page opinion article. The Melby-Lervåg and Hulme (2013) and Shipstead et al. (2012a) papers are metaanalyses. The authors of the former paper came to the conclusion that there is no far-transfer training effect on intelligence, whereas the authors of the latter concluded that the literature review provided insufficient evidence that training works, and that the main problem is that training influences cannot be attributed to changes in WM capacity, which is the main prerequisite for a far transfer effect (Table 3.1). Another interesting coincidence in the Redick (2015) paper is that the two more recent metaanalyses that are more favorable with respect to the possibility of a far transfer effect of training on intelligence (e.g., Au et al., 2015; Karbach and Verhaeghen, 2014), are mentioned in relation to the following statement:

> [...] meta-analyses of the WM training literature will therefore be influenced by studies such as the five reviewed here, positively biasing the meta-analytic estimate of the efficacy of WM training to increase intelligence (Redick, 2015, p. 18).

Nevertheless, even the greatest hardliners among the critiques of the possibility to raise intelligence with training must admit that it is puzzling that such a short intervention lasting only 20 h shows any effect at all. In comparison to ontogeny, 20 h represents the blink of an eye. For instance, we need about 6 years to develop a number concept, and roughly 12 years to grasp abstract thinking (Piaget and Inhelder, 1974). Moreover, some will never reach the stage of epistemic cognition (e.g., Kitchner, 1983); to know that there are some questions that have no correct answer—such as how to write a bestseller?

3.3.1 Changes in Brain Structure and Function Induced With Training

Several studies, additionally to analyzing training influences on IQ test performance, investigated possible training-induced changes in patterns of brain activity and structure (Table 3.1). As stressed by Klingberg (2010, p. 318), there is nothing magic about WM and the same holds true for intelligence, both are the result of synaptic connections governed by the same principles of brain plasticity as other behaviors. Some of the most often cited examples of this stem from the finding that repeated performance of activities (e.g., practice of skills in one's professional career) can induce lasting changes in brain structure: London taxi drivers who have abundant navigation experiences were found to have larger hippocampal gray matter volumes than matched controls (Maguire et al., 2006), professional typists showed increased gray matter volume in brain regions related to motor performance like the supplementary motor area, prefrontal cortex, and cerebellum (Cannonieri et al., 2007), string players (e.g., violinists) who use the second to the fifth digits for string manipulation have larger cortical representation of these digits in the primary somatosensory cortex (Elbert et al., 1995). Moreover, it was shown that even short lasting training can induce changes in brain structure in areas related to training. For example, 3 months of juggling training induced changes in brain areas associated with processing and storage of complex visual motion (Boyke et al., 2008; Draganski et al., 2004). This effect was also found for shorter training periods of 7 days (Driemeyer et al., 2008), and in white matter structure (Scholz et al., 2009). Changes in brain structure were also observed for training of other motor tasks like the Morse code (Schmidt-Wilcke et al., 2010) and a complex motor skill-learning task (Taubert et al., 2010). These findings on the plasticity of our brain during adulthood lend further support to the possibility that intelligence could be increased with training. Even if we take into account that intelligence is heritable and genetically determined, this relation is not fixed, but as several studies have indicated, it depends on different factors, some of which can be influenced (e.g., Chiang et al., 2009, 2011; Harden et al., 2007; Turkheimer et al., 2003).

In an early review paper, Klingberg (2010) analyzed 10 studies reporting changes in brain activation patterns (fMRI) related to working memory training. Differences were mainly observed in the prefrontal and parietal areas, with two studies reporting differences also in the basal ganglia in the caudate nucleus. These findings are in line with the assumed function of the multimodal fronto-parietal network. Another characteristic of the analyzed scans was that they did not show the same patterns of activity. Shorter trainings (less than 3 h) resulted in decreased brain activity, whereas longer periods of training showed mixed patterns of increased and decreased brain activity. A possible explanation was that increases in WM capacity are positively related to activity in the fronto-parietal network, yet this effect cooccurs with deactivations linked to learned strategy refinement, priming effects, and decreased time needed to perform the task (Klingberg, 2010).

In a second review paper by Buschkuehl et al. (2012), 18 studies were analyzed. The analysis suggested four possible scenarios of brain activation patterns that could be induced by training, three of which were related to the activity in the same brain areas and the fourth in different brain areas: (1) decreased activity after training, (2) increased activity after training, (3) a combination of the first two patterns, and (4) a change in activation patterns in different brain areas which would indicate a reorganization of brain networks related to the trained characteristic. As discussed in the previous chapter, decreased brain activity is associated with neural efficiency in the neural efficiency theory put forward by Haier et al. (1988). Therefore, it is not surprising that eight studies reported activation decreases after practicing on different visuospatial working memory tasks and visual and auditory *n*-back tasks. The increases/decreases in brain activity related to training duration reported by Klingberg (2010) were not observed in this study. On the other hand, there is much less evidence for activation increases after WM training which was only observed in three studies. One of these increases was reported in a study with just three subjects (Westerberg and Klingberg, 2007). Increases in bran activity are usually found after sensory or motor practice reflecting an increase in cortical recruitment due to an increase in the size of cortical representations (Pascual-Leone et al., 2005). A mix of increases and decreases in brain activation patterns related to training suggests that some processes become more automated causing less brain activation (neural efficiency) due to training, whereas in areas supporting task-specific functions, activation increases. One would expect that frontal brain areas related to attentional control would become less involved in task performance throughout the training process therefore a frontal decrease in activity would be observed. However, these activation patterns were confirmed only in a few studies (Buschkuehl et al., 2012). The fourth possibility suggesting a reorganization of brain networks after WM training was not observed in this review (Buschkuehl et al., 2012). Due to the few studies published on this topic and the different brain activation patterns observed after training, the authors concluded that it was not possible to identify a specific neural mechanism that would fit within one single framework.

In a recent review by Taya et al. (2015), 22 studies using different cardiovascular (fMRI, NIRS, ASL) and neuroelectric (EEG) indicators of brain activity, as well as 12 studies showing training-related changes in brain structure, were analyzed. The conclusion was that training tasks can improve higher cognitive functions regardless of sensory modality, and that these improvements are reflected in different brain activation patterns as well as changes in gray and white matter structure that can be captured by a variety of neuroimaging techniques. Taya et al. (2015) further suggested that the brain connectome approach would be a useful framework for analyzing different brain states related to training interventions and their influence on higher cognitive functions like working memory and intelligence. Furthermore, the analysis of 10 studies on training-induced changes in functional connectivity suggested that biomarkers based on the brain connectome approach could provide information for the optimization of cognitive training interventions aiming to increase intelligence.

An example of such an approach is the study by Langer et al. (2013). In a pre–posttest experimental design, 66 individuals were assigned to active control and experimental groups; the latter trained on three working memory tasks. The tasks were designed to trigger different processes: (1) storage and processing related to simultaneous maintenance and manipulation of information (a complex memory span task was used), (2) executive processes defined as selective activation of relevant and suppression of irrelevant information (a task switching paradigm was used), and (3) relation integration involved in combining bits of information into structures (a self constructed task—the tower of fame was used). High-density (256 electrodes) eyes-closed EEG segments (prior and posttraining) were used for the analysis of neural correlates induced by the WM training. Pre–posttraining differences were significant only in the theta frequency band, showing posttraining power increases in relation to memory performance. The graph-theoretical network analysis showed a positive correlation for the clustering coefficient and a negative relationship for the path length. Both parameters are indicators of small-world networks characterized by high clustering and short path length. The former is a marker of local connectedness, whereas the latter points to the efficiency of information transfer in a network. The nodes identified as responsible for the observed training effects corresponded mainly to the frontoparietal network, showing increased dissociation in frontal areas and more associations in parietal areas. This finding is in good agreement with the neural efficiency theory suggesting that fewer neural resources, primarily in frontal brain regions, are related to better task performance in more intelligent individuals (Neubauer and Fink, 2009).

Furthermore, a similar frontoparietal dissociation during rest was also observed in an fMRI/relative cerebral blood flow (rCBF) study by Takeuchi et al. (2013) in relation to WM training-induced changes in brain activation patterns. The results for the posttest showed: (1) decreased functional connectivity between the medial prefrontal cortex, the right posterior parietal cortex, and the right lateral prefrontal cortex, (2) increased functional connectivity between the medial prefrontal cortex and the precuneus, and (3) an increase in cerebral blood flow in the right lateral prefrontal cortex. This pattern would suggest dissociation between the default mode network (medial prefrontal cortex, precuneus) and the external attentional network (lateral prefrontal cortex, posterior parietal cortex), accompanied by increased activity in the default mode network and the right lateral prefrontal cortex. This pattern could be regarded as indicative for superior WM performance, because exactly opposite activation patterns have been observed in many conditions with reduced WM performance, such as Alzheimer's disease, schizophrenia, autism spectrum disorders, ADHD, and aging (Baddeley, 2003; Broyd et al., 2009; Goldman-Rakic, 1994; Sambataro et al., 2010; Steele et al., 2007; Westerberg et al., 2004; Wingfield et al., 1988).

Changes in the frontoparietal network related to cognitive training were also observed in diffusion tensor imaging (DTI) studies measuring white matter integrity. Mackey et al. (2012) reported that trained individuals displayed decreased radial diffusivity in white matter connecting frontal cortical areas and decreased mean diffusivity in left frontal and right parietal areas. In a second study, Wolf et al. (2014) found that long-term transfer of logical reasoning training was associated with increased structural integrity in corpus and genu of the corpus callosum.

All together this brief review demonstrated that cognitive training (e.g., WM, *n*-back, reasoning) not only shows transfer effects on performance but also significantly changes brain structure and activity. These changes were observed mainly in the frontoparietal network, which corresponds to brain areas that are central in most neurobiological explanations of intelligence (e.g., P-FIT, MD system). Moreover, the reported training-related decreases in frontal brain activity accompanied by increased parietal activity correspond to the neural efficiency theory suggesting that less brain activity during task performance (frontal brain regions) is typically observed in highly intelligent individuals (Neubauer and Fink, 2009). These findings are also in line with the dual-process theory (Chein and Schneider, 2005), which assumes that training increases automated task performance reducing the need for attentional control related to frontal brain regions, but increases activity in task-specific areas (parietal cortex). Another point that must be mentioned is the diversity of the observed findings, which suffer from similar methodological shortcomings as behavioral studies do (small sample sizes, no control group or no-contact control group, lack of randomized subject recruitment).

3.4 THE JAEGGI ET AL. STUDY

An adaptive dual *n*-back was the only training task used by Jaeggi et al. (2008) that led them to conclude that after all, fluid intelligence can be improved. The previous sections revealed that the *n*-back task (either single or dual) is one of the most frequently used tasks in WM training research, and that this task most often produces small far transfer effects on fluid intelligence (Table 3.1). Multifactorial training approaches targeting multiple cognitive skills also used versions of the *n*-back (e.g., Jaušovec and Jaušovec, 2012b; Schmiedek et al., 2010) therefore I will provide a synopsis of cognitive and neurocognitive research elucidating the characteristics of the *n*-back task.

3.4.1 The *n*-Back: Cognitive and Neurocognitive Characteristics

Fig. 3.3 shows an example of the dual *n*-back training task used by Jaeggi. Two sequences of stimuli were presented simultaneously: (1) visually displayed squares in one of eight different locations, and (2) one of eight consonants presented through headphones. The participants were required to indicate whenever one of the presented stimuli matched the one presented *n* positions back in the sequence. They had to press a letter—"A" for visual correspondence and "L" for auditory correspondence—on the keyboard. Each requirement alone corresponds to a visual or auditory single *n*-back. All stimuli were displayed for 500 ms with an interstimulus interval of 2.5 s. One training block consisted of 20 + *n* trials and in a daily training session 20 blocks were delivered.

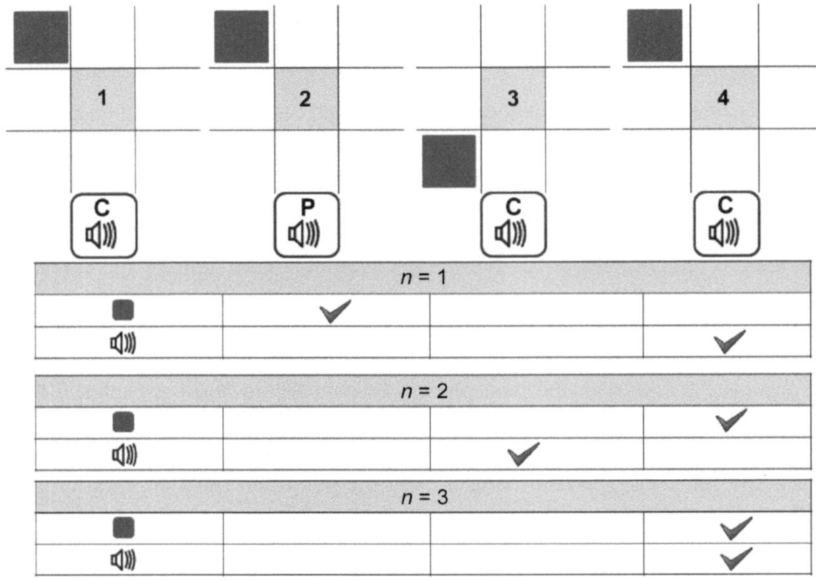

FIGURE 3.3 Example of the dual *n*-back task used in the study by Jaeggi et al. (2008). The spatial *(squares)* and auditory *(letters)* stimuli were presented simultaneously for a duration of 500 ms with an interstimulus interval of 2500 ms. Respondents were required to indicate by pressing a letter (A or L) on the keyboard whenever one of the presented stimuli matched the one presented *n* positions back in the sequence. The three tables show correct responses in an $n = 1$; $n = 2$ and $n = 3$ task paradigm (see the main text for a more exhaustive account).

From a theoretical viewpoint, two questions that are to some extent related should be addressed: first, given that the *n*-back is classified as a WM task, its relation to the construct of WM should be clarified. Second, far transfer training effects are based on the assumption that the constructs of intelligence and WM share the same domains, hence the overlapping processes between *n*-back task performance and *g* should be examined.

As stressed by Kane et al. (2007), the *n*-back has received little empirical justification as a WM measure. Early research by Roberts and Gibson (2002) reported a nonsignificant negative correlation between the *n*-back and a composite complex span score ($r = -0.09$), and significant positive correlations with the backward digit-span task ($r = 0.43$; $p < 0.05$) and a composite sentence memory score ($r = 0.48$; $p < 0.01$). However, the data were obtained on just 26 individuals. A slightly different pattern of correlations was reported by Oberauer (2005) obtained on a sample of 120 young individuals. A verbal *n*-back task (*n* size 1–3) adjusted for lures[f] correlated positively ($r = 0.22$ to $r = 0.27$; $p < 0.05$)

f. Example of a 1-back lure in a 2-back task: the second *A* in the sequence *B–A–A* is a 1-back lure (Kane et al., 2007).

with a verbal and spatial memory-updating task. Both tasks were according to Oberauer et al. (2000) classified as memory tasks that require mainly storage and transformation (complex span tasks). On the other hand, no significant correlations were observed for a reading span task, also classified as a memory task requiring mainly storage and transformation, and a spatial short-term memory task classified as requiring mainly integration (simple span task). Similarly, low correlations with the operation span task (OSPAN[g]) were reported by Kane et al. (2007) and Jaeggi et al. (2010). It was further shown that the *n*-back as well as the complex memory span task correlate with the Raven's matrices test. An interesting finding was that, even though *n*-back and OSPAN both predicted variance in Raven's Advanced Progressive Matrices (RAPM), they did so independently. It seems that the performance on the two tasks relates to different sources of variance. Both authors (Kane et al., 2007; Jaeggi et al., 2010) suggested that the main difference between the *n*-back and the OSPAN or similar WM capacity measures is that the former requires recognition and the latter serial recall. Recognition to a great extent minimizes interference, which is often responsible for the failure to recall anything (Kane et al., 2007). To overcome this problem, Shelton et al. (2010) used a modified version of the *n*-back that required recall. In their version of the task, the subjects were presented with lists of six-to-eight words and were asked to recall the word presented 1-, 2-, or 3-back. This version of the *n*-back correlated significantly with complex WM capacity measures. Furthermore, the recall version of the *n*-back (2- and 3-back) and the complex WM capacity tasks (operation and capacity span) both predicted common variance in a composite intelligence measure (matrix reasoning, RAPM, and block design). However, this cannot explain the shared domains between the dual/single *n*-back and intelligence. The recall *n*-back is with respect to the cognitive processes that are required for its performance more similar to a backward memory span task than to the original *n*-back. The main characteristic of the dual *n*-back is fast forward and backward switching between two streams of stimuli—visually displayed squares in one of eight different locations and consonants presented in the auditory domain (Jaeggi et al., 2010). Even the single *n*-back task demands swift shifting of elements in and out of the focus of attention. These processes are, at least from a theoretical viewpoint, rather similar to the solution requirements of a matrix reasoning test like the RAPM, in which switching between different educed relations and correlates is important (Jaeggi et al., 2010).

The question of shared variance between the *n*-back and intelligence must be seen in a broader context of the relationship between WM processes and intelligence, which is still to some extent mysterious (Colom et al., 2008). Different WM features were suggested as being central for intelligent reasoning. Cowan et al. (2005) proposed that the scope of attention that does not include

g. A complex measure of WM capacity requiring participants to solve mathematical operations [e.g., $(6/3) - 2 = 4$?] between to-be-remembered unrelated words or digits for immediate recall at the end of a series (2–5 equation–word pairs in the Kane et al. (2007) study).

any processing component showed high correlation with intelligence. In the same direction points a large-scale study by Colom et al. (2008), which revealed that simple short-term storage largely accounted for the relation between intelligence and working memory. In contrast, Engle et al. (1999) suggested that executive functioning, especially the control of attention over interference and conflict is the central process contributing to the shared variance with intelligence. Their idea has been further developed suggesting that when subjects are confronted with a complex WM test in which they have to solve a mathematical equation between the presentation of the to-be-remembered items (e.g., OSPAN), they have to displace the items from primary to secondary memory. The same happens for long lists in a memory span task (e.g., more than four items). The relation with intelligence derives from these two processes: the ability to retain information in primary memory, and to recover it from secondary memory (Unsworth and Engle, 2007). Even more extreme is the position that only the ability to retrieve information from secondary memory contributes to the shared variance between WM and intelligence (Mogle et al., 2008). However, this finding was not replicated (Shelton et al., 2010). Yet another process identified as crucial for the intelligence-WM overlap was relational integration, defined as the ability to build new relations between elements thereby creating structural representations (Oberauer et al., 2008).

A central problem of research into the relationship between intelligence and WM is the immense diversity of tasks used to measure WM. Some of these tasks resemble those used in intelligence tests, which would mean that the criterion is predicted by another example of the criterion (e.g., the tasks of relational integration used by Oberauer et al., 2008). Furthermore, the same tasks have been administered in different ways. For example, the display duration of stimuli in the *n*-back task ranges from 500 ms (Kane et al., 2007; Jaeggi et al., 2008) to 2500 ms (Roberts and Gibson, 2002) and even includes self-paced paradigms (Oberauer, 2005). Similar differences can be observed in the interstimulus intervals spanning from 200 ms (Oberauer, 2005), 500 ms (Roberts and Gibson, 2002), 2000 ms (Kane et al., 2007) to 2500 ms (Jaeggi et al., 2008). The main characteristics of the WM construct is the temporary storage of information, hence it can be assumed that these temporal differences in administration influenced study outcomes, making comparisons difficult. The same can be said of the recall version of the *n*-back task (Shelton et al., 2010), which actually resembles a backward memory span task, thus classifying it as *n*-back is misleading.

That even a small task modification can alter study outcomes was demonstrated by an analysis performed in our lab (unpublished data). Seventy-two female students participated in the study in which they solved four memory tasks requiring recall (forward and backward Corsi block-tapping and digit-span tasks) and four WM tasks requiring recognition (verbal and figural versions of a visual-array comparison task and of a single *n*-back task). The verbal form of the visual-array comparison task (Luck and Vogel, 1997) consisted of three letter words (e.g., dog), which were presented in the same way as the figural stimuli. In every trial, an array of colored squares (or words) was presented for 400 ms, followed by an interstimulus interval of 1000 ms, and a second array that was similar to the first one (for 2000 ms). A square in the second array was encircled and the respondent had to press "2" on the response pad if the square differed in color (different word) compared to the square in the previously presented array, or press "1" if it did not differ in color (same word). Three series with 4, 6, and 8 simultaneously presented squares were used. Each series consisted of 30 trials. A memory capacity score for the verbal and figural forms was determined for each respondent (Pashler, 1988) along with average reaction times for each type of task.

The stimuli in the figural *n*-back tasks were colored squares, while in the verbal version three-letter words were presented for 2000 ms with a 500 ms interstimulus interval. Three series of 20 + *n* items (1-, 2-, and 3-back) were presented. For each span task, a memory span score was determined. For each series of the *n*-back the number of correct responses (corrected for wrong responses) was determined. Figural and verbal *n*-backs (2- and 3-back) were collapsed into a figural and verbal *n*-back score. Since the correlation between the two forms was high ($r = 0.90$) we collapsed the forms into a total *n*-back score, which was used as the dependent variable in three regression analyses. Descriptive statistics and correlations among all measures are presented in Table 3.2.

TABLE 3.2 Descriptive Statistics and Pearson's Correlations Among All Memory Tasks Used in the Study

Measure	1	2	3	4	5	6	7	8	9
1. *n*-back T	—	0.13	0.06	0.12	0.26[a]	0.37[b]	0.24[a]	−0.26[a]	−0.12
2. DsFw		—	0.53[a]	0.33[b]	0.15	0.10	0.08	0.08	0.08
3. DsBw			—	0.27[a]	0.07	0.15	0.08	0.05	0.12
4. CbFw				—	0.31[b]	0.20	0.20	−0.14	0.02
5. CbBw					—	0.13	0.01	−0.01	−0.01
6. VacF						—	0.48[b]	−0.04	0.19
7. VacV							—	0.02	0.16
8. RTF								—	0.79[b]
9. RTV									—
Mean	81.90	5.33	5.83	5.77	5.43	4.16	2.98	722.56	840.55
SD	12.59	1.09	1.49	0.95	1.01	1.06	0.90	146.32	175.50

CbBw, Corsi block span backward; CbFw, Corsi block span forward; DsBw, digit-span backward; DsFw, digit-span forward; *n*-back T, average of All 2- and 3-back tasks; RTF, figural reaction time; RTV, verbal reaction time; VacF, capacity score figural visual-array comparison task; VacV, capacity score verbal visual-array comparison task.
[a]$p < 0.05$
[b]$p < 0.01$

To further explore the relations between *n*-back and other measures of working memory capacity, we conducted three stepwise regression analyses. In the first analysis, we included all six WM tasks and the average reaction time (RT) measures obtained during performance on the figural and verbal forms of the visual-array comparison task (Luck and Vogel, 1997). The aim was to determine which of these WM tasks accounted for common variance in the *n*-back. A second question was if differences in task modality (e.g., verbal vs. figural) affect this relation. For that purpose, we conducted two additional regression analyses, both of which included all four recall memory span tasks, but were separated based on task modality: one was based on the figural visual-array comparison task with the corresponding average reaction time, whereas the other was based on the verbal version. The results of the stepwise regression analysis are summarized in Table 3.3.

As can be seen in Table 3.3, when the figural form was used, speed, RT, and performance on the visual-array comparison task accounted for the variance on the *n*-back task. In contrast, when the verbal form was used, performance on the backward Corsi block-tapping task accounted for the variance on the *n*-back task. The two outcomes of the analyses differed significantly and would probably also have implications for the conclusions drawn. This once again highlights the problems related to the interpretation of WM training effects on intelligence using single/dual *n*-back tasks.

The regression analysis based on the whole set of WM tasks emphasizes yet another point to be considered in future research related to *n*-back training. As revealed in Table 3.3, fast and correct performance on the visual-array comparison

TABLE 3.3 Summary of Stepwise Regression Analyses [a]

Variable	*B*	*SE*	β	*t*	R^2
Stepwise regression analysis (1)					
Step 1: VacF	4.37	1.33	0.37	3.30**	0.13
Step 2: RTF	−0.02	0.01	−0.25	−2.30*	0.20
Step 3: CbBw	2.71	1.33	0.22	2.03*	0.24
Stepwise regression analysis (2)					
Step 1: VacF	0.84	0.23	0.38	3.61**	0.14
Step 2: RTF	−0.02	0.01	−0.25	−2.35*	0.19
Stepwise regression analysis (3)					
Step 1: CbBw	3.27	1.44	0.26	2.27*	0.07

* $p < 0.05$; ** $p < 0.01$
[a]*Using (1) all memory tasks (CbBw, Corsi block span backward; CbFw, Corsi block span forward; DsBw, digit-span backward; DsFw, digit-span forward; RTF, figural reaction time; RTV, verbal reaction time; VacF, capacity score figural visual-array comparison task; VacV, capacity score verbal visual-array comparison task), (2) four recall memory tasks and VacF, and (3) four recall memory tasks and VacV to predict* n-back*T (average 2- and 3-back tasks)*

task as well as performance on the backward Corsi block-tapping task accounted for 24% of variance on the *n*-back task. According to Cowan et al. (2005), the visual-array task is a measure of the scope of attention limited to three-to-four objects. This limitation does not stem from a failure to modulate attentional control, but from a capacity limit—in the sense of the neurocognitive model of discrete slots, each with all-or-none properties (e.g., Luck and Vogel, 2013).

Analyzing the characteristics of forward and backward versions of the Corsi block task in relation to Baddeley and Hitch's (1974) multicomponent model of WM, indicated a clear relation with the visuospatial sketchpad and for longer sequences also the involvement of the central executive. Furthermore, there was evidence that in the backward presentation some reliance on the phonological loop was present for longer sequences. It was suggested that as the task becomes more difficult, subjects may call on verbal support to cope with the situation (Vandierendonck et al., 2004). In general, the backward presentation was more difficult than the forward one, not only requiring visuospatial representation but also stronger order representations (Higo et al., 2014). All these findings suggest that the shared domain between WM processes and the *n*-back task mainly relates to capacity (inherent limitation not related to attentional control) of visuospatial and order representations. This would also explain why *n*-back training had no influence on WM measured with simple or complex WM tasks, such as the reading-span, digit-span, or OSPAN (Jaeggi et al., 2008, 2010).

From a neurocognitive perspective, the overlap seems even more determined. Todd and Marois's (2004) fMRI study showed that performance on the visual-array comparison task was associated with activity in the intraoccipital and intraparietal sulci. Recent findings support that the superior intraparietal sulcus plays a central role in visual short-term storage (Bettencourt and Xu, 2016). For performance on the Corsi block tasks a lesion mapping analysis of stroke patients revealed a robust correlation with damage to the right dorsolateral prefrontal cortex and the posterior parietal cortex (Asselen et al., 2006). These findings were confirmed and extended in a recent study by Chechlacz et al. (2014). Poor performance on the Corsi block task was related to bilateral lesions within the middle occipital gyrus, the right supramarginal gyrus, angular gyrus, and disconnection within the right posterior segment of the perisylvian network. The impairment was greater for right hemisphere lesions and strongly correlated with lesion size. Neurocognitive research on brain activity during performance on the *n*-back is even more extensive. A metaanalysis by Owen et al. (2005) including 668 foci from 24 studies revealed six cortical regions related to performance on the *n*-back mainly including parietal (precuneus and inferior parietal lobules BA7, 40), and frontal areas (premotor and supplementary motor cortex, BA6, 8, 32; prefrontal cortex BA9, 10, 46, 45, 46). A region in the medial cerebellum was also identified. Frontoparietal involvement in relation to the *n*-back was also confirmed by a recent lesion mapping study (Barbey et al., 2014), an analysis based on network theory (Braun et al., 2015), and in a fNIRS study (León-Domínguez et al., 2015).

This brief review has shown that there is cognitive and neurocognitive evidence in favor of the suggested theoretical framework for the relation among *n*-back training, WM, and fluid intelligence that can explain far transfer effects (e.g., Jaeggi et al., 2008, 2010).

3.4.2 The *n*-Back in Action

The positive influence of dual *n*-back training on fluid intelligence reported in Jaeggi's first study (Jaeggi et al., 2008) was also observed in a group of undergraduates from the National Taiwan Normal University in Taipei (Jaeggi et al., 2010). In this study, two groups of students were trained for 4 weeks on either a single or a dual *n*-back task. Both groups, in comparison with a no-contact control group, showed significant improvements on two tests of fluid intelligence. An interesting finding was that single *n*-back training was equally effective as dual *n*-back training.

Similar improvements were also observed in a group of elementary and middle school children (Jaeggi et al., 2011). In this study, children in the experimental group trained on an adaptive spatial single *n*-back instead of a dual *n*-back task. By contrast, children in the active control group learned how to answer questions related to general knowledge and vocabulary. This intervention was assumed to promote skills related to crystallized intelligence. Effects on trained tasks were only observed for the experimental group; however, there were no transfer effects on measures of fluid intelligence. In an additional analysis, the experimental group was split in relation to training performance (below/above average training gains). Far transfer effects on fluid intelligence were only present for students showing pronounced training gains on the *n*-back, whereas no effects were found for the control group, and the students in the experimental group that displayed low *n*-back training gains. Furthermore, far transfer effects remained stable even after a 3-month hiatus from training. The authors concluded that far transfer effects can only be expected in the presence of training gains on the trained tasks. A questionnaire revealed that although the training was adaptive, some children who had improved the least reported that they found the tasks difficult and effortful.

To further analyze training characteristics that may improve far transfer effects Jaeggi et al. (Katz et al., 2014) tested seven versions of a game-like *n*-back training paradigm. All training versions were adaptive and designed to measure the influence of five motivational features: points (a bar at the bottom of the screen displayed the actual score); theme changes (several different themes were included using different characters, such as a frog or a cat); lives and levels (a level indicating *n* was shown, as well as lives, indicating the number of errors that could be made before dropping to a lower *n* level); prizes were offered after training in exchange for points, and end-of-session certificates were awarded showing the level of *n* a person reached. A surprising finding was that none of the provided incentives affected training improvement. On the contrary, children performed better when no points, levels, or lives were shown compared to a group of children that were exposed to all five motivational features. These findings are counterintuitive and hard to explain. The authors suggested that given the demanding nature of the *n*-back task, additional information might have distracted the attention of trainees.

In yet another study, Jaeggi et al. (2014) investigated whether individual differences of trainees (their need for cognition and their implicit theories about intelligence) influence transfer training effects. Students were randomly assigned to three training groups: a dual *n*-back; a single *n*-back where subjects trained on just the auditory part of the dual *n*-back task, and an active control group where subjects learned how to answer questions related to general knowledge and vocabulary. The results replicated previous findings indicating that both training groups showed far transfer effects to a composite score representing five visuospatial reasoning measures. Of particular interest is the finding that training on the *n*-back task presented in the auditory stream resulted in transfer to measures of visuospatial reasoning, which implies that the transfer is modality independent. The moderator variable *need for cognition,* which indicated how much one enjoys cognitively challenging tasks, did not significantly influence transfer effects. On the other hand, individuals who thought that intelligence can be changed (is malleable) showed more transfer to visuospatial ability. Besides replicating previous findings, this study is important because it resolved several methodological shortcomings that were present in previous studies conducted by the same research group.

Another question that attracted the researchers that had used the *n*-back as a training task was what happens in the brain when training improves performance on a trained task? To answer this question Buschkuehl et al. (2014) used a technique called arterial spin labeling (ASL) which is from a methodological viewpoint more suited for longitudinal studies (such as cognitive training studies) than the more often used BOLD measure that is prone to scanner drifts (e.g., Hernandez-Garcia and Buschkuehl, 2012). The participants in the experimental group trained on an adaptive visuospatial *n*-back task whereas the control group answered vocabulary and general knowledge questions. Before and after training, the participants solved an easy 1-back and a more difficult 4-back task in a MRI scanner. The MRI analysis used perfusion as a surrogate for neural activity indicating increases in activity (contrast difference 4-back −1-back) in several brain areas. The results showed that higher task proficiency was related to increased resting perfusion in frontal brain regions accompanied with decreased activity in one parietal cluster. A combined analysis of all factors (e.g., rest vs. *n*-back proficiency and control vs. training) revealed increased activity after the intervention in the BA6 area located in the left precentral gyrus/frontal middle gyrus/superior frontal gyrus, a finding that is in line with previous research (Section 3.3.1). The authors concluded that the MRI analysis revealed that training on the *n*-back increased the physical fitness of the brain—a condition that might also explain other transfer effects.

3.5 COGITO

The COGITO study (Schmiedek et al., 2010, 2013, 2014; Shing et al., 2012) was carried out at the Max Planck Institute for Human Development in Berlin as part of the German Socio-Economic Panel (SOEP). SOEP is a wide-ranging representative longitudinal study of private households located at the German Institute for Economic Research, DIW Berlin.

In total, 101 younger (20–31 years) and 103 older adults (65–80 years) participated in the study. About half of the individuals in both groups were females. On average, the participants completed 101 practice sessions (1.0–1.5 h per session). Participants in the no-contact control group were 44 younger and 39 older adults who were matched on age, sex, initial cognitive status, and education with the individuals in the training group. A pre–posttest study design was adopted. The testing was carried out in 10 sessions (2.0–2.5 h) consisting of cognitive tests and self-reported questionnaires. Far transfer effects were assessed with the BIS (Berlin Intelligence Structure test), the Raven's Matrices test and established complex WM span tasks (e.g., counting, reading, and rotation span). Near transfer was assessed with parallel forms of the training tasks. Test retest assessment was separated by 188–197 days. The remuneration for participating was 1450–1950 EUR in the intervention group and 460 EUR (exchange rate: approximately 1 EUR = 1.1 USD) in the control group. It depended on the number of completed sessions and the rate at which the participants completed the whole training. I mention this for two reasons; first, the compensation is quite high in comparison to other studies, and second, several findings have suggested that extrinsic motivation, such as money, can reduce intrinsic motivation for training (e.g., Au et al., 2015, 2016; Jaeggi et al., 2014). Furthermore, it is possible that the offered payment may also bias recruitment of individuals, attracting those who are more interested in money than in self-improvement (e.g., Duckworth et al., 2007). The metaanalysis of *n*-back training studies by Au et al. (2015) actually confirmed this hypothesis indicating a negative correlation between the amount of remuneration and effect size, although when outliers were excluded the relation was no longer significant. The relation became interesting again in the combination with several other factors addressed in the metaanalysis: differences between the studies conducted in the United States and Europe (the former showing less training effects), differences in the type of control groups employed (less training effects when active control groups were used), and the finding that the type of control group (active, no-contact) had no influence on training gains of experimental groups (Section 3.3 for more detail). This led the authors to suggest that studies conducted outside the United States offered less remuneration for participating in experiments, which might have had a positive influence on the motivation of participants thereby increasing training effects in European studies (Au et al., 2016).

The participants in the COGITO study trained individually in lab rooms with up to six computers. The training consisted of 12 different tasks classified as perceptual speed (six tasks), WM (three tasks), and episodic memory (three tasks). The memory tasks were adaptive in terms of presentation time, which was adjusted in relation to individual performance on trained memory tasks. The perceptual speed tasks were further divided into choice reaction and comparison tasks. All task types were presented in a numerical, verbal, and figural layout, requesting participants to, for example, judge as quickly as possible if two strings of numbers presented on the computer monitor were of the same length (numerical form of the comparison task). In the three episodic memory-training tasks, the participants had to remember lists of words, pairs of numbers and words, and the position of 12 colored real world objects in a 6×6 grid. The presentation duration for all to-be-remembered items varied between 1, 2, and 4 s, whereas the interstimulus interval was always 1 s. In the number–noun pairs task, for instance, 12 pairs of two digit numbers together with a noun in plural form (e.g., 22 dogs) were presented sequentially each second, for a duration of either 1, 2, or 4 s. When the sequence ended the nouns appeared randomly on the monitor and the participants had to add the correct numbers. The WM tasks included (1) a 3-back task using figural stimuli appearing in different locations in a 4×4 grid, (2) alpha span that required the participants to determine the correspondence between simultaneously presented consonants and numbers with respect to the position in the alphabet and the already presented letters, and (3) numerical memory updating in which the participants had to perform addition and subtraction operations from four numbers simultaneously presented in a grid for a duration of 4 s, after which the requested operations (±) and numbers appeared in a second row. The net training gains of WM and fluid intelligence compared with those of the no-contact control group were tested with a model that allowed for a latent variable analysis. The results can be summarized as follows. Improvements in trained tasks were observed for all tasks except that older participants showed no gains for the episodic memory task where lists of words had to be remembered. At the latent factor level significant far transfer effects to fluid intelligence and to episodic memory were only observed for younger individuals, while for older individuals a near transfer effect to WM was established. Latent factor analysis for the other constructs (perceptual speed and far transfer to WM) was not reliable. On the level of individual tasks, 17 comparisons for each age group in relation to the control group were performed. In the younger group, seven comparisons showed significant positive training gains and one (figural perceptual speed) was negative. With respect to fluid intelligence, significant increases were present for the figural/spatial and numerical reasoning tasks, whereas for Raven's matrices a significant increase just on trend level was present ($p < 0.06$). There were no significant increases for the verbal reasoning test. Further, no gains were observed for far transfer on memory tasks, though there was a pronounced positive far transfer effect for episodic memory (verbal, spatial figural, and numerical). By contrast, in the group of older individuals only 4 out of 17 comparisons showed a significant training effect. Significant gains were reported for fluid intelligence (performance on Raven's matrices), for working memory on reading and rotation span, and verbal episodic memory, while no significant gains were observed for the perceptual speed tasks.

The authors also addressed the methodological issue related to no-contact control groups. Namely, they attempted to rule out this issue with a statistical analysis showing that in the training group, transfer effects on specific latent factors were more related to the specific tasks practiced during training than to tasks that trained other abilities, which was interpreted as an ability specific transfer effect (Schmiedek et al., 2010). In my opinion, an active control group would be a far more efficient way to rule out all possible motivational and placebo influences.

The importance of the obtained results was further confirmed in a recent follow-up analysis performed 2 years after the training had ended (Schmiedek et al., 2014). The analysis was performed on 80 younger individuals of the intervention group, and on 32 individuals of the no-contact control group. The results at the latent factor level at 2-year follow-up showed that the gains in episodic memory and fluid intelligence were retained. However, for episodic memory, this gain was reduced significantly at follow up compared to net gains at posttest, whereas for the latent factor of reasoning, no such effect was observed. Furthermore, analysis at task level for three reasoning tasks (except Raven's matrices) still showed significant transfer effects at follow-up, whereas for the memory tasks, these effects were no longer present. An interesting observation in relation to the difference between episodic memory and reasoning, not addressed by the authors, was that latent means for reasoning in the control group at follow-up dropped to pretest level, whereas for episodic memory they increased, which was opposite to the trend observed in the trained group. It is further worth mentioning that at follow-up, the near transfer effect for the 3-back task was also significant.

The data obtained in the COGITO study were also analyzed with respect to several age-related differences in training gains, e.g., showing that for older adults, longer presentation times were more beneficial than when they practiced the tasks at a faster pace (Shing et al., 2012). Another interesting finding reported by Schmiedek et al. (2013) was based on the intuitively reported impression that there are days when everything goes wrong, whereas on other days everything seems perfect. The study showed that variability in cognitive performance does indeed exist. However, the impression that the whole day goes wrong is false. Most performance fluctuations occurred within shorter periods of time. Surprisingly, older adults showed less day-to-day performance variability in all tasks compared to younger adults. The reasons were multifaceted, but mainly due to learned strategies to solve tasks, one could say experience, high motivation, balanced daily routine, and stable mood (Schmiedek et al., 2013).

On the whole, the COGITO study is an example of well-designed research, except for the lack of an active control group, which to some extent reduces the validity of observed findings. One point that should be mentioned is the high remuneration given to trainees but also to the controls, an aspect that makes the explanation by Au et al. (2016)—that differences in training gains obtained in US and non-US studies might be attributed to the negative effect of higher payments in the former, and by that reducing intrinsic motivation—less plausible. Hence the search for a moderating variable explaining the discrepancy in relation to training gains (far transfer/no far transfer), region where the study was conducted (the United States/Europe), and the type of control group (active/passive) continues.

3.6 RESEARCH FROM OUR LAB

The main objectives of our study were to investigate whether WM training can improve performance on tests of fluid intelligence, and whether these influences are observable in brain activity. In order to gain deeper insight into brain activity under cognitive load, a multimodal brain imaging approach was used, combining EEG methodology (based on electrophysiological principles) with NIRS imaging technique (based on hemodynamic principles). This multimodal approach compensated for the poor spatial resolution of EEG and allowed for more reliable testing of the main research questions (Jaušovec and Jaušovec, 2012b). Further, the experiment was designed to eliminate some of the criticisms (Sternberg, 2008; Moody, 2009) of the Jaeggi et al. (2008) study, and integrate suggestions for improvements put forward in several reviews (e.g., Buschkuehl and Jaeggi, 2010; Klingberg, 2010; Morrison and Chein, 2011) on WM training (e.g., an active control group was included, different types of training tasks as well as different tests of fluid intelligence were used). We used a multifactorial training approach targeting multiple cognitive skills (von Bastian and Oberauer, 2013). Besides the diversity of tasks used, the training involved different instructional approaches and platforms: (1) computerized WM training tasks which the trainees performed individually at home, (2) lecture-style instruction by the experimenter with tasks presented on a projection screen to the whole training group, but individually solved by paper-and-pencil testing, and (3) social games performed in small groups (three-to-five students).

The tasks used in the WM training were theoretically based on the multicomponent model of WM proposed by Baddeley et al. (2011). This model provided a framework for the distinction between different inputs related to the two slave systems (e.g., visuospatial sketchpad: visually presented objects of different form, color, content, and spatial location; phonological-loop: auditory and visually presented information, including speech and music). An additional component was the requirement for motor performance, which was achieved by paper-and-pencil testing during the lab training sessions. For all tasks, the participants were given printed tables in which they had to mark the correct answers. This was rather easy for the memory maintenance tasks and also for the matrices, but more demanding for the single/dual *n*-back, requiring the respondent to shift attention not only in relation to the task presented on the projection screen, but also to timely mark the correct answer (+ equal; − not equal) in the table placed on the desk in front of the participant. The motor act required an additional shift in attention, but also a change from visual input to complex motor action (e.g., finding the right column in the table for the ± mark in a narrow timeframe). Theoretically, the educational approach was based on the behaviorist paradigm of Keller's (1968) mastery learning: providing small discreet course units, with observable criteria for achievement, self-pacing, and feedback for each participant in form of weekly updated graphically displayed average percentages of correctly solved tasks. On the other hand, the two social games performed in between the computerized WM training were based on Vygotsky's (1977) social cognitive learning theory, allowing for social interaction and communication between trainees while performing the tasks.

The computerized WM training included three task types: matrices, memory updating, and *n*-back. Matrices consisted of 4 × 4 matrices with eight pairs of elements; 4 × 5 matrices with 10 pairs and 5 × 6 matrices with 15 pairs of elements (Fig. 3.4. showing an example of an eight pair matrix). The matrices were, depending on the number of pairs, displayed between 45–90 s on a 2 × 2 m projection screen. After this time period, the matrix that showed all elements was replaced with a matrix in which just one element was displayed (right matrix in Fig. 3.4). The respondents were asked to mark the position of the corresponding pair on an answer sheet. The task continued until all pairs were identified. We gradually

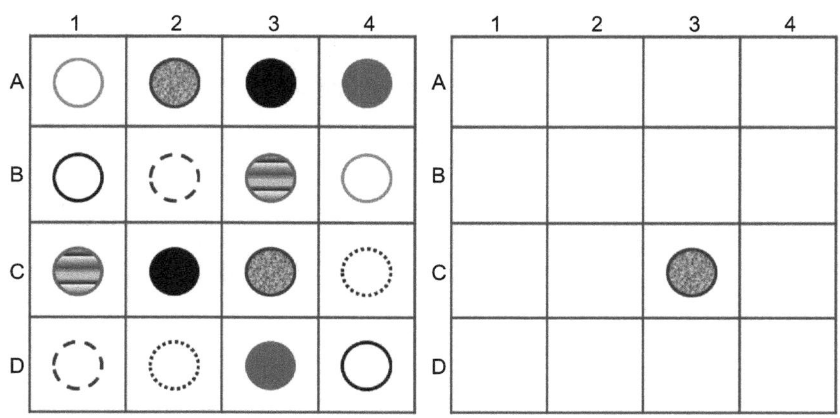

FIGURE 3.4 The matrices tasks required subjects to remember the position of the eight pairs presented in the left matrix (exposed for 75 s). For example, the circle in A1 corresponds to the circle in B4; A2 corresponds to C3, and so forth. After this exposition period, the left matrix was replaced by the right matrix showing one element of the previously displayed pairs (the circle in C3). The respondents were asked to mark the position of the corresponding pair, which was for the present example—A2. After respondents marked their answer, the left matrix was presented again, followed by a matrix showing just one of the circles (easy version). In the more difficult version, respondents were just shown another element and asked to identify its pair. This procedure continued until all eight pairs were identified.

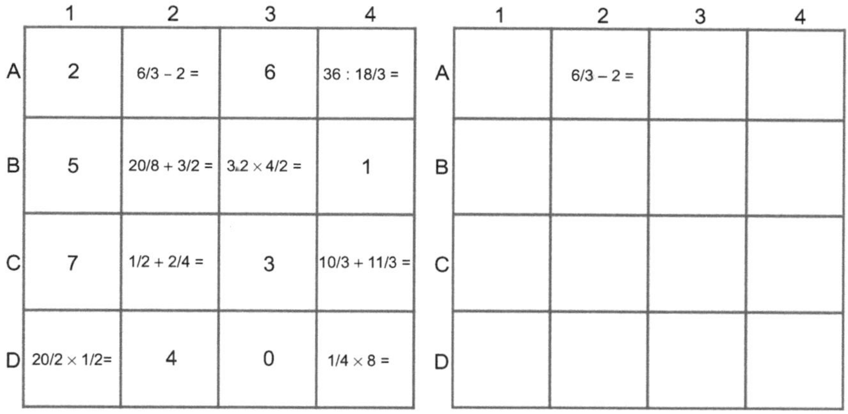

FIGURE 3.5 **Example of the matrices tasks showing eight equations and their corresponding solutions (e.g., equation in A2: 6/2 − 3 = the corresponding solution is 0 in D3; see the main text for a more exhaustive account).**

increased the difficulty level of the tasks by increasing the number of pairs, their similarity, and by reducing the number of times the initial matrix with all elements was presented (after each answer was marked) from four to just one presentation. The elements in the matrices were presented visually (e.g., pictures, words), or aurally (e.g., piano tones, beep tones, short music clips). The content of the elements was figural, symbolic, or semantic. The difficulty level was further increased by changing the content of the target and initially presented elements (e.g., combining equations with corresponding solutions, Fig. 3.5).

Working memory maintenance tasks were similar to the task used by Crone et al. (2006). They contained different visual and auditory elements. At each session all of the elements used in the session (10–50) were presented first on the projection screen, then a blank screen was presented for 500 ms followed by a set of three-to-five items (800–1500 ms presentation per item and 500 ms between items). Next, the instruction “forward” or “backward” appeared on the screen, followed by a time delay of 6000–8000 ms. Finally, one of the target elements was displayed for 2000 ms, during which the participants wrote the number that indicated the correct order of the presented target element with respect to the instruction provided (Fig. 3.6).

The number of elements presented (three to five), as well as rules (forward vs. backward) was alternated randomly between trials. As in the matrices tasks, the content of the elements was figural, symbolic, or semantic. In some tasks, the content or the modality of presentation (visual vs. auditory) of the three-to-five set of elements differed from the target element (e.g., visually presented picture of a train combined with an aurally presented sound of a train). These properties of the elements were used to gradually increase the level of difficulty. Fig. 3.7, for instance, shows a Stroop-like task.

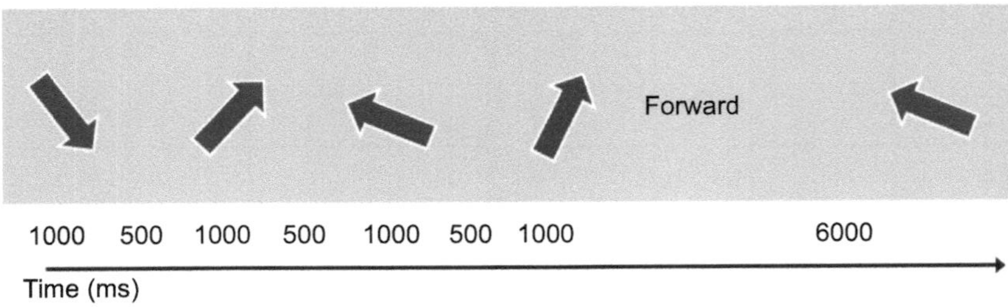

FIGURE 3.6 Example of the memory maintenance tasks that consisted of sets of three-to-five items (800–1500 ms presentation per item and 500 ms between items). Next, the instruction "forward" or "backward" appeared on the screen, followed by a time delay of 6000–8000 ms after which one of the target elements was displayed for 2000 ms. At that time, participants wrote down the number that indicated the correct order of the presented target element with respect to the instruction provided. In the example, the correct answer would be 3 for the forward instruction and 2 for the backward instruction.

n-Back tasks were similar to the tasks used in the Jaeggi et al. (2008) study. The task consisted of blocks of 20 + *n* items presented visually, aurally or in a combination of both (1000–2000 ms presentation per item and 500 ms between items). The participants were asked to indicate if the currently presented item on the screen was identical or not to the previously presented item (1-back), the item presented before the previously presented item (2-back), or the item presented before the penultimate item (3-back). Beside the *n*-back condition, the complexity (single/dual) of the presented items was used to gradually increase the difficulty level.

The memory and sentence construction tasks resembled a game and were played in groups of three-to-five participants. The memory game consisted of 72 cards (36 pairs). At the beginning, all cards were placed backside-up and were spread out on the table. One of the participants in the group uncovered two of the cards, if identical he/she kept the pair and had the opportunity to uncover the next pair, if not, the cards were returned backside-up to the set on the table, and the next player proceeded. The game was played until no cards were left on the table. The sentence construction task required the first participant in

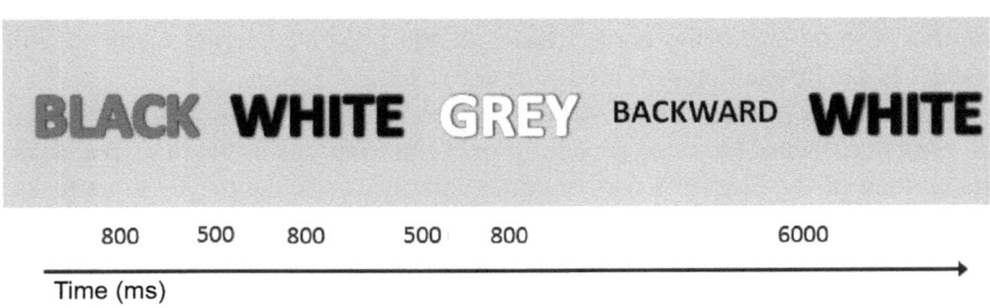

FIGURE 3.7 Example of the memory maintenance task presented in a Stroop-like format (correct answer, 2; see the main text for a more exhaustive account).

the group to say a word, the next one had to repeat this word and add another word and so on, until one of the respondents could not correctly repeat the word chain.

The training session started with one of the tasks presented on the screen, and was then alternated by one of the tasks played in groups. Each of the tasks was administered for about 15–20 min. The training lasted for 12 weeks with one training session per week (75–100 min). The matrices, *n*-back and memory maintenance tasks were uploaded to the university server and the participants were instructed to practice the WM tasks twice a week for 30 min. If they failed to do so (the e-learning program provided information about individual training sessions), they were reminded by e-mail to continue training. On average each of the participants completed 30 h of training. The participants were provided with information about their training results on a weekly basis.

The active control group was trained in communication and social skills mainly based on Gordon's communication training (Gordon, 1974). It was assumed that this training was not directly related to fluid intelligence. However, to resemble the training conditions of the WM group, similar forms of presentation, interaction between participants and duration of the training sessions were applied. The training took place during the same time interval as the WM training. The same number of lessons as for the WM group was provided. The tasks and videos used were presented on a screen, whereas role-playing and interactive exercises were practiced in groups of two-to-three participants. Participants were further asked to practice weekly the acquired techniques with their friends and colleagues and to report about the training sessions by presenting a written or video report (all together 9 h). The training consisted of 12 h (1 h weekly) of a theoretical introduction accompanied by examples (video clips). An experimenter pointed out different communication skills, obstacles in communication, active and passive listening, examples of effective communication, mirroring, identifying emotions in others, and in themselves. After each theoretical introduction, about 45 min (all together 9 h) were devoted to applying the learned skills in provided scenarios. On average each of the participants completed 30 h of training. One of the scenarios would read:

A friend asks you for advice. His/her company has just announced that about 30% of the people will lose their jobs sometime in the next 6 months. The final decisions are still being made, and it will be at least 4 weeks before there is any announcement about the cut. What should he/she do?

Thirty students participated in the study (mean age 20 years and 3 months; 4 male and 26 female participants—one female student of the WM group did not complete the training and was therefore not invited for the posttest session). They were assigned to two groups (WM, working memory group, $n = 14$, $M_{IQ} = 105.38$, $SD = 9.25$; and AC, active control group, $n = 15$, $M_{IQ} = 105.40$, $SD = 8.93$) that were equalized with respect to gender and intelligence (WAIS-R, Wechsler, 1981). The participants were informed about the goal of the experiment—to improve intelligence with training. They were further told that the training will aim at improving working memory (WM group), or components of emotional intelligence (AC group). The participants were given a partial course credit (40%) for participating in the research.

A pre–posttest experimental design was employed. The participants solved four test-batteries; the same procedure, and the same test-batteries were used for pre- and posttesting. The digit-span subtest (WAIS-R) was administered separately. The subjects were tested on the whole set of seven number strings (three to nine numbers read aloud by the experimenter) even if they failed on two items of a set of the same length. This was done because it was shown that longer number strings (>4) are a much better measure of WM capacity corresponding with complex memory span measures (e.g., Unsworth and Engle, 2007). The RAPM was based on a modified version of Raven's progressive matrices (Raven, 1990) and included 50 test items—25 easy (Advanced Progressive Matrices Set I–12 items and the B Set of the Colored Progressive Matrices), and 25 difficult items (Advanced Progressive Matrices Set II, items 12–36). A figural matrix was presented in which the lower right entry was missing and the participants had to determine which of the four options fit into the missing space. The tasks were presented on a computer screen (positioned about 80–100 cm in front of the participant), at fixed 10 or 14 s interstimulus intervals. They were exposed for 6 s (easy) or 10 s (difficult) following a 2-s interval, when a cross was presented. During this time the participants were instructed to press a button on a response pad (1–4), which indicated their answer. The verbal analogy test was based on a well-established Slovene intelligence test-BTI (Mihelič, 1972). The participants were presented with a verbal analogy with one word missing and were asked to chose one out of four options presented (e.g., FISH: WATER, BIRD: ?; 1 = AIR; 2 = BRANCH; 3 = SHORE; 4 = WAVE). The presentation was the same as for the RAPM, except that the 25 analogies were exposed for 7 s. The spatial rotation test was based on the Paper Folding and Cutting—PF&C subtest of the Stanford–Binet IQ test (Rideout and Laubach, 1996), a measure of visual–spatial reasoning. The participants were asked to decide which of the four figures on the right corresponded to the figure in the left frame (last row in Fig. 3.1). The figure in the left frame showed how a piece of paper was folded and cut, whereas in the right frame four unfolded papers were shown, one of which corresponded to the folded paper in the left frame. The presentation of the 25 items was the same as for the verbal analogy test. Based on these four test batteries a composite score of fluid intelligence was determined (M= 100; SD = 10).

Upon arrival at the lab, the participants were asked to relax with their eyes closed during which baseline EEG and NIRS recordings were made (duration = 5 min). The tasks were administered in a fixed order: RAPM, verbal analogies, and PF&C. Prior to each test the participants solved four-to-five probe items to get acquainted with the response pad. The posttest was administered 6 months after the pretest, within 10 days after the training had stopped. This 10-day time span was necessary to allow for the female respondents to be tested at about the same time (±2 days) of their menstrual cycle as on the pretest.

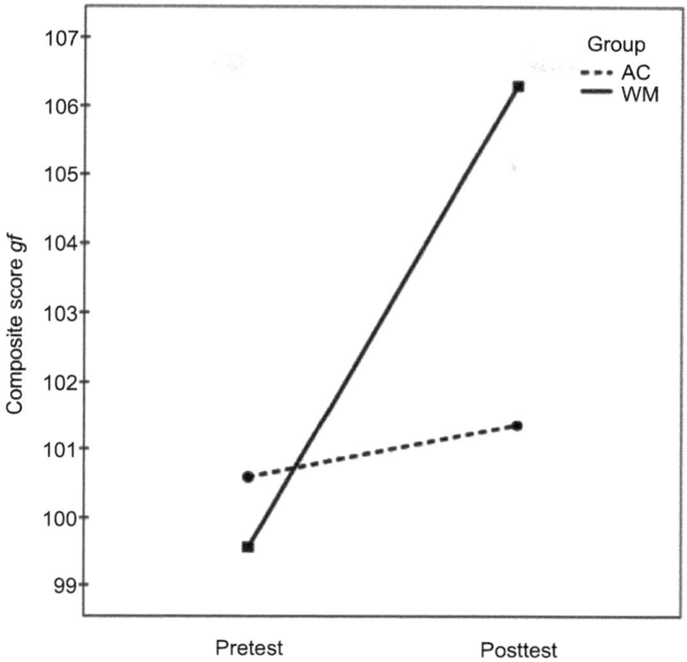

FIGURE 3.8 Line graphs of performance changes on a composite measure of *gf* (active control and working memory group) observed between pre- and posttests.

The training showed a robust positive effect on the composite *gf* measure which was present in the WM group but not in the controls. This was indicated by a general linear model for repeated measures with the within subject factor pre–posttest and the between subject factor group (WM vs. active control). The interaction for these factors was significant [F (1, 27) = 7.40; $p < 0.01$; $\eta^2 = 0.22$]. As can be seen in Fig. 3.8, participants in the active control group also increased performance on posttest, however the difference was not significant as revealed by a paired sample t-test [t (14) = 0.47; $p < 0.65$, $d = 0.12$]. By contrast, the WM group significantly increased on posttest performance indicated by the composite *gf* measure [t (13) = 4.75; $p < 0.0004$, $d = 1.27$].

Of interest (Fig. 3.9) was also the finding that the WM group increased significantly in performance on the digit-span task [t (13) = 3.05; $p < 0.009$, $d = 0.82$], while on the other hand the increase of the controls from pre- to posttest was not significant [t (14) = 0.98; $p < 0.34$, $d = 0.25$]. This finding is important since the test was administered with the whole set of forward and backward number sequences, and was not terminated when subjects failed on two successive number strings. In that way, the longer (e.g., supraspan) sets were also tested. It is assumed that if a memory task exceeds the capacity limit

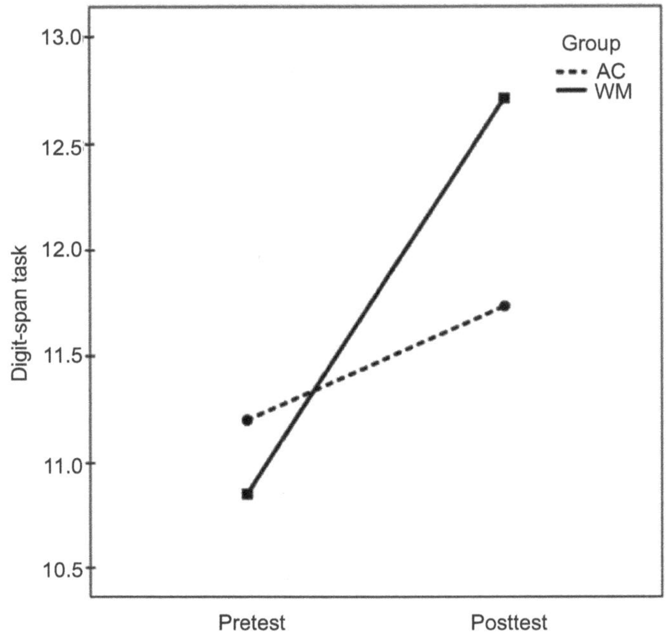

FIGURE 3.9 Line graphs of performance changes on the digit-span task (active control and working memory group) observed between pre- and posttests.

of primary memory (similar to scope of attention) individuals move the to-be-remembered items to secondary memory (Shelton et al., 2010). As mentioned in Section 3.4.1, some authors believe that the shared variance between intelligence and WM can be explained by the ability to retain information in primary memory and to recover it from secondary memory (Unsworth and Engle, 2007), or just by the ability to retrieve information from secondary memory (Mogle et al., 2008).

Training-induced changes in performance on tests of fluid intelligence were accompanied by changes in neuroelectric and hemodynamic patterns of brain activity. The characteristics of the impact of WM training on neuroelectric brain activity were time-related synchronization (ERS) in the theta and lower-1 alpha bands and a more topographically significant (location by hemisphere) increase in lower-2 and upper alpha desynchronization (ERD). The increase in theta synchronization was most pronounced in frontal and parietooccipital brain areas (Fig. 3.10). The frontoparietal topography of theta synchronization (for a review see Klimesch et al., 2008; Sauseng et al., 2010) is an often reported characteristic of theta during encoding and retrieval (parietal), and during the performance of central executive and attentional control (frontal). Theta synchronization was accompanied by lower-1 alpha synchronization in the occipital brain areas and an early left parietal desynchronization (ERD). Given the attentional function of

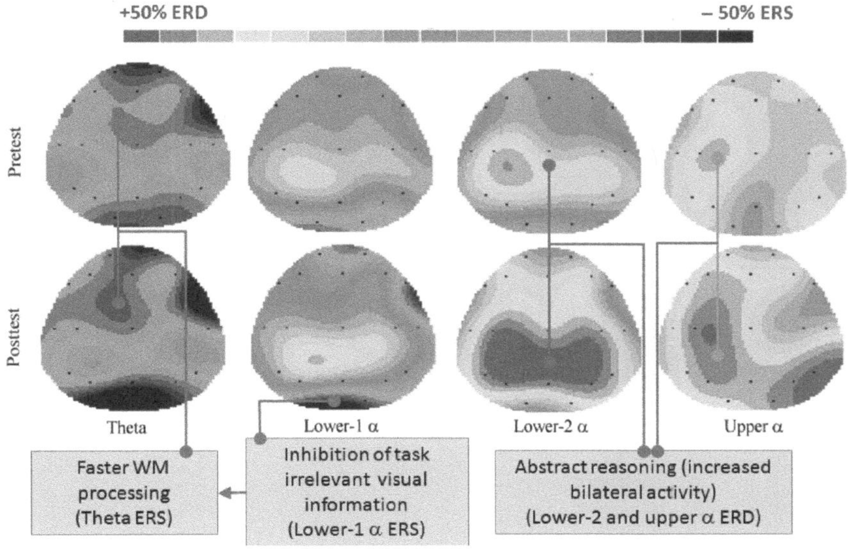

FIGURE 3.10 Topographic ERD/ERS maps of respondents' (working memory group) neuroelectric activity in the theta and the three α frequency bands averaged for all three test batteries during pre- and posttest (see the main text for a more exhaustive account).

the lower-1 alpha, the ERD/ERS topography would suggest inhibition processes in the occipital area probably suppressing task irrelevant information in the visual cortex and in that way allowing for faster WM processing, as was indicated by earlier occurring patterns of theta synchronization that were topographically similar to the ones observed in individuals prior to WM training. This is in line with research relating the level of intelligence with brain activity (Jaušovec and Jaušovec, 2004). Specifically, it was found that during performance on a WM task, high intelligent individuals displayed greater theta synchronization in the early phases of task completion as compared to low intelligent individuals. Interestingly, WM training influenced the two areas of the brain that are assumed to constitute the central executive network (CEN), a highly important network involved in complex information processing (e.g., van den Heuvel and Sporns, 2013). According to the parieto-frontal integration theory by Jung and Haier (2007), the fidelity of neuronal transmissions in these areas is a key characteristic distinguishing high versus low intelligent individuals. A similar involvement of brain areas in intelligent reasoning was also suggested in Duncan's (2010) multidemand system theory.

Neuroelectric activity in the lower-2 and upper alpha band after WM training showed increased desynchronization (ERD). In the lower-2 alpha band, this desynchronization after WM training occurred earlier and was more intense, located in the left and right parietal areas, as opposed to a more left hemispheric parietal ERD prior to training. In the upper alpha band, after WM training an

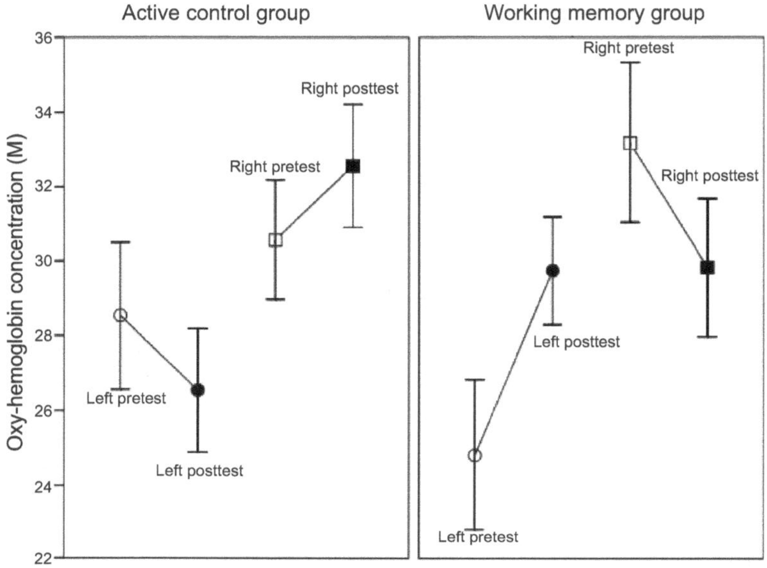

FIGURE 3.11 Oxy-hemoglobin concentrations in the right and left frontal brain areas of respondents (active control and working memory group) averaged for all three test-batteries during pre- and posttesting.

early right parietal ERD (1000 ms) intensified and spread to a left parietal and bipolar frontal ERD pattern (6000 ms). Prior to training, less intense and more scattered ERD patterns in the upper alpha band were observed (Fig. 3.10).

Even though research on the relations between NIRS and EEG measures (Koch et al., 2006) suggests that electrophysiological resonance between an individual's alpha frequency and stimulation is not mirrored by the vascular response (probably because there is no straightforward translation of neuronal excitation and inhibition into respective increases or decreases in vascular response), the training induced changes in upper alpha ERD/ERS were to some extent similar to the distribution of oxygenated hemoglobin changes in frontal brain areas. After WM training about equal concentrations of oxyhemoglobin in the left and right frontal areas could be observed (Fig. 3.11). This equalization in oxyhemoglobin concentrations was due to increases in left hemisphere and decreases in right hemisphere oxyhemoglobin concentrations. A similar mixture of increases and decreases in frontoparietal brain areas after different types of WM training was also reported by studies using functional magnetic resonance imaging (e.g., Moore et al., 2006; Olesen et al., 2003; for a review see, Klingberg, 2010; Section 3.3.1 this chapter).

The only significant change in neuroelectric activity observed in the control participants was an increase in theta synchronization. Because it is well documented that with increasing WM load, theta power increases (Klimesch, 1999;

Gevins and Smith, 2000) this could point to more intense WM processes in these participants. Similar findings were obtained for changes in frontal oxyhemoglobin concentrations. After training, the same but more intensified pattern in right/left frontal brain activity was observed compared to the hemodynamic patterns observed prior to training. This would further suggest that the participants in the control group tried hard on the tests but could not solve more tasks than before training.

The neuroelectric as well as hemodynamic patterns observed in individuals after WM training suggest that the training influenced working memory capacity as well as processing components guided by the central executive. Training-induced changes in brain activity were not only observed in same brain areas but also in different brain areas (e.g., the left hemispheric ERD patterns in upper alpha frequency bands), which would indicate a reorganization of brain networks related to the trained characteristic (Buschkuehl et al., 2012). The changes in the upper alpha band ERD would further suggest that processes related to LTM were influenced. This was expected given that both WM and LTM systems interact on several levels of processing. This relation was elaborated in Cowan's embedded processes model of WM, in which only the control function of the central executive is assigned to WM, whereas all other processes are part of LTM (Cowan, 2000). The model corresponds well with the neuroelectric and hemodynamic patterns of brain activity observed in our study (Jaušovec and Jaušovec, 2012b).

Although the behavioral changes in fluid intelligence induced by training are rather robust and correspond with neuroelectric as well as hemodynamic changes in brain activity, there are some limitations that restrict the generalization of conclusions. First, the investigated sample was small and not representative of the full population of 20-year-old individuals. Second, a more essential constraint was the predominantly female structure of the sample. Given the immense body of evidence with regard to gender-related differences in general as well as specific abilities (e.g., Johnson and Bouchard, 2007; Nyborg, 2005, see also Section 2.5 in Chapter 2), the sample structure could have affected the results. It should be further mentioned that the use of an identical test battery both for pre- and posttests, in combination with the small and predominantly female sample, could have influenced the obtained gains on the composite measure of fluid intelligence to a certain extent.

3.6.1 Training of Specific Abilities

In Chapter 2, we discussed sex-related differences in intelligence and brain function demonstrating that the most pronounced sex differences have been reported for mental rotation (Mackintosh and Bennett, 2005; Masters and Sanders, 1993; Voyer et al., 1995). It is likely that men and women apply different strategies while solving mental rotation tasks (e.g., Butler et al., 2006; Jordan et al., 2002). Some have suggested that women perform mental rotation tasks in a "top-down" way, whereas men use a "bottom-up" strategy. Top–down processing has been linked to activity in the prefrontal cortex, whereas bottom–up processing tends to be associated with less prefrontal involvement, and a greater involvement of primary sensory brain regions (Butler et al., 2006).

In a study conducted in our lab (Jaušovec and Jaušovec, 2012a), we examined whether training could reduce the deficit of mental rotation performance in females, and change their brain activity patterns. A second objective was to determine whether sex-differences in brain activity patterns observed during mental rotation could be attributed to fixed differences in cerebral organization or to sex-specific experiences.

Support for the latter hypothesis has been provided by recent studies showing that mental rotation performance can be enhanced through practice, and that these enhancements are sex-specific (Feng et al., 2007; Quaiser-Pohl and Lehmann, 2002). Research exploring the influence of training on brain activity has mostly dealt with general aspects of learning and has for the most part ignored the interaction between training, spatial ability, and sex (Haier et al., 1992, 2009; Jaušovec et al., 2006; Neubauer et al., 2004). One exception involves the study by Neubauer et al. (2010). In a pretest–training–posttest design, the authors found a weak influence of training (decrease of brain activation after training) among females, which was found only for 3D presentations in frontal brain areas. One finding common to most of the studies researching the influence of learning on brain activity either with hemodynamic (Haier et al., 1992, 2009) or neuroelectric brain imaging techniques (Jaušovec et al., 2006; Neubauer et al., 2004), is a decrease in brain activation after training. Moreover, in all of these training studies, a computer-based training approach was used and in most of them the same (or similar) tasks were employed for training and pre- and posttests (e.g., the game Tetris; Haier et al., 1992, 2009). The decrease in brain activity observed after training could therefore partially result from repetition effects, familiarity with the testing situation, as well as increased computer expertise. To overcome these shortcomings, we based our training on a Japanese paper folding technique, origami, which was presented in written form (an instruction booklet), and with oral instructions. A pretest–posttest study design with an active control group (communications and social skills training, similar to the one described in Jaušovec and Jaušovec, 2012b) was used. The groups (40 female students assigned to two groups) were equalized with respect to general intelligence and mental rotation ability (WAIS; block design subtest). Brain activity was measured with EEG and near infrared spectroscopy (NIRS)—a technique that measures absolute oxyhemoglobin concentration (HbO_2) in left and right orbitofrontal brain areas. The posttest was administered 6 months after pretesting, within 10 days after the training stopped. Mental rotation ability was measured with the PF&C subtest of the Stanford–Binet IQ test (Rideout and Laubach, 1996).

On average, each student in the origami group completed 18 h of training. Origami is the art of folding paper into a sculpture without the aid of scissors or glue (Hull, 1994). Modern advances in the complexity of origami have

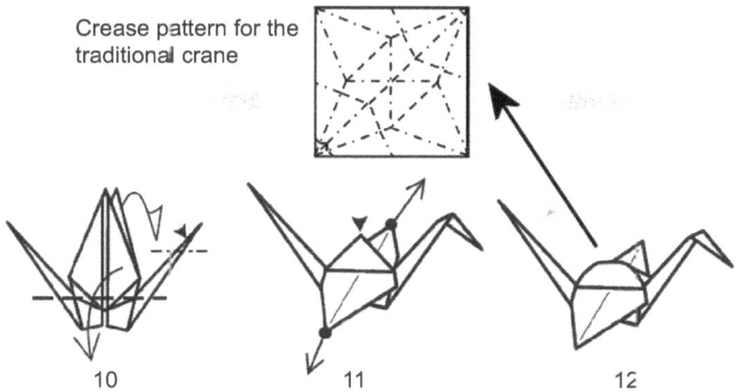

FIGURE 3.12 Example of the training instruction (booklet) given to respondents in the origami group. Respondents were asked to fold a crane, following steps demonstrating each fold separately. They were also asked to make a sculpture based on the crease patterns provided.

revealed a rich geometric structure governing the possibilities of paper folding (Herrlich, 2006), as well as its usefulness in education (Tubis and Mills, 2006). Participants were provided with origami folding paper and instruction booklets showing how to fold paper into different pieces of sculpture. The difficulty level of the folding gradually increased from just 14 folds for the "waterbomb" to 83 folds for the "dragon in flight". Students were also asked to build models based on crease patterns (Fig. 3.12). The training sessions were supervised by an experimenter who provided additional instruction on how to fold the paper, and demonstrated the folds described in the instruction booklets. Interaction between students during training sessions was allowed.

No significant increase in performance (pre–posttest comparison) among participants in the active control group was observed. After communication training, they correctly solved an approximately equal number of task items as before the training (Fig. 3.13). By contrast, after the training, the participants in the origami group showed a significant increase in PF&C performance scores (d= 0.70).

Even more notable were the changes in brain activation patterns displayed by the origami group after training. After 18 h of origami training, females changed their brain activation patterns so that they became almost identical to those of highly intelligent males with high spatial ability (Fig. 3.14). The same patterns of brain activity were also observed for induced event-related coherence values. A nonparametric Wilcoxon test with a FDR correction showed that the required $p < 0.05$ was reached by 14 ErCoh values. As revealed in Fig. 3.15, after origami training females showed inter- and intrahemispheric decoupling in frontal brain areas, accompanied by intra- and interhemispheric coupling of parietal areas with frontal and central brain areas, as well as intrahemispheric coupling in the left parietooccipital brain area.

The neuroelectric activation patterns were accompanied by similar patterns of cardiovascular brain activity measured with NIRS. As can be seen in

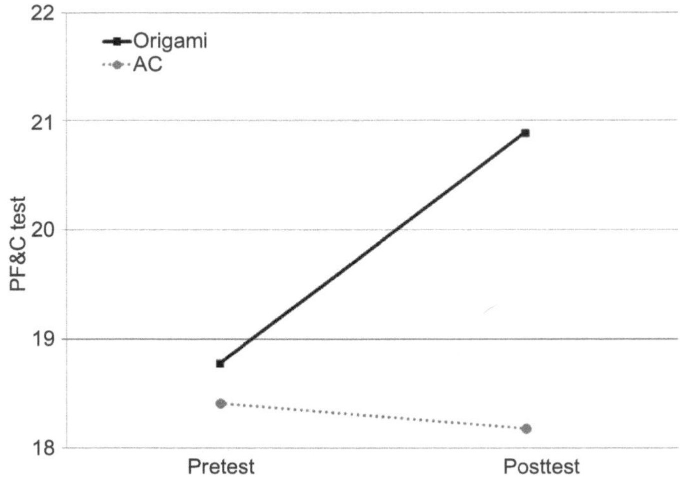

FIGURE 3.13 Line graphs of performance changes on the PF&C task (active control and origami group) observed between pre- and posttests.

Fig. 3.16, after origami training a decrease in females' frontal brain HbO_2 concentration was observed, a pattern which corresponded with the decrease in frontal theta synchronization. On the other hand, after the communication training administered in the control group, women showed no differences in left frontal HbO_2 concentrations, whereas an increase in right hemisphere HbO_2 concentrations was observed.

The robust differences observed in brain activity and in performance measures after just 18 h of training are surprising. In our opinion, this influence is best explained by the characteristics of the origami training that involved motor coordination and transformation of a 2D figural instruction into a 3D sculpture.

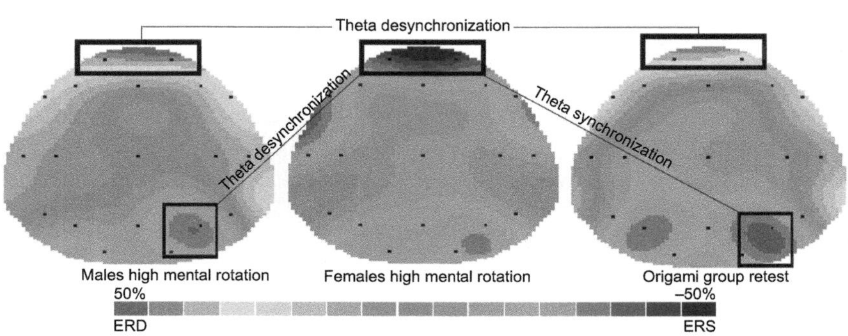

FIGURE 3.14 Topographic theta ERD/ERS maps for males and females who were highly intelligent and had high spatial ability while solving a mental rotation task and topographic theta ERD/ERS maps for females of the origami group while solving a mental rotation task after training (retest).

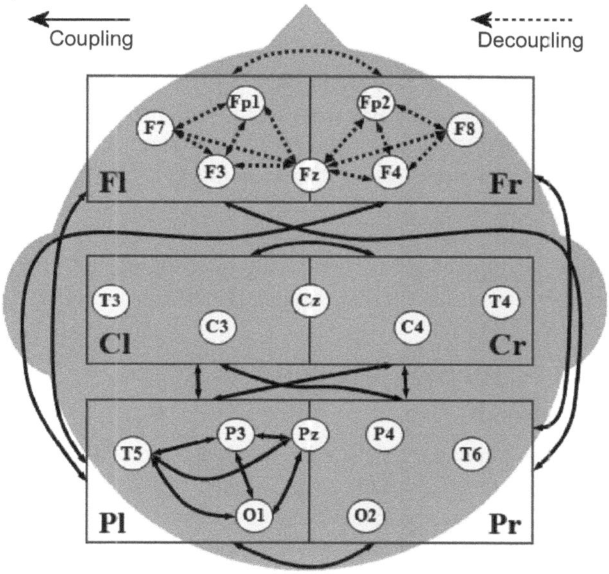

FIGURE 3.15 Topographic map of significant difference in ErCoh% between pre- and posttest for respondents in the origami group (FDR corrected). *Solid lines* represent functional coupling (increase in pre–posttest ErCoh), *dashed lines* represent functional decoupling (decrease in pre–posttest ErCoh). Cl, Central left; Cr, central right; Fl, frontal left; Fr, frontal right; Pl, occipitoparietal left; Pr, occipitoparietal right.

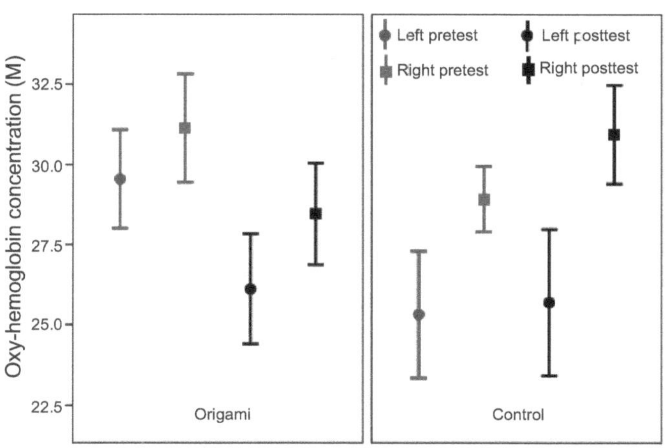

FIGURE 3.16 Oxy-hemoglobin concentrations HbO_2 in the right and left frontal brain areas of respondents (control and origami group) during pre- and posttests.

Support for this assumption comes from several recent studies showing a relationship between volumes of the caudate nucleus (a subcortical structure of the basal ganglia system) and fluid intelligence (Grazioplene et al., 2015; Rhein et al., 2014). The basal ganglia are associated with various functions, such as motor control and simple forms of learning. Moreover, it has been suggested that the caudate nucleus plays a central role in implicit memory and expressive language (Kolb and Whishaw, 2009). A similar finding was reported for pianists who, while listening to piano pieces, activated their motor cortex corresponding to the fingers most involved in the playing (e.g., thumb or little finger). One could say that they listened with their fingers (Haueisen and Knösche, 2001).

3.7 VIDEO GAMES: WITH FUN TO MORE INTELLIGENT BRAINS?

Computerized training approaches like Cogmed, Lumosity, Cognifit, Jungle Memory, and others (e.g., Lampit et al., 2014; Schwaighofer et al., 2015; Table 3.1) have been discussed in the overview of metaanalyses in Section 3.3, therefore I will just briefly summarize some findings and critical remarks, and to a greater extent, focus on video games that were distributed originally for entertainment before they were investigated in terms of cognitive training.

Cogmed was the most prevalent software analyzed in reviews of computerized training interventions (e.g., Lampit et al., 2014; Shipstead et al., 2012a; von Bastian and Oberauer, 2013), and was found to have a more pronounced effect on WM than on other cognitive abilities (Melby-Lervåg and Hulme, 2013; Schwaighofer et al., 2015; Weicker et al., 2016). In an analysis specifically dealing with Pearson's Cogmed WM training, Shipstead et al. (2012b) concluded that many claims made by Cogmed are based on findings that have not been replicated and studied profoundly enough to provide firm scientific evidence for transfer effects. Furthermore, with respect to far transfer effects, the findings are disappointing and the only indisputable statement about Cogmed is that it will improve performance on tasks that resemble training. The same conclusion was put forward by Hulme and Melby-Lervåg (2012). A more favorable opinion was proposed by Gathercole et al. (2012), arguing that large-scale studies that would meet all necessary methodological criteria allowing for solid conclusions on possible far transfer effects of Cogmed training are still missing. A main problem is that such studies are work-intensive, time consuming, and expensive. However, in their opinion, even the studies that do not meet all methodological criteria (randomized assignment, active control group, etc.) can still provide valuable information, and should not be a priori rejected to avoid throwing the baby out with the bathwater.

David Marino-Nachison honored Ralph Baer (1922–2014) as the "godfather of video games" in *The Washington Post* (12/9/2014). The console, the *Magnavox Odyssey*, was available in shops in 1972. The ancestor of the Sony PlayStation, the Nintendo Wii, and the Microsoft Xbox, also called the "brown box" because it was covered with adhesive tape that resembled wood, was by modern standards quite primitive as it offered a limited number of games like the popular

Pong arcade game. However, it has developed to a global market valued at 70.8 billion USD in 2014. It is estimated that 155 million Americans play videogames [h], four out of five US households own a device used to play video games and almost half of Americans play a videogame at least 3 h a week (source: Essential Facts About the Computer and Video Game Industry, 2015). Today, video games can be played on computers, tablets, game consoles, or smart phones, almost everywhere, even on the bus station where Ralph Baer some 50 years ago had his eureka experience and the video game odyssey started.

As stressed by Latham et al. (2013), although this new technology was accepted by the public, its usefulness was not without objection. In the past decades, a large number of reports have accumulated that have analyzed video game playing and the influence it might have on human cognition (e.g., Green and Bavelier, 2012). In the past, these studies considered video games as a unique entity, whereas today the effects of one particular subgenre of games known as "action video games" is of particular interest. These games include complex visual scenes and a wide variety of goals at different timescales (Bavelier et al., 2012; Green and Bavelier, 2012; McDermott et al., 2014; Powers et al., 2013). Another aspect that distinguishes them from other types of video games like strategy or role-playing games, is the speed of the moving objects on screen, how fast they enter it and leave it. Furthermore, players receive feedback about their success throughout the game, which includes a rather complex rewarding system (Green and Bavelier, 2012). All these characteristics of action video games led Bavelier et al. (2012) to suggest that they teach individuals how to learn-to-learn—to quickly learn to perform new tasks.

Two strands of research try to tackle the question of the potential influence of action video game playing on cognitive skills. On the one hand, a comparison is made between habitual action video game players (VGP) and those who do not play action video games (NVGP, nonvideo game players). The assumption is that if video game playing is beneficial, then VGPs should exceed NVGPs on tests measuring a variety of cognitive skills. However, it is possible that those who score high on certain cognitive skills are more attracted by video games than those who score low on these skills thus the comparison might be biased. In order to avoid this, some studies investigated the effects of video game playing on cognitive skills in NVGPs.

Several differences between nonplayers and players of action video games have been revealed by metaanalyses, mainly with respect to attentional characteristics (McDermott et al., 2014; Bavelier et al., 2012; Green and Bavelier, 2012; Powers et al., 2013). Some metaanalyses used a rather fine-grained distinction between different types of attention that can most often be best explained by the tasks that were used to measure it, which is a circular explanation and represents a methodological shortcoming (Table 3.4). Moreover, the classification of these attentional components differs between analyses (e.g.,

h. Worlwide, it is estimated that more than 1.2 billion people play video games, of those 700 million play online games (2013 data).

TABLE 3.4 Overview of Metaanalyses of Differences Between Action Video Game Players (VGP) and Nonaction Video Game Players (NVGP)

Green and Bavelier (2012)

Differences VGP>NVGP:

- Selective attention defined as the ability to focus attention on a target and ignore distracting information:
  - Selective attention over space—the minimal distance between objects at which discrimination can be made (not limited to static objects but also to moving targets).
  - Selective attention in time—to select relevant objects over time
  - Selective attention to objects—faster tracking of more objects
- Attentional control (more efficient top-down allocation of resources): VGP better overcome attentional capture; suppression of unattended, distracting information
- Sustained attention and impulsivity—for example, fast answers to target stimuli (location A) versus no answers to no targets (location B)

Conclusion: Greatest influence on tasks where performance is limited by top-down attention. VGPs have more efficient processes for attentional control and resource management. Also enhanced are processes, such as task switching, multi-tasking, and visual short-term memory tasks.

Bavelier et al. (2012)

Differences VGP > NVGP:

- Vision
  - Spatial resolution—better discrimination of small differences
  - Temporal resolution—reductions in backward masking (the influence of the time interval of following stimuli)
  - Contrast sensitivity, or the ability to detect small changes in levels of gray—improving eyesight
- Cognitive functions
  - Visual short-term memory
  - Multitasking
  - Executive function
  - Task switching
  - Mental rotation abilities
- Decision making—accumulation of information over time
- Reaction time and speed-accuracy trade-off—the ability of quick and accurate responses to environmental changes, faster reaction times in the visual and auditory domain
- Attention—top-down attentional control, such as selective attention, divided attention, and sustained attention

Conclusion: Action game play increases the ability to extract regularity in the outer world—fostering learn-to-learn ability.

TABLE 3.4 *(cont.)*

Latham et al. (2013)
Differences VGP > NVGP:
• Hand-eye coordination and reaction time (enhanced visuomotor reaction times to a wide variety of stimuli) • Spatial visualization (visual pursuit, spatial orientation, spatial relations, mental rotation). • Visuospatial attention (attentional capacity, attentional blink, divided attention, visual short-term memory capacity) • Visual anticipation and visual search strategies—the ability to anticipate the stimulus onset (accuracy of the timing and location of stimuli in motion; efficient top-down search strategies because of broader search patterns) • Temporal dynamics of sensory attention (judgments of multisensory temporal order; reduced effect of backward masking) • Endogenous attention—no influence of distracter stimuli on performance • Task switching Conclusions: Video game playing enhances cognition beyond the context of the video game played. All studies lack adequate (active) control groups.

Bavelier et al., 2012; Green and Bavelier, 2012; Latham et al., 2013), which makes it difficult to compare the results. Furthermore, the analyses are packed with ad hoc constructs that are sometimes nothing more than "cheap talk". For example, the study by Karle et al. (2010), which was by Latham et al. (2013) classified as an example of task switching superiority of VGPs over NVGPs, explained this superiority as the result of the capacity to activate relevant task networks and inhibiting irrelevant ones (those that were previously activated). Explaining the difference in terms of selective attention would probably be much more adequate and understandable. However, in their review, attention was divided according to modality into visuospatial attention, and then further fine-grained in relation to time, capacity, exogenous/endogenous cues, and strategy (Latham et al., 2013). One cannot get rid of the impression that all these distinctions were made to increase the number of beneficial influences of videogame playing. Most of the metaanalyses would benefit from a more theoretically based classification, like the one provided by Powers et al. (2013). An additional problem is that several reviewers have been involved in the commercial exploitation of video games (e.g., Daphne Bavelier, Shawn Green, Alexandre Pouget), which presents a conflict of interest. Table 3.4. provides a summary of differences in cognitive processes between VGPs and NVGPs reported in three metaanalyses.

A different perspective on the effects of video game playing on cognition was provided by Baniqued et al. (2013). Their study focused on the relation between performance on video games and different measures of intelligence, such as fluid intelligence, perceptual speed, episodic memory, and vocabulary, and measures of executive function like attention, inhibition, working memory, and task switching. Two-hundred-nineteen participants played 20 casual video games (in five sessions) that were classified into four groups according to the cognitive process they presumably triggered (reasoning: Silversphere, Sushi-Go-Round; working memory: Memotri, Oddball; attention: Filler, Enigmata; perceptual speed: Phage Wars, Crashdown). The highest correlations were observed between participants' performance on working memory and reasoning games and WM and fluid intelligence test scores ($r = 0.44$ to $r = 0.65$). The authors suggested that the findings provide information for the validity of using certain video games as training devices for specific cognitive abilities. Their main idea was that the overlap in cognitive processes engaged by the casual game and assessed by cognitive tests is the main prerequisite for successful transfer.

In a metaanalysis by Powers et al. (2013), a distinction was made between quasiexperimental and true-experimental studies. The former compared habitual gamers to nongamers or the skill level in playing a particular game. In this type of comparison it is not possible to determine whether the differences between VGPs and NVGPs are due to the influence of game playing or because the VGPs already had superior capabilities that made video game playing more attractive for them. The latter compared participants who received video game training to a control group that did not receive video game training in a within-subjects design, with pre- and posttests assessing the effects of the training. The analysis classified the games according to five cognitive processes: auditory processing, executive functions, motor skills, spatial skills, and visual processing. Variables that were also included were video game genre (action/violent—shooter games, such as Medal of Honor and Unreal Tournament, mimetic—the player mimics the action on screen such as Wii Sports and Fit, no action such as educational games, sports games, and simulation games like Word Whomp, Mario Kart, and The Sims 2, puzzles, such as Tetris and its variations, or nonspecific), the type of control group, and the length of training (more or less than 10 h of training) as well as the moderators age and sex. The analysis of quasiexperiments involved 72 studies (318 comparisons), whereas the analysis of true experiments involved 46 studies (251 comparisons). The analysis of quasiexperiments showed a general medium-to-large size effect showing that video game playing had a significant influence on cognitive performance. Cognitive processing also had a significant influence on effect size and was most prominent in visual and auditory processing domains. The processing domains of visual imagery and executive control were less important. Major influences were observed in players of mimetic games followed by action games and small effects for all the other game types. Most effects were observed in the group of young adults (18–22 years). The analysis of true experiments showed a similar albeit less significant general influence of video game playing on cognition as compared to the quasiexperiments. With respect to cognitive processing, the most pronounced effects were observed for motor skills, small effects were reported for auditory processing, spatial imagery, and visual processing, and negligible effects for executive functions. With respect to the genre of the game, the most pronounced effects were observed for mimetic game playing followed by action games and puzzles. A subsequent analysis of the executive function processing domain indicated only a small effect size for inhibition tasks. More promising was the same analysis of the quasiexperiments showing moderate effects for dual/multitasking and task switching, and small effects for the other subskills (inhibition, intelligence, and working/short-term memory).

An interesting study was conducted by Oei and Patterson (2013) who investigated whether playing different video games on a mobile device (iPhone) influences cognitive performance. The study was conducted over a period of 4 weeks and involved 75 nongamers who played video games on their mobile devices for 20 h. Action games had the strongest influence on cognition: they eliminated the attentional blink [i], and increased performance in cognitive control, and multiple-object tracking. Other game types like spatial memory and hidden object improved visual search performance and spatial working memory. The authors concluded that cognitive performance is increased by video game playing in domains that share common underlying cognitive processing demands.

A less favorable result was reported for the influence of video game playing on ADHD behaviors (Dovis et al., 2015). Compared to controls, the trained group showed improvements in measures of visuospatial short-term memory, WM, inhibitory performance, and interference control, however, no far transfer effects were observed for cognitive-flexibility, verbal WM, complex-reasoning; child, teacher, or parent ratings of ADHD and executive function; motivational and general problem behaviors.

Several studies investigated the influence of video game playing on cognitive performance in older individuals. As put forward by Rose et al. (2015), the ability to remember and successfully execute our planned activities (prospective memory) is a key component of successful everyday living. This ability declines in the normal aging process. The question is whether video game playing could have a beneficial effect on this activity. In the study, a group of older individuals (60–79 years) played a video game called the Virtual Week game board. The respondents rolled a die and moved around the board performing different tasks—one circuit was one day from 7:00 a.m. till 10:15 p.m. They then had to perform appropriate everyday tasks at different times (e.g. have lunch, go shopping). In comparison to two active control groups, the respondents' prospective memory performance significantly improved. Furthermore, a far transfer effect to real-world activities of daily living was observed. The differences were also reflected in ERP components that were associated with WM retrieval. Similar beneficial

i. In a rapidly presented stream of letters, participants can identify a white letter among black letters, which creates a blink in attention causing subjects to be unaware of the next few items following the white letter.

influences on cognitive performance of older individuals (60–85 years) for the video game NeuroRacer were reported by Anguera et al. (2013). The game required participants to drive a car in a virtual environment. The training resulted in performance benefits on the trained task but also on tasks of sustained attention and working memory even 6 months after the end of the training.

This brief analysis of video game training on cognition shows that improvements are observed usually for tasks that correspond to the trained characteristics of the game (near-transfer), but are much less visible in tasks that measure cognitive processes that are not directly trained by the game (far-transfer).

Yet another aspect related to computerized training was addressed recently by Sternberg et al. (2013) for the Lumosity platform and dataset. The dataset comprises 36,140,947 users with 6,661,302 online neuropsychological assessments and demographic information (Lumosity's Human Cognition Project), which presents immense possibilities for large-scale research of cognitive performance. The dataset can be used by external researchers (http://hcp.lumosity.com). The first example of this type of research was provided by Sternberg et al. (2013), who investigated the relation between cognitive performance and general health and lifestyle factors, and the effect of age on performance on ability tests. However, the use of this type of database opens up another more complex ethical question, namely what such use and insight into one's personal data means for the individual's livelihood, privacy, and liberty (Purcell and Rommelfanger, 2015)?

REFERENCES

Adaman, J.E., Blaney, P.H., 1995. The effects of musical mood induction on creativity. J. Creat. Behav. 29, 95–108.

Anderson, D.E., Vogel, E.K., Awh, E., 2011. Precision in visual working memory reaches a stable plateau when individual item limits are exceeded. J. Neurosci. 31, 1128–1138.

Anguera, J.A., Boccanfuso, J., Rintoul, J.L., Al-Hashimi, O., Faraji, F., Janowich, J., Gazzaley, A., 2013. Video game training enhances cognitive control in older adults. Nature 501 (7465), 97–101.

Anokhin, A.P., Lutzenberger, W., Birbaumer, N., 1999. Spatiotemporal organization of brain dynamics and intelligence: an EEG study in adolescents. Int. J. Psychophysiol. 33, 259–273.

Asselen, M., Kessels, R.P., Neggers, S.F., Kappelle, L.J., Frijns, C.J., Postma, A., 2006. Brain areas involved in spatial working memory. Neuropsychologia 44 (7), 1185–1194.

Au, J., Buschkuehl, M., Duncan, G.J., Jaeggi, S.M., 2016. There is no convincing evidence that working memory training is NOT effective: a reply to Melby-Lervåg and Hulme (2015). Psychon. Bull. Rev. 23 (1), 331–337.

Au, J., Sheehan, E., Tsai, N., Duncan, G.J., Buschkuehl, M., Jaeggi, S.M., 2015. Improving fluid intelligence with training on working memory: a meta-analysis. Psychon. Bull. Rev. 22 (2), 366–377.

Baddeley, A., 2012. Working memory: theories, models, and controversies. Ann. Rev. Psychol. 63 (1), 1–29.

Baddeley, A.D., Allen, R.J., Hitch, G.J., 2011. Binding in visual working memory: the role of the episodic buffer. Neuropsychologia 49 (6), 1393–1400.

Baddeley, A.D., Hitch, G.J., 1974. Working memory. In: Bower, G.A. (Ed.), The Psychology of Learning and Motivation: Advances in Research and Theory. Academic press, New York, pp. 47–89.

Baddeley, A., 2003. Working memory: looking back and looking forward. Nat. Rev. Neurosci. 4 (10), 829–839.

Baniqued, P.L., Lee, H., Voss, M.W., Basak, C., Cosman, J.D., DeSouza, S., Kramer, A.F., 2013. Selling points: what cognitive abilities are tapped by casual video games? Acta Psychol. 142 (1), 74–86.

Barbey, A.K., Colom, R., Paul, E.J., Grafman, J., 2014. Architecture of fluid intelligence and working memory revealed by lesion mapping. Brain Struct. Funct. 219 (2), 485–494.

Bavelier, D., Green, C.S., Pouget, A., Schrater, P., 2012. Brain plasticity through the life span: learning to learn and action video games. Annu. Rev. Neurosci. 35 (1), 391–416, https://doi.org/10.1146/annurev-neuro-060909-152832.

Bettencourt, K.C., Xu, Y., 2016. Decoding the content of visual short-term memory under distraction in occipital and parietal areas. Nat. Neurosci. 19 (1), 150–157.

Bhattacharya, J., Petsche, H., 2001. Universality in the brain while listening to music. Proc. R. Soc. Lond. 268, 2423–2433.

Bhattacharya, J., Petsche, H., Pereda, E., 2001a. Long-range synchrony in the rgamma band: role in music perception. J. Neurosci. 21, 6329–6337, 2001a.

Bhattacharya, J., Petsche, H., Feldmann, U., Rescher, B., 2001b. EEG gamma-band phase synchronization between posterior and frontal cortex during mental rotation in humans. Neurosci. Lett. 311, 29–32.

Bodner, M., Muftuler, L.T., Nalcioglu, O., Shaw, G.L., 2001. fMRI study relevant to the Mozart effect: brain areas involved in spatial-temporal reasoning. Neurolog. Res. 23, 683–690.

Bogg, T., Lasecki, L., 2015. Reliable gains? Evidence for substantially underpowered designs in studies of working memory training transfer to fluid intelligence. Front. Psychol. 5, 1589.

Boyke, J., Driemeyer, J., Gaser, C., Büchel, C., May, A., 2008. Training-induced brain structure changes in the elderly. J. Neurosci. 28, 7031–7035.

Braun, U., Schäfer, A., Walter, H., Erk, S., Romanczuk-Seiferth, N., Haddad, L., …, Bassett, D.S., 2015. Dynamic reconfiguration of frontal brain networks during executive cognition in humans. Proc. Natl. Acad. Sci. USA 112 (37), 11678–11683, https://doi.org/10.1073/pnas.1422487112.

Broyd, S.J., Demanuele, C., Debener, S., Helps, S.K., James, C.J., Sonuga-Barke EJS, 2009. Default-mode brain dysfunction in mental disorders: a systematic review. Neurosci. Biobehav. Rev. 33 (3), 279–296.

Bryck, R.L., Fisher, P.A., 2012. Training the brain: practical applications of neural plasticity from the intersection of cognitive neuroscience, developmental psychology, and prevention science. Am. Psychol. 67 (2), 87–100.

Buschkuehl, M., Hernandez-Garcia, L., Jaeggi, S.M., Bernard, J.A., Jonides, J., 2014. Neural effects of short-term training on working memory. Cogn. Affect. Behav. Neurosci. 14 (1), 147–160.

Buschkuehl, M., Jaeggi, S.M., 2010. Improving intelligence: a literature review. Swiss Med. Wkly. 140 (19–20), 266–272.

Buschkuehl, M., Jaeggi, S.M., Jonides, J., 2012. Neuronal effects following working memory training. Dev. Cogn. Neurosci. 2, S167–S179.

Butler, T., Imperato-McGinley, J., Pan, H., Voyer, D., Cordero, J., Zhu, Y.-S., Silbersweig, D., 2006. Sex differences in mental rotation: top–down versus bottom–up processing. NeuroImage 32 (1), 445–456.

Cabeza, R., Nyberg, L., 2000. Imaging cognition II: an empirical review of 275 PET and fMRI studies. J. Cogn. Neurosci. 12, 1–47.

Cannonieri, G.C., Bonilha, L., Fernandes, P.T., Cendes, F., Li, L.M., 2007. Practice and perfect: length of training and structural brain changes in experienced typists. Neuroreport. 18, 1063–1066.

Caprara, V., Barbaranello, C., Borgogni, L., Bucik, V., Boben, D., 2002. Model velikih pet, Cenetr za psihodiagnostična sredstva. Ljubljana.

Carretti, B., Borella, E., De Beni, R., 2007. Does strategic memory training improve the working memory performance of younger and older adults? Exp. Psychol. 54, 311–320.

Chase, W.G., Ericsson, K.A., 1982. Skill and working memory. Bower, G.H. (Ed.), The Psychology of Learning and Motivation, vol. 16, Academic Press, New York, NY, pp. 1–58.

Chechlacz, M., Rotshtein, P., Demeyere, N., Bickerton, W.L., Humphreys, G.W., 2014. The frequency and severity of extinction after stroke affecting different vascular territories. Neuropsychologia 54, 11–17.

Chein, J.M., Schneider, W, 2005. Neuroimaging studies of practicerelated change: FMRI and meta-analytic evidence of a domaingeneral control network for learning. Cogn. Brain Res. 25, 607–623.

Chiang, A.K.I., Rennie, C.J., Robinson, P.A., van Albada, S.J., Kerr, C.C., 2011. Age trends and sex differences of alpha rhythms including split alpha peaks. Clin. Neurophysiol. 122 (8), 1505–1517.

Chiang, M.-C., Barysheva, M., Shattuck, D.W., Lee, A.D., Madsen, S.K., Avedissian, C., Thompson, P.M., 2009. Genetics of brain fiber architecture and intellectual performance. J. Neurosci. 29 (7), 2212–2224.

Colom, R., Abad, F.J., Quiroga, M.Á., Shih, P.C., Flores-Mendoza, C., 2008. Working memory and intelligence are highly related constructs, but why? Intelligence 36 (6), 584–606.

Colom, R., Quiroga, M.Á., Shih, P.C., Martínez, K., Burgaleta, M., Martínez-Molina, A., Ramírez, I., 2010. Improvement in working memory is not related to increased intelligence scores. Intelligence 38 (5), 497–505.

Cohen, J., 1988. Statistical Power Analysis for the Behavioral Sciences, second ed. Erlbaum, Hillsdale, NJ.

Cowan, N., 2000. The magical number 4 in short-term memory: a reconsideration of mental storage capacity. Behav. Brain Sci. 24 (1), 87–114.

Cowan, N., Elliott, E.M., Scott Saults, J., Morey, C.C., Mattox, S., Hismjatullina, A., Conway, A.R.A., 2005. On the capacity of attention: its estimation and its role in working memory and cognitive aptitudes. Cogn. Psychol. 51 (1), 42–100.

Crone, E.A., Wendelken, C., Donohue, S., van Leijenhorst, L., Bunge, S.A., 2006. Neurocognitive development of the ability to manipulate information in working memory. Proc. Natl. Acad. Sci. USA 103, 9315–9320.

D'Esposito, M., Postle, B.R., 2015. The cognitive neuroscience of working memory. Ann. Rev. Psychol. 66 (1), 115–142.

Danielsson, H., Zottarel, V., Palmqvist, L., Lanfranchi, S., 2015. The effectiveness of working memory training with individuals with intellectual disabilities—a meta-analytic review. Front. Psychol. 6, 1230, http://doi.org/10.3389/fpsyg.2015.01230.

Donkin, C., Nosofsky, R.M., Gold, J.M., Shiffrin, R.M., 2013. Discrete-slots models of visual working-memory response times. Psychol. Rev. 120 (4), 873–902.

Doppelmayr, M., Klimesch, W., Hödlmoser, K., Sauseng, P., Gruber, W., 2005. Intelligence related upper alpha desynchronization in a semantic memory task. Brain Res. Bull. 66 (2), 171–177.

Dovis, S., Van der Oord, S., Wiers, R.W., Prins, P.J.M., 2015. Improving executive functioning in children with ADHD: training multiple executive functions within the context of a computer game. A randomized double-blind placebo controlled trial. PLoS One 10 (4), e0121651.

Draganski, B., Gaser, C., Busch, V., Schuierer, G., Bogdahn, U., May, A., 2004. Neuroplasticity: changes in grey matter induced by training. Nature 427, 311–312.

Driemeyer, J., Boyke, J., Gaser, C., Büchel, C., May, A., 2008. Changes in gray matter induced by learning—revisited. PLoS One 3, e2669.

Duckworth, A.L., Peterson, C., Matthews, M.D., Kelly, D.R., 2007. Grit: perseverance and passion for long-term goals. J. Person. Soc. Psychol. 92 (6), 1087–1101.

Duncan, J., 2010. The multiple-demand (MD) system of the primate brain: mental programs for intelligent behaviour. Trends Cogn. Sci. 14 (4), 172–179.

Elbert, T., Pantev, C., Wienbruch, C., Rockstroh, B., Taub, E., 1995. Increased cortical representation of the fingers of the left hand in string players. Science 270, 305–307

Engle, R.W., Tuholski, S.W., Laughlin, J.E., Conway, A.R.A., 1999. Working memory, shortterm memory and general fluid intelligence: a latent variable approach. J. Exp.Psychol. Gen. 128, 309–331.

Eriksson, J., Vogel, E.K., Lansner, A., Bergström, F., Nyberg, L., 2015. Neurocognitive architecture of working memory. Neuron 88 (1), 33–46.

Entertainment Software Association (ESA), 2015. Essential facts about the computer and video game industry. http://www.theesa.com/wp-content/uploads/2015/04/ESA-Essential-Facts-2015.pdf

Fedorenko, E., McDermott, J.H., Norman-Haignere, S., Kanwisher, N., 2012. Sensitivity to musical structure in the human brain. J. Neurophysiol. 108 (12), 3289–3300.

Feng, J., Spence, I., Pratt, J., 2007. Playing an action video game reduces gender differences in spatial cognition. Psychol. Sci. 18, 850–855.

Gathercole, S.E., Dunning, D.L., Holmes, J., 2012. Cogmed training: let's be realistic about intervention research. J. App. Res. Mem. Cogn. 1 (3), 201–203.

George, E.M., Coch, D., 2011. Music training and working memory: an ERP study. Neuropsychologia 49 (5), 1083–1094, https://doi.org/10.1016/j.neuropsychologia.2011.02.001.

Gevins, A., Smith, M.E., 2000. Neurophysiological measures of working memory and individual differences in cognitive ability and cognitive style. Cereb. Cortex 10 (9), 829–839.

Goldman-Rakic, P.S., 1994. Working memory dysfunction in schizophrenia. J. Neuropsych. Clin. Neurosci. 6 (4), 348–357.

Gordon, T., 1974. Teacher Effectiveness Training. Peter H. Wyden Publisher, New York.

Grazioplene, R.G., G Ryman, S., Gray, J.R., Rustichini, A., Jung, R.E., DeYoung, C.G., 2015. Subcortical intelligence: caudate volume predicts IQ in healthy adults: caudate volume and intelligence. Hum. Brain Mapp. 36 (4), 1407–1416.

Green, A.J., Easton, R.D., LaShel, L.S.R., 2001. Visual-auditory events: cross-modal perceptual priming and recognition memory. Conscious. Cogn. 10, 425–435.

Green, C.S., Bavelier, D., 2012. Learning, attentional control, and action video games. Curr. Biol. 22 (6), R197–R206.

Guisan, M.-C., 2005. Universities and research expenditure in Europe and the USA, 1993-2003: an analysis of countries and regions. Reg. Sector. Econ. Stud. 5 (2), 35–46.

Haier, R.J., Neuchterlein, K.H., Hazlett, E., Wu, J.C., Paek, J., Browning, H.L., Buchsbaum, M.S., 1988. Cortical glucose metabolic rate correlates of abstract reasoning and attention studied with positron emission tomography. Intelligence 12, 199–217.

Haier, R.J., Siegel, B., Tang, C., Abel, L., Buchsbaum, M.S., 1992. Intelligence and changes in regional cerebral glucose mezabolic rate following learning. Intelligence 16, 415–426.

Haier, R.J., Karama, S., Leyba, L., Jung, R.E., 2009. MRI assessment of cortical thickness and functional activity changes in adolescent girls following three months of practice on a visual-spatial task. BMC Res. Notes 2 (1), 174.

Haier, R.J., 2014. Increased intelligence is a myth (so far). Front. Syst. Neurosci. 8, 34, http://doi.org/10.3389/fnsys.2014.00034.

Harden, K.P., Turkheimer, E., Loehlin, J.C., 2007. Genotype by environment interaction in adolescents' cognitive aptitude. Behav. Genet. 37 (2), 273–283.

Haueisen, J., Knösche, T.R., 2001. Involuntary motor activity in pianists evoked by music perception. J. Cogn. Neurosci. 13 (6), 786–792.
Hebb, D.O., 1949. Organization of Behavior. John Wiley, New York.
Hernandez-Garcia, L., Buschkuehl, M., 2012. Advances in longitudinal MRI diagnostic tests. Exp. Opin. Med. Diagn. 6 (4), 1–11.
Herrlich, F., 2006. Characteristic origamis. The Geometry of Riemann Surfaces and Abelian Varieties. Contemporary Mathematics, vol. 397. American Mathmatical Society, Providence, RI, pp. 133–144.
Higo, K., Minamoto, T., Ikeda, T., Osaka, M., 2014. Robust order representation is required for backward recall in the Corsi blocks task. Front. Psychol. 5, 1285, http://doi.org/10.3389/fpsyg.2014.01285.
Hughes, J.R., 2002. The Mozart effect: additional data. Epilepsy Behav. 3, 182–184.
Hughes, J.R., Daaboul, Y., Fino, J.J., Shaw, G.L., 1998. The "Mozart effect" on epilepticform activity. Clin. Electroencephalogr. 29, 109–119.
Hull, T., 1994. On the mathematics of flat origamis. Congres. Numer. 100, 215–224.
Hulme, C., Melby-Lervåg, M., 2012. Current evidence does not support the claims made for CogMed working memory training. J. App. Res. Mem. Cogn. 1 (3), 197–200.
Husain, G., ThompsonWF, Schellenberg, E.G., 2002. Effects of musical tempo and mode on arousal, mood, and spatial abilities. Music Percept. 20, 151–171.
Jaeggi, S.M., Buschkuehl, M., Jonides, J., Perrig, W.J., 2008. Improving fluid intelligence with training on working memory. Proc. Natl. Acad. Sci. 105 (19), 6829–6833.
Jaeggi, S.M., Buschkuehl, M., Jonides, J., Shah, P., 2011. Short- and long-term benefits of cognitive training. Proc. Natl. Acad. Sci. 108 (25), 10081–10086.
Jaeggi, S.M., Buschkuehl, M., Shah, P., Jonides, J., 2014. The role of individual differences in cognitive training and transfer. Mem. Cogn. 42 (3), 464–480.
Jaeggi, S.M., Studer-Luethi, B., Buschkuehl, M., Su, Y.-F., Jonides, J., Perrig, W.J., 2010. The relationship between *n*-back performance and matrix reasoning—implications for training and transfer. Intelligence 38 (6), 625–635.
Janata, P., Grafton, S.T., 2003. Swinging in the brain: shared neural substrates for behaviors related to sequencing music. Nat. Neurosci. 6, 682–687.
Jaušovec, N., Habe, K., 2004. The influence of auditory background stimulation (Mozart's sonata K 448) on visual brain activity. Int. J. Psychophys. 51, 261–271.
Jaušovec, N., Jaušovec, K., 2000. Differences in event-related and induced brain oscillations in the theta and alpha frequency bands related to human intelligence. Neurosci. Lett. 293, 191–194.
Jaušovec, N., Habe, K., 2005. The influence of Mozart's sonata K 448 on brain activity during the performance of spatial rotation and numerical tasks. Brain Topogr. 17, 207–218.
Jaušovec, N., Jaušovec, K., 2004. Differences in induced brain activity during the performance of learning and working-memory tasks related to intelligence. Brain Cogn. 54 (1), 65–74.
Jaušovec, N., Jaušovec, K., 2005. Differences in induced gamma and upper alpha oscillations in the human brain related to verbal/performance and emotional intelligence. Int. J. Psychophysiol. 56 (3), 223–235, https://doi.org/10.1016/j.ijpsycho.2004.12.005.
Jaušovec, N., 1998. Are gifted individuals less chaotic thinkers? Pers. Individ. Dif. 25, 253–267.
Jaušovec, N., 1996. Differences in EEG alpha activity related to giftedness. Intelligence 3, 159–173.
Jaušovec, N., 1994. Flexible Thinking: An Explanation for Individual Differences in Ability. Hampton Press Inc, Cresskill.
Jaušovec, N., Jaušovec, K., 2012a. Sex differences in mental rotation and cortical activation patterns: can training change them? Intelligence 40 (2), 151–162.
Jaušovec, N., Jaušovec, K., 2012b. Working memory training: improving intelligence—changing brain activity. Brain Cogn. 79 (2), 96–106.

Jausovec, N., Jausovec, K., Gerlic, I., 2006. The influence of Mozart's music on brain activity in the process of learning. Clin. Neurophysiol. 117 (12), 2703–2714.

Jensen, A.R., 2011. The theory of intelligence and its measurement. Intelligence 39 (4), 171–177.

Johnson, W., Bouchard, T.J., 2007. Sex differences in mental ability: a proposed means to link them to brain structure and function. Intelligence 35 (3), 197–209.

Jonides, J., Lewis, R.L., Nee, D.E., Lustig, C.A., Berman, M.G., Moore, K.S., 2008. The mind and brain of short-term memory. Ann. Rev. Psychol. 59, 193–224.

Jordan, K., Wüstenberg, T., Heinze, H.J., Peters, M., Jäncke, L., 2002. Women and men exhibit different cortical activation patterns during mental rotation tasks. Neuropsychologia 40 (13), 2397–2408.

Jung, R.E., Haier, R.J., 2007. The parieto-frontal integration theory (P-FIT) of intelligence: converging neuroimaging evidence. Behav. Brain Sci. 30 (02), 135.

Kane, M.J., Conway, A.R.A., Miura, T.K., Colflesh, G.J.H., 2007. Working memory, attention control, and the n-back task: a question of construct validity. J. Exp. Psychol. Learn. Mem. Cogn. 33 (3), 615–622.

Karbach, J., Verhaeghen, P., 2014. Making working memory work: a meta-analysis of executive-control and working memory training in older adults. Psychol. Sci. 25 (11), 2027–2037.

Karle, J.W., Watter, S., Shedden, J.M., 2010. Task switching in videogame players: benefits of selective attention but not resistance to proactive interference. Acta Psychol. 134, 70–78.

Katz, B., Jaeggi, S., Buschkuehl, M., Stegman, A., Shah, P., 2014. Differential effect of motivational features on training improvements in school-based cognitive training. Front. Hum. Neurosci. 8.

Keller, F.S., 1968. "Good-bye, teacher..."1. J. Appl. Behav. Anal. 1 (1), 79–89.

Kitchner, K.S., 1983. Cognition, metacognition, and epistemic cognition. Hum. Dev. 26 (4), 222–232.

Klimesch, W., 1999. EEG alpha and theta oscillations reflect cognitive and memory performance: a review and analysis. Brain Res. Brain Res. Rev. 29 (2–3), 169–195.

Klimesch, W., Doppelmayr, M., Pachinger, T., Russegger, H., 1997. Event-related desynchronization in the alpha band and the processing of semantic information. Cogn. Brain Res. 6, 83–94.

Klimesch, W., Freunberger, R., Sauseng, P., Gruber, W., 2008. A short review of slow phase synchronization and memory: evidence for control processes in different memory systems? Brain Res. 1235, 31–44.

Klingberg, T., Fernell, E., Olesen, P., Johnson, M., Gustafsson, P., Dahlström, K., et al., 2005. Computerized training of working memory in children with ADHD—a randomized, controlled, trial. J Am. Acad. Child Adolesc. Psych. 44, 177–186.

Klingberg, T., Forssberg, H., Westerberg, H., 2002. Training of working memory in children with ADHD. J Clin. Exp. Neuropsychol. 24, 781–791.

Klingberg, T., 2010. Training and plasticity of working memory. Trends Cogn. Sci. 14 (7), 317–324.

Koch, S.P., Steinbrink, J., Villringer, A., Obrig, H., 2006. Synchronization between background activity and visually evoked potential is not mirrored by focal hyperoxygenation: Implications for the interpretation of vascular brain imaging. J. Neurosci. 18, 4940–4948.

Koenigs, M., Barbey, A.K., Postle, B.R., Grafman, J., 2009. Superior parietal cortex is critical for the manipulation of information in working memory. J. Neurosci. 29, 14980–14986.

Kolb, B., Whishaw, I.Q., 2009. Fundamentals of Human Neuropsychology, sixth ed. Worth Publishers, New York, NY.

Lampit, A., Hallock, H., Valenzuela, M., 2014. Computerized cognitive training in cognitively healthy older adults: a systematic review and meta-analysis of effect modifiers. PLoS Med. 11 (11), e1001756.

Langer, N., von Bastian, C.C., Wirz, H., Oberauer, K., Jäncke, L., 2013. The effects of working memory training on functional brain network efficiency. Cortex 49 (9), 2424–2438.

Latham, A.J., Patston, L.L.M., Tippett, L.J., 2013. The virtual brain: 30 years of video-game play and cognitive abilities. Front. Psychol. 4, 629.
Leng, X., Shaw, G.L., 1991. Toward a neural theory of higher brain function using music as a window. Conc. Neurosci. 2, 229–258.
León-Domínguez, U., Martín-Rodríguez, J.F., León-Carrión, J., 2015. Executive n-back tasks for the neuropsychological assessment of working memory. Behav. Brain Res. 292, 167–173.
Lisman, J.E., Idiart, M.A.P., 1995. Storage of 7 +/– 2 short-term memories in oscillatory subcycles. Science 267, 1512–1515.
Lozanov, G., 1978. Sugestology and Outlines of Suggestopedy. Gordon and BreachPL London.
Luck, S.J., Vogel, E.K., 1997. The capacity of visual working memory for features and conjunctions. Nature 390 (6657), 279–281.
Luck, S.J., Vogel, E.K., 2013. Visual working memory capacity: from psychophysics and neurobiology to individual differences. Trends Cogn. Sci. 17 (8), 391–400.
Lustig, C., Shah, P., Seidler, R., Reuter-Lorenz, P.A., 2009. Aging, training, and the brain: a review and future directions. Neuropsychol. Rev. 19 (4), 504–522.
Ma, W.J., Husain, M., Bays, P.M., 2014. Changing concepts of working memory. Nat. Neurosci. 17 (3), 347–356.
Mackey, A.P., Whitaker, K.J., Bunge, S.A., 2012. Experience-dependent plasticity in white matter microstructure: reasoning training alters structural connectivity. Front. Neuroanat. 6, 32.
Mackintosh, N.J., Bennett, E.S., 2005. What do Raven's Matrices measure? An analysis in terms of sex differences. Intelligence 33 (6), 663–674.
Maguire, E.A., Woollett, K., Spiers, H.J., 2006. London taxi drivers and bus drivers: a structural MRI and neuropsychological analysis. Hippocampus 16, 1091–1101.
Martin, D.M., Liu, R., Alonzo, A., Green, M., Loo, C.K., 2014. Use of transcranial direct current stimulation (tDCS) to enhance cognitive training: effect of timing of stimulation. Exp. Brain Res. 232 (10), 3345–3351.
Masters, M.S., Sanders, B., 1993. Is the gender difference in mental rotation disappearing? Behav. Genet. 23 (4), 337–341.
Mayer, J.D., Salovey, P., Caruso, D.R., 2002. Mayer-Salovey-Caruso Emotional Intelligence Test (MSCEIT). MHS, Toronto.
McCarney, R., Warner, J., Iliffe, S., van Haselen, R., Griffin, M., Fisher, P., 2007. The Hawthorne Effect: a randomised, controlled trial. BMC Med. Res. Methodol. 7, 30.
McDermott, A.F., Bavelier, D., Green, C.S., 2014. Memory abilities in action video game players. Comp. Hum. Behav. 34, 69–78.
McGrann, J.V., Shaw, G.L., Shenoy, K.V., Leng, X., Mathews, R.B., 1994. Computation by symmetry operations in a structured model of the brain. Phys. Rev. 49, 5830–5839.
McKelvie, P., Low, J., 2002. Listening to Mozart does not improve children's spatial ability: final curtains for the Mozart effect. Br. J. Dev. Psychol. 20, 241–285.
McNamara, D., Scott, J., 2001. Working memory capacity and strategy use. Mem. Cognit. 29 (1), 10–17.
Melby-Lervåg, M., Hulme, C., 2013. Is working memory training effective? A meta-analytic review. Dev. Psychol. 49 (2), 270–291.
Melby-Lervåg, M., Hulme, C., 2016. There is no convincing evidence that working memory training is effective: a reply to Au et al. (2014) and Karbach and Verhaeghen (2014). Psychon. Bull. Rev. 23 (1), 324–330.
Mihelič, A., 1972. Baterija testov sposobnosti BTI Priročnik. Ljubljana, Zavod SR Slovenije za produktivnost dela.

Mogle, J.A., Lovett, B.J., Stawski, R.S., Sliwinski, M.J., 2008. What's so special about working memory? An examination of the relationships among working memory, secondary memory, and fluid intelligence. Psychol. Sci. 19, 1071–1077.

Mölle, M., Marshall, L., Wolf, B., Fehm, H., Born, J., 1999. EEG complexity and performance measures of creative thinking. Psychophysiology 36, 95–104.

Moody, D.E., 2009. Can intelligence be increased by training on a task of working memory? Intelligence 37 (4), 327–328, https://doi.org/10.1016/j.intell.2009.04.005.

Moore, C.D., Cohen, M.X., Ranganath, C., 2006. Neural mechanisms of expert skills in visual working memory. J. Neurosci. 26, 11187–11196.

Moreau, D., Conway, A.R A., 2014. The case for an ecological approach to cognitive training. Trends Cogn. Sci. 18 (7), 334–336.

Moreno, S., Bialystok, E., Barac, R., Schellenberg, E.G., Cepeda, N.J., Chau, T., 2011. Short-term music training enhances verbal intelligence and executive function. Psychol. Sci. 22 (11), 1425–1433.

Morra, S., Borella, E., 2015. Working memory training: from metaphors to models. Front. Psychol. 6, 1097, http://doi.org/10.3389/fpsyg.2015.01097.

Morrison, A.B., Chein, J.M., 2011. Does working memory training work? The promise and challenges of enhancing cognition by training working memory. Psychon Bull. Rev. 18 (1), 46–60.

Nantanis, K.M., Schellenberg, E.G., 1999. The Mozart effect: an artifact of preference. Psychol. Sci. 10, 370–373.

Neubauer, A.C., Fink, A., 2003. Fluid intelligence and neural efficiency: effects of task complexity and sex. Pers. Individ. Dif. 35 (4), 811–827.

Neubauer, A.C., Fink, A., 2009. Intelligence and neural efficiency. Neurosci. Biobehav. Rev. 33 (7), 1004–1023.

Neubauer, A.C., Fink, A., Schrausser, D.G., 2002. Intelligence and neural efficiency: the influence of task content and sex on the brain–IQ relationship. Intelligence 30, 515–536.

Neubauer, V., Freudenthaler, H.H., Pfurtscheller, G., 1995. Intelligence and spatiotemporal patterns of event-related desynchronization. Intelligence 3, 249–266.

Neubauer, A.C., Sange, G., Pfurtscheller, G., 1999. Psychometric intelligence and event-related desynchronization during performance of a letter matching task. In: Pfurtscheller, G., Lopes da Silva, F.H. (Eds.), Handbook of Electroencephalography and Clinical Neuropsychology, Vol 6: Eventrelated Desynchronization. Elsevier, Amsterdam, pp. 219–232.

Neubauer, A.C., Bergner, S., Schatz, M., 2010. Two- vs. three-dimensional presentation of mental rotation tasks: sex differences and effects of training on performance and brain activation. Intelligence 38 (5), 529–539.

Neubauer, A.C., Grabner, R.H., Freudenthaler, H.H., Beckmann, J.F., Guthke, J., 2004. Intelligence and individual differences in becoming neurally efficient. Acta Psychol. 116 (1), 55–74.

Nyborg, H., 2005. Sex-related differences in general intelligence g, brain size, and social status. Pers. Individ. Dif. 39, 497–509.

Oberauer, K., 2002. Access to information in working memory: Exploring the focus of attention. J. Exp. Psychol. Learn. Mem. Cogn. 28 (3), 411–421.

Oberauer, K., 2005. Binding and inhibition in working memory: individual and age differences in short-term recognition. J. Exp. Psychol. Gen. 134 (3), 368–387.

Oberauer, K., Sü\s s, H.-M., Schulze, R., Wilhelm, O., Wittmann, W.W., 2000. Working memory capacity—facets of a cognitive ability construct. Pers. Individ. Dif. 29 (6), 1017–1045.

Oberauer, K., Süβ, H.-M., Wilhelm, O., Wittmann, W.W., 2008. Which working memory functions predict intelligence? Intelligence 36 (6), 641–652, https://doi.org/10.1016/j.intell.2008.01.007.

Oei, A.C., Patterson, M.D., 2013. Enhancing Ccognition with video games: a multiple game training study. PLoS One 8 (3), e58546.
Olesen, P., Westerberg, H., Klingberg, T., 2003. Increased prefrontal and parietal brain activity after training of working memory. Nat. Neurosci. 7, 75–79.
Owen, A.M., McMillan, K.M., Laird, A.R., Bullmore, E., 2005. N-back working memory paradigm: a meta-analysis of normative functional neuroimaging studies. Hum. Brain Mapp. 25 (1), 46–59.
Pahor, A., Jaušovec, N., 2016. The role of theta and gamma oscillations in working memory capacity. Andres, C., Eugenio, V. (Eds.), Horizons in Neuroscience Research, vol. 25, Nova Science publishers Inc, Hauppauge, NY, pp. 27–48.
Panksepp, J., 1998. Affective Neuroscience: the Foundations of Human and Animal. Emotions Oxford University Press, New York.
Panksepp, J., Bernatzky, G., 2002. Emotional sounds and the brain: the neuroaffective foundations of musical appreciation. Behav. Proc. 60, 133–155.
Pascual-Leone, A., Amedi, A., Fregni, F., Merabet, L.B., 2005. The plastic human brain cortex. Annu. Rev. Neurosci. 28, 377–401.
Pashler, H., 1988. Familiarity and visual change detection. Percep. Psychophys. 44 (4), 369–378.
Patel, A.D., 2003. Language, music, syntax and the brain. Nat. Neurosci. 6, 674–681.
Peterson, M., 2000. Spatial-temporal animation reasoning (STAR) b. Academic Press, NewYork.
Piaget, J., Inhelder, B., 1974. Gedächtnis und Intelligenz. Walter, Olten.
Pietschnig, J., Voracek, M., Formann, A.K., 2010. Mozart effect–Shmozart effect: a meta-analysis. Intelligence 38 (3), 314–323.
Pollack, I., Johnson, L.B., Knaff, P.R., 1959. Running memory span. J. Exp Psychol. 57, 137–146.
Postle, B.R., 2015. The cognitive neuroscience of visual short-term memory. Curr. Opin. Behav. Sci. 1, 40–46.
Powers, K.L., Brooks, P.J., Aldrich, N.J., Palladino, M.A., Alfieri, L., 2013. Effects of video-game play on information processing: a meta-analytic investigation. Psychon. Bull. Rev. 20 (6), 1055–1079.
Purcell, R.H., Rommelfanger, K.S., 2015. Internet-based brain training games, citizen scientists, and big data: ethical issues in unprecedented virtual territories. Neuron 86 (2), 356–359.
Quaiser-Pohl, C., Lehmann, W., 2002. Girls' spatial abilities: Charting the contributions of experiences and attitudes in different academic groups. Br. J. Educ. Psych. 72, 245–260.
Raffone, A., Wolters, G., 2001. A cortical mechanism for binding in visual working memory. J. Cogn. Neurosci. 13, 766–785.
Rapport, M.D., Orban, S.A., Kofler, M.J., Friedman, L.M., 2013. Do programs designed to train working memory, other executive functions, and attention benefit children with ADHD? A meta-analytic review of cognitive, academic, and behavioral outcomes. Clin. Psychol. Rev. 33 (8), 1237–1252.
Rauscher, F.H., Shaw, G.L., Ky, K.N., 1993. Music and spatial task performance. Nature 365, 611.
Rauscher, F.H., Shaw, G.L., Ky, K.N., 1995. Listening to Mozart enhances spatial temporal reasoning: towards a neurophysiological basis. Neurosci. Lett. 195, 44–47.
Raven, J.C., 1990. Advanced Progressive Matrices: Sets I, II. Oxford Univ Press, Oxford.
Redick, T.S., 2015. Working memory training and interpreting interactions in intelligence interventions. Intelligence 50, 14–20.
Rhein, C., Mühle, C., Richter-Schmidinger, T., Alexopoulos, P., Doerfler, A., Kornhuber, J., 2014. Neuroanatomical correlates of intelligence in healthy young adults: the role of basal ganglia volume. PLoS One 9 (4), e93623.
Rideout, B.E., Laubach, C.M., 1996. EEG correlates of enhanced spatial performance following exposure to music. Percept. Mot. Skill 82, 427–432.

Rideout, B.E., Taylor, J., 1997. Enhanced spatial performance following 10 minutes exposure to music: a replication. Percept. Mot. Skill 85, 112–114.

Rideout, B.E., Dougherty, S., Wernert, L., 1998. Effects of music on spatial task performance: a test of generality. Percept. Mot. Skill 86, 512–514.

Rigotti, M., Barak, O., Warden, M.R., Wang, X.-J., Daw, N.D., Miller, E.K., Fusi, S., 2013. The importance of mixed selectivity in complex cognitive tasks. Nature 497, 585–590.

Rimland, B., Edelson, S.M., 1995. A pilot study of auditory integration training in autism. J. Aut. Dev. Disord. 25, 61–70.

Roberts, R., Gibson, E., 2002. Individual differences in sentence memory. J. Psycholinguist. Res. 31, 573–598.

Rose, N.S., Rendell, P.G., Hering, A., Kliegel, M., Bidelman, G.M., Craik, F.I.M., 2015. Cognitive and neural plasticity in older adults' prospective memory following training with the Virtual Week computer game. Front. Hum. Neurosci. 9, 592, http://doi.org/10.3389/fnhum.2015.00592.

Rouder, J.N., Morey, R.D., Cowan, N., Zwilling, C.E., Morey, C.C., Pratte, M.S., 2008. An assessment of fixed-capacity models of visual working memory. Proc. Natl. Acad. Sci. USA 105, 5975–5979.

Sambataro, F., Murty, V.P., Callicott, J.H., Tan, H.Y., Das, S., Weinberger, D.R., et al., 2010. Age-related alterations in default mode network: impact on working memory performance. Neurobiol. Aging 31 (5), 839–852.

Sauseng, P., Griesmayr, B., Freunberger, R., Klimesch, W., 2010. Control mechanisms in working memory: a possible function of EEG theta oscillations. Neurosci. Biobehav. Rev. 34, 1015–1022.

Schmidt-Wilcke, T., Rosengarth, K., Luerding, R., Bogdahn, U., Greenlee, M.W., 2010. Distinct patterns of functional and structural neuroplasticity associated with learning Morse code. Neuroimage 51, 1234–1241.

Schmiedek, F., Lövden, M., Lindenberger, U., 2010. Hundred days of cognitive training enhance broad cognitive abilities in adulthood: findings from the COGITO study. Front. Aging Neurosci. 2, 1–10.

Schmiedek, F., Lovden, M., Lindenberger, U., 2013. Keeping it steady: older adults perform more consistently on cognitive tasks than younger adults. Psychol. Sci. 24 (9), 1747–1754.

Schmiedek, F., Lövdén, M., Lindenberger, U., 2014. Younger adults show long-term effects of cognitive training on broad cognitive abilities over 2 years. Dev. Psychol. 50 (9), 2304–2310.

Scholz, J., Klein, M.C., Behrens, T.E., Johansen-Berg, H., 2009. Training induces changes in white-matter architecture. Nat. Neurosci. 12, 1370–1371.

Schwaighofer, M., Fischer, F., Bühner, M., 2015. Does working memory training transfer? A meta-analysis including training conditions as moderators. Educ. Psychologist 50 (2), 138–166.

Shaw, G.L., Silverman, D.J., Pearson, J.C., 1985. Model of cortical organization embodying a basis for a theory of information processing and memory recall. Proc. Natl. Acad. Sci. USA 82, 2364–2368.

Shelton, J.T., Elliott, E.M., Matthews, R.A., Hill, B.D., Gouvier, W.D., 2010. The relationships of working memory, secondary memory, and general fluid intelligence: Working memory is special. J. Exp. Psychol. Learn. Mem. Cogn. 36 (3), 813–820.

Shenoy, K.V., Kaufman, J., McGrann, J.V., Shaw, L.G., 1993. Learning by selection in the Trion model of cortical organization. Cereb. Cortex 3, 239–248.

Shipstead, Z., Redick, T.S., Engle, R.W., 2012a. Is working memory training effective? Psychological Bulletin 138 (4), 628–654, https://doi.org/10.1037/a0027473.

Singer, W., Gray, C.M., 1995. Visual feature integration and the temporal correlation hypothesis. Rev. Neurosci. 18, 555–586.

Shing, Y.L., Schmiedek, F., Lövdén, M., Lindenberger, U., 2012. Memory updating practice across 100 days in the COGITO study. Psychol. Aging 27 (2), 451–461.

Shipstead, Z., Hicks, K.L., Engle, R.W., 2012b. Cogmed working memory training: Does the evidence support the claims? J. App. Res. Mem. Cogn. 1 (3), 185–193.

Shipstead, Z., Redick, T., Engle, R., 2010. Does working memory training generalize? Psychologica Belgica 50 (3–4), 245–276, Retrieved from http://www.psychologicabelgica.com/articles/10.5334/pb-50-3-4-245/?toggle_hypothesis=on.

Skoe, E., Kraus, N., 2012. A little goes a long way: how the adult brain is shaped by musical training in childhood. J. Neurosci. 32 (34), 11507–11510.

Sreenivasan, K.K., Curtis, C.E., D'Esposito, M., 2014. Revisiting the role of persistent neural activity during working memory. Trends Cogn. Sci. 18 (2), 82–89.

St Clair-Thompson, H., Stevens, R., Hunt, A., Bolder, E., 2010. Improving children's working memory and classroom performance. Educ. Psychol. 30 (2), 203–219.

Steele, S.D., Minshew, N.J., Luna, B., Sweeney, J.A., 2007. Spatial working memory deficits in autism. J. Autism Dev. Dis. 37 (4), 605–612.

Sternberg, R.J., 2008. Increasing fluid intelligence is possible after all. Proc. Natl. Acad. Sci. 105, 6791–6792.

Sternberg, D.A., Ballard, K., Hardy, J.L., Katz, B., Doraiswamy, P.M., Scanlon, M., 2013. The largest human cognitive performance dataset reveals insights into the effects of lifestyle factors and aging. Front. Hum. Neurosci. 7, 292.

Takeuchi, H., Taki, Y., Kawashima, R., 2010. Effects of working memory training on cognitive functions and neural systems. Rev. Neurosci. 21, 427–449.

Takeuchi, H., Taki, Y., Nouchi, R., Hashizume, H., Sekiguchi, A., Kotozaki, Y., Kawashima, R., 2013. Effects of working memory training on functional connectivity and cerebral blood flow during rest. Cortex 49 (8), 2106–2125.

Tallon-Baudry, C., Bertrand, O., 1999. Oscillatory gamma activity in humans and its role in object representation. Trends Neurosci. 19, 151–162.

Taubert, M., Draganski, B., Anwander, A., Müller, K., Horstmann, A., Villringer, A., et al., 2010. Dynamic properties of human brain structure: learning-related changes in cortical areas and associated fiber connections. J. Neurosci. 30, 11670–11677.

Taya, F., Sun, Y., Babiloni, F., Thakor, N., Bezerianos, A., 2015. Brain enhancement through cognitive training: a new insight from brain connectome. Front. Syst. Neurosci. 9, 44.

Thompson, W.F., Schellenberg, E.G., Husain, G., 2001. Arousal, mood, and the Mozart effect. Psychol. Sci. 12, 248–251.

Todd, J.J., Marois, R., 2004. Capacity limit of visual short-term memory in human posterior parietal cortex. Nature 428, 751–754.

Tomatis, A.A., 1996. The Ear and Language. Moulin Press, Norval.

Tubis, A., Mills, C., 2006. Unfolding Mathematics With Origami Boxes. Key Curriculum Press, Emeryville, CA.

Turkheimer, E., Haley, A., Waldron, M., D'Onofrio, B., Gottesman, I.I., 2003. Socioeconomic status modifies heritability of IQ in young children. Psychol. Sci. 14 (6), 623–628.

Unsworth, N., Engle, R.W., 2007. The nature of individual differences in working memory capacity: active maintenance in primary memory and controlled search from secondary memory. Psychol. Rev. 114, 104–132.

van den Berg, R., Shin, H., Chou, W.C., George, R., Ma, W.J., 2012. Variability in encoding precision accounts for visual short-term memory limitations. Proc. Natl. Acad. Sci.USA 109, 8780–8785.

van den Heuvel, M.P., Sporns, O., 2013. Network hubs in the human brain. Trends Cogn. Sci. 17 (12), 683–696.

Vandierendonck, A., Kemps, E., Fastame, M.C., Szmalec, A., 2004. Working memory components of the Corsi blocks task. Br. J. Psychol. 95 (1), 57–79.

Vogel, E.K., Machizawa, M.G., 2004. Neural activity predicts individual differences in visual working memory capacity. Nature 428, 748–751.

von Bastian, C.C., Oberauer, K., 2013. Effects and mechanisms of working memory training: a review. Psychol. Res. 78 (6), 803–820, https://doi.org/10.1007/s00426-013-0524-6.

Voyer, D., Voyer, S., Bryden, M.P., 1995. Magnitude of sex differences in spatial ability: a meta-analysis and consideration of critical variables. Psychol. Bull. 117, 250–270.

Vygotsky, L.S., 1977. Mišljenje I govor. Nolit, Beograd.

Wechsler, D., 1981. Wechsler Adult Intelligence Scale—Revised. Psychologica Corporation, San Antonio TX.

Weicker, J., Villringer, A., Thöne-Otto, A., 2016. Can impaired working memory functioning be improved by training? A meta-analysis with a special focus on brain injured patients. Neuropsychology 30 (2), 190–212.

Westerberg, H., Klingberg, T., 2007. Changes in cortical activity after training of working memory—a single-subject analysis. Physiol. Behav. 92, 186–192.

Westerberg, H., Hirvikoski T., Forssberg, H., Klingberg T, 2004. Visuospatial working memory span: a sensitive measure of cognitive deficits in children with ADHD. Child Neuropsychol. 10 (3), 155–161.

Wingfield, A., Stine, E.A.L., Lahar, C.J., Aberdeen, J.S., 1988. Does the capacity of working memory change with age? Exp. Aging Res. 14 (2), 103–107.

Wolf, D., Fischer, F.U., Fesenbeckh, J., Yakushev, I., Lelieveld, I.M., Scheurich, A., Fellgiebel, A., 2014. Structural integrity of the corpus callosum predicts long-term transfer of fluid intelligence-related training gains in normal aging: callosal integrity predicts training transfer. Hum. Brain Mapp. 35 (1), 309–318.

Chapter 4

Changing Brain Activity, Increasing Intelligence: Transcranial Electrical and Magnetic Stimulation

The "Nürnberger Trichter," a magic funnel used to pour knowledge, expertise and wisdom into students in virtually no time, was first mentioned in the book "German Arithmetic" in 1545, by Michael Stifels. However, it gained worldwide popularity in a book with a long title about a poetic funnel that teaches German poetry without the use of Latin language in 6 h, written by the Nürnberg nobleman and senator Georg Philipp Harsdörffer, in 1647. The oldest picture of the process of wisdom pouring that can be found on several postcards of Nürnberg is based on a copper engraving from the 17th century kept in the City Library of Nürnberg. The engraving shows three men and a student lying on the ground. The three men, probably teachers, are pouring all the wisdom through a big funnel into the student's head. The picture is titled: "Look dear people, here is the man who can pour in all art."

This short historical introduction shows that the idea of effortless learning and the power of intelligence was trendy almost 500 years ago. Still, the items that are depicted in the funnel (a pair of compasses, books, and musical notations) in the engraving suggest that increasing crystallized intelligence might have been the main objective in the past. At that time, the book was a bestseller and even had to be reprinted after three years. At the present time, transcranial electrical brain stimulation (TES[a]) could be regarded as a contemporary replacement for the magic funnel. The lay public has displayed great interest in the device, as demonstrated by numerous websites that describe how to build your own "electric brain booster." The idea that applying electricity to your head can enhance intelligence became so widespread that on August 24, 2014, Melissa Hogenboom provided a warning about electrical brain stimulation on

a. TES refers to the generic term that is used for any transcranial electrical stimulation method and should be differentiated from the same term TES that refers to applying high-voltage electric stimuli through electrodes on the scalp, pioneered by Merton and Morton (1980).

Increasing Intelligence. http://dx.doi.org/10.1016/B978-0-12-809407-5.00004-X

BBC Science news. In the TV clip, experts like Oxford Professor Roy Cohen Kadosh explained the limitations and side effects of electrical brain stimulation, discouraging lay public to try it without expert guidance. Nevertheless, most researchers view electrical and magnetic brain stimulation as a promising tool in neurocognitive research. TES and TMS can be used to explore correlations between brain activity and behavior uncovered with neuroimaging, moreover, they can be used to test causality. Influencing cognitive processes could also improve behavioral processes in health and disease, presenting a possible treatment for a variety of medical conditions (Filmer et al., 2014; Shin et al., 2015; Veniero et al., 2015; Wagner et al., 2007; Woods et al., 2016).

In this chapter I will present certain theoretical and technical considerations regarding electrical and magnetic brain stimulation, providing tentative evidence for the possibility of influencing performance on IQ tests and increasing working memory performance.

4.1 TRANSCRANIAL ELECTRICAL BRAIN STIMULATION

According to Wagner et al. (2007) the first account of electrical brain stimulation dates back to 43 AD. At that time, Scribonious Largus, the court physician of the Roman emperor Claudius (Bunson, 2002; p. 301), composed a list of 271 medical prescriptions (Compositiones Medicamentorum) among which was a cure for headache describing electrical stimulation. The idea was to expose affected body parts to electrical currents produced by torpedo fish,[b] or to put them into a pool of water with these fish. This approach was especially popular for the treatment of gouty arthritis. Contemporary electrical brain stimulation, called electrosleep and electroanesthesia, was initiated circa 1900, with the work of Louise Robinovitch (1914) describing a voltage-controlled device with monophasic stimulation below 100 Hz with a 1 ms pulse width (for a detailed review, see Guleyupoglu et al., 2013). However, pioneer work in short duration TES (transcranial direct current stimulation, tDCS) was described in a paper by Priori et al. (1998) and, shortly after, Nitsche and Paulus used prolonged tDCS that produced lasting aftereffects (c. 30 min to 1 h) in cortical excitability (Nitsche and Paulus, 2000).

Actually, electric brain stimulation is the superordinate concept for at least three noninvasive brain stimulation techniques: (1) direct current (DC), (2) alternating current (AC), and (3) random noise (RN) stimulation (Herrmann et al., 2013). The applied currents can be constant over a time (tDCS), or they can alternate at a specific frequency (tACS) or in a random frequency window (tRNS). The oscillations and frequencies can be also combined into oscillatory tDCS (otDCS). Although the AC oscillations are usually sinusoidal, they can also have rectangular or more complex shapes (Herrmann et al., 2013;

---

b. Torpedo fish is a member of the family of electric rays, flattened cartilaginous fish. They are known for being capable of producing an electric discharge ranging from 8 to 220 V.

Groppa et al., 2010). Some authors also categorize TES techniques with respect to the number and shape of stimulation electrodes used, distinguishing between approaches with two big electrodes (tDCS; anode/cathode, 25 cm^2) and high-definition transcranial direct current stimulation HD-tDCS. In the latter method bigger electrodes are replaced with smaller ones (e.g., Pi electrodes, 3.1 cm^2) that are optimized for a specific target brain area to be stimulated. For example, a 4×1 ring montage with a central electrode (anode/cathode) and four return electrodes can be used in HD-tDCS (Dmochowski et al., 2012).

These transcranial electrical stimulation methods share several common characteristics. They are neuromodulatory techniques that do not induce activity in resting neuronal networks, but influence or alternate spontaneous nerve activity. Hence the amount and direction of effects depend on the previous physiological state of the stimulated neural structures. The stimulation intensities of TES are not high enough to trigger discharge in resting neurons or axons and do not cause massive synchronized discharge of action potentials as TMS does (Paulus, 2011; Woods et al., 2016). Furthermore, these methods share basic electrode montage and low-intensity stimulation features (1–2 mA) and their application is generally regarded as safe. The electrodes do not directly contact the skin. Usually an electrolyte is used as a buffer between the electrode and the skin. The experimental procedures require blinding of participants (single-blind) and blinding of experimenters (double-blind) is even more desirable. In order to do so, sham and verum experimental conditions are required, either performed on the same individuals or on matched samples. In the sham condition that can be thought of as a placebo condition, the same ramping stimulation up and down as in the verum condition is applied, but the stimulation with the target intensity lasts only a few seconds. In that way participants will feel the initial itching sensation, but the stimulation duration is too short (about 30 s) to induce after effects (Cohen Kadosh, 2015; Paulus, 2011; Woods et al., 2016; Wagner et al., 2007).

4.1.1 tDCS

Transcranial direct current stimulation (tDCS) has the longest tradition, and is the most well-known and frequently used method of electrical brain stimulation. In this approach a constant current is passed from one electrode (the positive anode) to the other (the negative cathode) over a period of time (usually 8–15 min). Stimulation induces polarity-specific modulations in cortical excitability, and in neurotransmitter and neuromodulator systems in the stimulated cortex. Anodal stimulation (where the current exits the electrode and enters the brain) causes the resting membrane potential to become more positive, whereas cathodal stimulation (where the current exits the brain and reenters the electrode) causes hyperpolarization (Bindman et al., 1964; Nitsche and Paulus, 2000). These tDCS-induced neural modulations make it more (anode) or less (cathode) likely that a stimulated neuron will produce an action potential (Filmer et al., 2014).

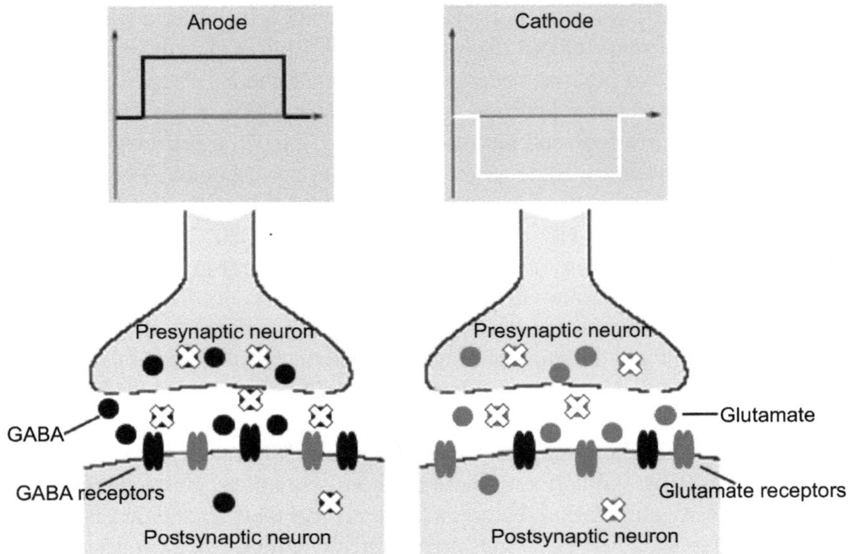

FIGURE 4.1 Anodal and cathodal stimulation. A simplified diagram showing a presynaptic and a postsynaptic GABAergic and glutamatergic neuron. Anodal stimulation inhibits GABA whereas cathodal stimulation inhibits glutamate.

As stressed by Brunoni et al. (2012), tDCS aftereffects can last up to 1 h, therefore, they cannot be solely attributed to changes in electrical neuronal membrane potential. Actually, it was shown (Stagg et al., 2009) that anodal stimulation inhibits neurotransmission via GABA (an inhibitory neurotransmitter), while cathodal stimulation inhibits neurotransmission via glutamate (an excitatory neurotransmitter; see Fig. 4.1). It was further suggested that tDCS might also influence neuronal function by inducing prolonged neurochemical changes (Nitsche et al., 2003; Rango et al., 2008). In addition, tDCS can induce indirect effects, such as connectivity-driven alterations of distant cortical and subcortical areas (Boros et al., 2008), influence several different tissues (vessels, connective tissue), cellular structures (cytoskeleton, mitochondria, membrane) and pathophysiologic mechanisms (inflammation, cell migration, vascular motility) (Brunoni et al., 2012).

The idea that by applying a low 1–2 mA current to the scalp can in one area increase activity (anode) and in another decrease activity (cathode) is rather appealing—too good to be true, as put forward by Cohen Kadosh (2015). In fact, a recent metaanalysis (Jacobson et al., 2012), including 52 tDCS studies (18 motor and 34 cognitive) showed that the relation is not as straightforward as suggested by the two pioneer studies by Nitsche and Paulus (2000) and Priori et al. (1998). The metaanalysis showed that anodal stimulation will in most cases show a corresponding excitation as measured by a relevant cognitive or perceptual task, with most robust findings for motor tasks measured with motor evoked potentials (MEP). On the other hand, the cathodal inhibition effect was often present only in motor studies, but rarely in cognitive studies (Jacobson et al., 2012). The explanation for this discrepancy may have several reasons. As already mentioned transcranial electrical stimulation effects depend on the previous physiological state of the stimulated neural structures, because they are too feeble to trigger discharge in resting neurons on their own. Brain areas related to the performance of a cognitive task are highly activated during the solution process (for example during an IQ task), hence cathodal stimulation cannot generate sufficient inhibition when the initial state is already high. On the other hand, during a motor task, because no voluntary activation is required, the motor areas are less active during stimulation, thus cathodal stimulation can inhibit activity (Jacobson et al., 2012). Another aspect that should be considered is the fact that motor performance has been mostly tested with MEP that only involves stimulated areas. In contrast, cognitive performance was measured with a variety of tasks and measures that increase the level of external noise. Yet another aspect put forward by Antal et al. (2004) was the possibility that cathodal tDCS might in fact improve performance in a certain cognitive task by decreasing neuronal competition.

Several other factors might also influence the impact of tDCS on motor, perceptual and cognitive performance. Among the most often studied factors are electrode size, number and positioning, intensity and duration of stimulation, number of sessions per day, interval between sessions, and subject's anatomy (Bikson et al., 2012; Brunoni et al., 2012; Cohen Kadosh, 2015). In tDCS two electrodes (anode/cathode) are typically used and their placement is determined according to the International EEG 10–20 System. For example, in motor cortex stimulation the electrodes are placed over C3 or C4, and for the visual system, electrodes are typically placed over O1 or O2 (for review, see Utz et al., 2010). Montages in which both electrodes are placed on the scalp are referred to as bipolar or bicephalic. If only one electrode is placed on the head and the other is not (e.g., it is placed below the neck), the montage is referred to as unipolar or monocephalic. Such placement, although theoretically supported (e.g., the reference electrode poses a confounding effect), can sometimes be problematic, because it was shown that, although bicephalic setups were effective, the monocephalic setup did not differ from sham stimulation (Mahmoudi et al., 2011). A possible way of overcoming these problems is to use larger reference electrodes that reduce the current density applied to the reference location (Filmer et al., 2014). Another option is to conduct control experiments with alternative reference locations (Filmer et al., 2013).

The central issue in tDCS is to apply an exact dose (electrode placement in combination with current magnitude) to a target brain area. Recently, computational forward models have been designed that enable customization and optimization of tDCS for specific applications and research questions. These models range in complexity from concentric sphere models to high-resolution models based on an individual's MRI (Bikson et al., 2012). All models follow similar patterns, starting with precise anatomical scans that allow demarcation of key tissues that are meshed together with the shape of electrodes to allow for finite element method calculations. The boundary conditions (energy of electrodes and applied ohms laws) thus make a prediction of possible resulting brain currents (Datta et al., 2008). As stressed by Bikson et al. (2012), increased model complexity is not necessarily related to greater accuracy or clinical value. There is evidence that the intensity of current flow in any specific brain region does not translate in a linear manner to the degree of brain activation or modulation. Furthermore, changes in excitability may not even be monotonic with stimulation dose, and may in fact reverse, depending on background activity (Datta et al., 2008; Salvador et al., 2010).

4.1.2 tACS

Transcranial alternating current stimulation (tACS) is a variant of transcranial electrical stimulation that is in most characteristics similar to tDCS. However, as stressed by Woods et al. (2016), it also differs in a few important features that have a major influence on the obtained changes in brain activity during and after stimulation with sinusoidal alternating currencies.

tACS is a relatively new stimulation technique in which weak oscillating electrical currents are applied to the head (Thut and Miniussi, 2009). The main differences in relation to tDCS is that there is no fixed anode/cathode position because polarity alternates during one half cycle of tACS oscillation. In that way one electrode will serve as anode and the other one as cathode in the first half cycle, whereas in the next half cycle the pattern will reverse—the former anode will now be the cathode, and vice-versa. The current strength will increase and decrease during this half sine wave (Woods et al., 2016). Because of that the neuron's membrane potential will not be affected, but tACS will entrain or synchronize with neuronal networks (Antal and Paulus, 2013) and in that way modulate brain oscillations in a frequency-specific manner (Herrmann et al., 2013). Fig. 4.2 schematically shows this alternating process between three tACS electrodes that, for descriptive purposes, are referred to as active (one channel is split and connected with two active electrodes) and a return electrode.

Evidence for tACS modulation of brain activity in a frequency-specific manner comes from an EEG study in which tACS applied at individual alpha frequency (IAF) increased individual alpha power, but did not affect lower and upper alpha power (Zaehle et al., 2010). Moreover, Neuling et al. (2013) reported that the induced alpha power enhancement remained present for as long as 30 min after stimulation onset, however, this effect depended on endogenous IAF power: tACS-induced aftereffects were present in the low IAF power condition (eyes open), whereas in the high IAF power condition (eyes closed) no aftereffects were observed. Similar findings were reported by a modeling study in which EEG scalp activity was simulated under tACS (4–16 Hz), and under no

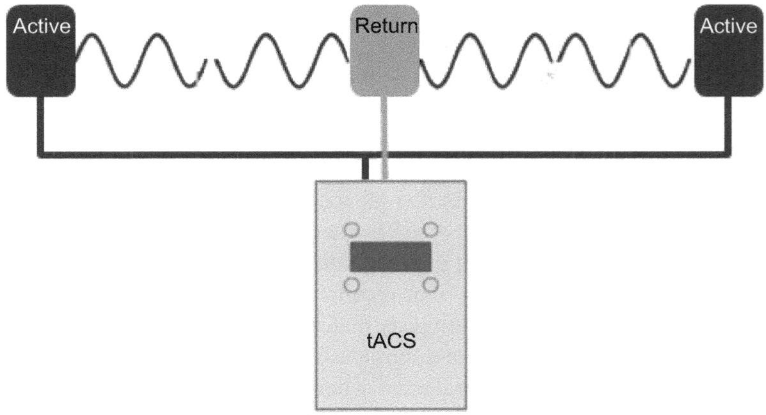

FIGURE 4.2 A schematic diagram showing the alternating process between three tACS electrodes which are for descriptive purposes referred to as active (one channel is split and connected with two active electrodes) and return electrodes. The sine wave alternates between the active and return electrodes that are, during one-half cycle of a tACS oscillation, the anode and, during the next half cycle, the cathode. The two sine waves applied at the active and return tACS electrodes have a phase difference of 180 degrees (out-of-phase), whereas the two sine waves at the two active electrodes (one split channel) have a 0 degree phase difference (in-phase). See the main text for a more exhaustive account.

stimulation (Merlet et al., 2013). The authors reported that only tACS frequencies ranging from 8 to 12 Hz affected alpha power. Specifically, 10 Hz stimulation resulted in an increase in alpha power in the simulated scalp EEG signals. In a recent study using pulsed current stimulation[c] in three frequency bands (1–5; 6–10, and 11–15 Hz), 6–10 Hz stimulation increased power and coherence for the whole alpha band in fronto–central brain areas, as compared with sham stimulation, whereas an opposite effect was observed for 11–15 Hz stimulation. The authors explained this difference in stimulation outcomes related to frequency suggesting that 6–10 Hz stimulation increased power and coherence values in the whole alpha band because it induced changes in naturally occurring alpha oscillatory activity, whereas stimulation with higher frequencies interfered with natural neural dynamics that in turn decreased the intrinsic alpha rhythm at rest (Morales-Quezada et al., 2015).

Based on a review of 22 tACS and otDSC studies, Veniero et al. (2015) concluded that only in three studies no aftereffects were reported (Antal et al., 2008; Krause et al., 2014; Strüber et al., 2015). The null effects reported in these three studies could be explained by insufficient stimulation intensity and duration and by the mental characteristics of subjects. While all other studies showed some aftereffects, they did not reveal a consistent pattern in aftereffects in relation to

c. Transcranial pulsed current stimulation (tPCS) delivers a quadratic biphasic alternate current in different frequencies; this is not sinusoidal as in tACS.

the stimulation frequency, especially when the frequency was not close to IAF or at least in the alpha band range. The authors suggested that entrainment or resonance of cortical networks cannot explain all aftereffects on brain oscillations. The results of a recent study by Vossen et al. (2015) also point in this direction. IAF tACS resulted in expected alpha power enhancements, however, the comparison between intermittent tACS protocols with a sham condition suggested that spike-timing dependent plasticity rather than entrainment is the main mechanism for the observed tACS aftereffects. It is assumed that tACS entrainment (periodic hyper- and depolarization of neural membranes) induces longer-lasting synaptic plasticity that results in frequency-specific changes in oscillatory activity. The model suggests that when neurons are stimulated at a frequency slightly slower than the dominant frequency (tACS $<$ IAF), the synapse is strengthened and it is weakened when presynaptic events follow postsynaptic events of the next cycle (tACS $>$ IAF).

Another key characteristic in alternating current stimulation is that the phase of the EEG can become phase-locked to the alternating current of tACS. In that way the phase at a specific brain location can be modulated hence some brain areas can be brought out-of-phase or in-phase. An aspect that has great relevance for research into the brain–intelligence relationship and follows recent trends in cognitive neuroscience describing brain structure and function as a dense connectome allowing for virtually unconstrained interactions among any pair of neurons in the brain, either through direct connections or via a few brain hubs (van den Heuvel and Sporns, 2013). An even more extreme position was put forward by Wokke et al. (2015), suggesting that intrinsic network dynamics in the brain are the key component in understanding the neural underpinnings of observed behavioral improvements due to transcranial electrical stimulation. For example, Polanía et al. (2012) showed that fronto–parietal theta tACS delivered in-phase (0 degree phase difference) significantly improved reaction times of a delayed letter discrimination task as compared to placebo stimulation. In contrast, fronto–parietal theta tACS delivered out-of-phase (180 degrees phase difference) deteriorated performance. As can be seen in Fig. 4.2, the two sine waves applied at the active and return tACS electrodes have a phase difference of 180 degrees (out-of-phase), whereas the two sine waves at the two active electrodes (one split channel) have a 0 degree phase difference (in-phase).

The main problem with drawing a firm generalization regarding the influence of tACS on neuroelectric patterns of brain activity is that the influence of tACS on rhythmic brain activity has been determined mainly by inference from perceptual, motor, and cognitive consequences of electrical stimulation or by analyzing tACS after-effects on brain activity and rarely with direct electrophysiological evidence (Thut et al., 2011; Zaehle et al., 2010). As put forward by Neuling et al. (2015), these offline analyses cannot make a fully convincing case of the impact of tACS on brain activity. tACS-induced electrophysiological brain dynamics can only be uncovered with approaches that measure brain activity during transcranial electrical stimulation. Such an approach would also separate online entrainment effects from aftereffects (Vossen et al., 2015). Recently, there have been several attempts towards monitoring the electrophysiological effects of tACS online in concurrent tACS-EEG experiments (Fehér and Morishima, 2016; Helfrich et al., 2014) and tACS-magnetoencephalography (MEG) experiments (Neuling et al., 2015).

The study by Helfrich et al. (2014) combined 10 Hz tACS at 1000 μA intensity with concurrent EEG recording. The tACS artifacts were removed in a two-step procedure first by using an artifact template with average subtraction and in the next step with the help of principal component analysis. The artifact-free EEG recordings revealed that tACS increased the oscillatory power in the alpha band by synchronization. Also observed was a transient shift of the individual alpha peak frequency towards the 10 Hz tACS frequency. All these stimulation effects were confined to the alpha band. Neuling et al. (2015) used MEG instead of EEG during tACS stimulation. This recording approach has several advantages. First, the recording electrodes are not attached to the highly conductive scalp, second it allows for increased spatial sampling of the signal and third, it is not based on an artifact reduction approach in which brain signals need to be separated from artifacts (Neuling et al., 2015). The respondents were stimulated with sham, strong and weak tACS at IAF. The results showed that strong tACS affected alpha power in eyes open and closed conditions; this could be an indicator of entrainment on the neuronal level.

4.1.2.1 Research From Our Lab

In a series of tACS experiments with the main goal of influencing higher cognitive functions (working memory and intelligence) conducted in our lab, we also investigated tACS aftereffects on eyes closed or open resting EEG neuroelectric patterns (Jaušovec and Jaušovec, 2014; Jaušovec et al., 2014; Pahor and Jaušovec, 2014; Pahor and Jaušovec, 2016). The analysis reported in this section mainly includes unpublished data.

The objective of the research conducted in 2014 was to test whether theta tACS can induce aftereffects on resting neuroelectric brain activity similar to those reported for alpha tACS or rTMS (Merlet et al., 2013; Morales-Quezada et al., 2015; Neuling et al., 2013; Thut et al., 2011; Zaehle et al., 2010). We explored two different tACS protocols with the placement of target electrodes over the left parietal (P3) and frontal (F3) brain areas. The return electrode was placed above the right eyebrow. The stimulating frequency was in the theta range based on individual alpha frequency (IAF). The influence of sham/verum tACS settings on brain activity was investigated by analyzing EEG amplitude and coherence in the theta to upper alpha frequency bands. Changes in source current density induced by tACS were determined using exact low–resolution electrotomography (eLORETA; Fuchs et al., 2002; Jurcak et al., 2007; Pascual-Marqui, 2007). Twenty-four students participated in the experiment, and they were divided into two groups (frontal/parietal tACS).

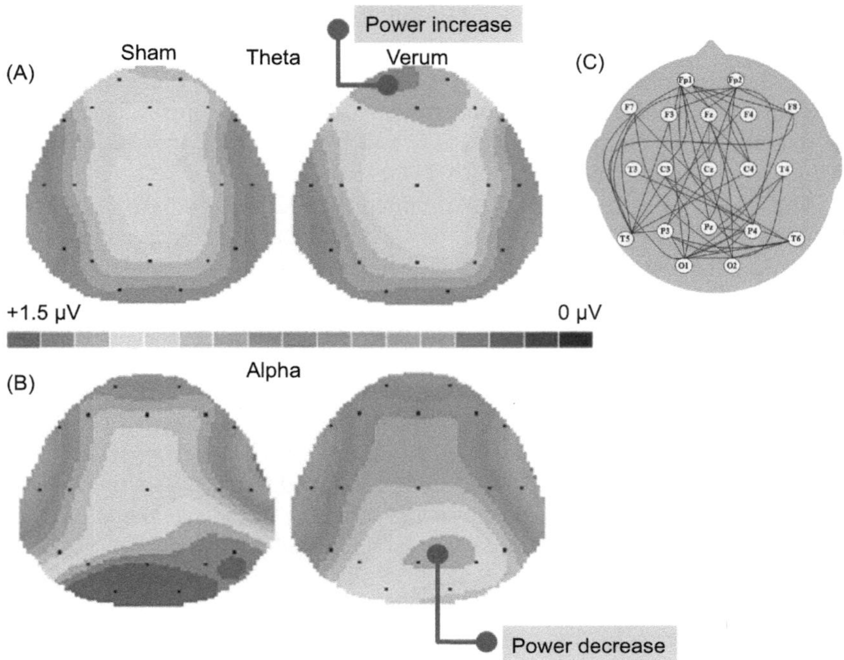

FIGURE 4.3 (A) Topographic power maps of sham and verum resting eyes–closed neuroelectric activity in the theta band and (B) alpha band (collapsed for the lower-1 and lower-2 alpha bands) after theta tACS target electrode positioned over the left parietal site (P3). (C) Topographic map of significant sham/verum difference in theta coherence. The arrows represent 51 significant ($p < 0.05$) FDR corrected differences (verum > sham).

The analysis showed that in the parietal group, theta tACS had no influence on the participants' IAF. On the other hand, in the frontal group theta tACS significantly increased the participants' IAF. On average the increase was about 0.5 Hz ($d = 0.64$). It was further observed that in the parietal group tACS significantly influenced EEG power in theta, lower-1 and lower-2 alpha frequency bands. As can be seen in Fig. 4.3A, participants in the verum setting showed increases in theta power mainly in left frontal brain areas compared to the sham setting. On the other hand, in the two alpha bands the participants in the verum setting showed decreases in alpha power mainly in posterior brain areas (Fig. 4.3B). Further, verum stimulation increased theta coherence as compared to sham stimulation (51 significant increases out of 171; $p < .05$ FDR corrected). Most of these increases were observed over posterior brain areas (75%) and were more numerous in the left hemisphere (35%) (Fig. 4.3C).

In the frontal group the only significant influence of theta tACS on power measures was observed in the lower-2 alpha band. A decrease in lower-2 alpha power was mainly observed in frontal brain areas (Fig. 4.4).

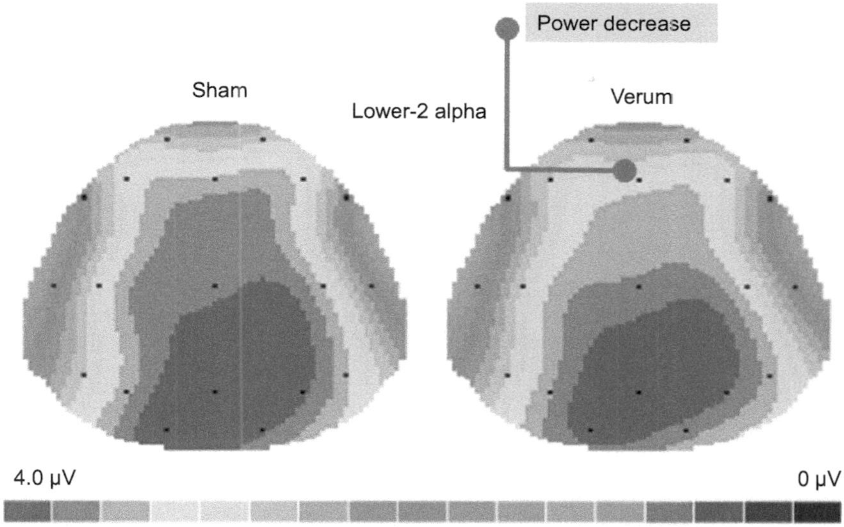

FIGURE 4.4 Topographic power map of sham and verum resting eyes-closed neuroelectric activity in the lower-2 alpha band after theta tACS—target electrode positioned over left frontal site (F3).

Overall, theta tACS decreased power in all three alpha bands. These alpha power decreases were related to the position of the active electrode. When the active electrode was located over the left frontal site, decreases in alpha power were more pronounced in frontal brain areas, whereas when the active electrode was positioned over the left parietal site, more pronounced decreases in alpha power were observed over parieto–occipital brain areas.

eLORETA analysis for the parietal group showed a significant difference ($p < 0.01$) between the sham and verum settings in current source density just in the lower-2 alpha band. As can be seen in Fig. 4.5, theta tACS decreased electric neuronal activity in the left occipital lobe (BA: 18, 19, 30).

In the frontal group significant differences ($p<0.01$) between the sham and verum settings in current source density were observed in the theta and lower-1 alpha bands. In the theta band (Fig. 4.6) these differences were mainly pronounced in the right frontal and temporal brain areas (BA: 38, 47, 45 11, 13). A similar pattern of differences was also observed in the lower-1 alpha band (BA: 32, 47).

The central finding of this study was that theta tACS significantly influenced oscillatory brain activity. This influence was mainly observed in oscillations below IAF. Theta tACS significantly decreased lower-1 and lower-2 alpha power in brain areas located close or beneath the site of stimulation. The observed effects on neuronal electric activity were present after stimulation had terminated. These data demonstrate the ability of tACS to specifically modulate oscillatory activity in particular brain areas—most likely by

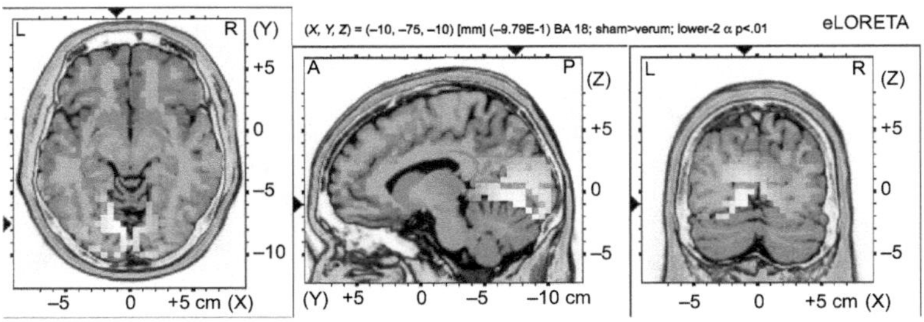

FIGURE 4.5 eLORETA statistical maps of lower-2 alpha oscillations. Dorsal, medial and frontal views of significant differences ($p < 0.01$; corrected for multiple testing) between sham and verum settings after tACS—target electrode positioned over the left parietal site (P3). Theta tACS decreased current density in the left occipital lobe (BA—18, lingual gyrus; BA—19, fusiform gyrus; BA—30, Cuneus).

interfering with the ongoing rhythmic activity of local pacemaker networks. However, it seems more likely that not entrainment, but other more complex biological mechanisms of neural plasticity (e.g., spike-timing dependent plasticity) mostly contribute to the outlasting power and current source density effects. This is in line with recent findings suggesting that tACS enhancement should be stronger when the stimulation frequency matches the spontaneous alpha frequency (Vossen et al., 2015).

Theta tACS also had a noticeable influence on coherence—it facilitated increased coupling between brain areas. However, this influence was far more pronounced for the parietal placement of the target electrode than for the frontal placement and was present just in the lower frequency bands of theta and lower-1 alpha. Increased coupling in the theta frequency band was mainly observed over posterior brain areas and between inter- and intrahemispheric long-distance fronto–parietal locations. In contrast, when the target electrode was placed over the left frontal site, a limited influence increasing coherence

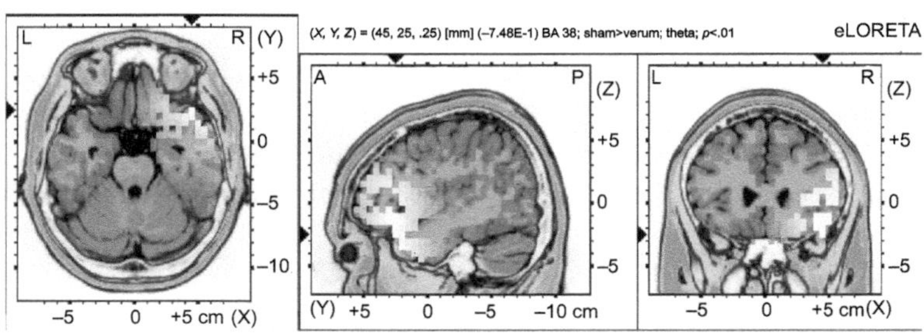

FIGURE 4.6 eLORETA statistical maps of theta oscillations. Dorsal, medial and frontal views of significant differences ($p < 0.01$; corrected for multiple testing) between sham and verum settings after tACS target electrode positioned over the left frontal site (F3). Theta tACS decreased current density in the right frontal and temporal brain areas (BA: 38, superior temporal gyrus; BA: 47, 45 inferior frontal gyrus; BA: 11, middle frontal gyrus; BA: 13, insula).

merely in the right frontal area was observed. A possible explanation for the difference between frontal and parietal electrode placement on coherence could be the interconnectivity of brain areas with a posterior–anterior direction from occipital visual input to frontal executive function (Luria, 1973). As suggested by Felleman and Van Essen (1991), prefrontal areas can be viewed as the end points of sensory streams originating in the occipito–parietal lobes. This also supports findings based on aggregated and synthesized human functional neuroimaging data (Anderson et al., 2013; Poldrack et al., 2009; Toro et al., 2008). Several of the identified networks are related to parietal areas, like the fronto–parietal central-executive network (CEN) or the cingulum–parietal default mode network (DMN), and could explain the displayed increase in theta coherence when the target electrode was placed over the left parietal area. It was further shown that maximal connectivity between the intra parietal sulcus (IPS) and other regions in the dorsal attention and cinguloinsular networks are topographically organized, suggesting that the IPS is involved in top–down control of multisensory attention (Anderson et al., 2013). This explanation is in line with contemporary theoretical explanations of neurobiological underpinnings of increased cognitive performance induced with transcranial electrical stimulation (Sale et al., 2015; Wokke et al., 2015).

In a recent study we combined EEG and NIRS recordings to analyze tACS effects prior, during, and after stimulation. NIRS is a noninvasive technique that can be used to investigate functional activation of the cerebral cortex. By illuminating a tissue at two different wavelengths, it is possible to detect changes in the concentration of deoxygenated (deoxy-Hb) and oxygenated hemoglobin (oxy-Hb) and hemoglobin volume (Ferrari et al., 2004). An activated cortical area is typically characterized by the following oxygenation response: a decrease in deoxy-Hb, accompanied by an increase in oxy-Hb (of two- to threefold magnitude), resulting in an increase in total hemoglobin volume. In contrast, a deactivated cortical area typically displays an increase in deoxy-Hb and a decrease in oxy-Hb (Obrig and Villringer, 2003). An advantage of using NIRS is that, unlike EEG, it can be recorded concurrently with tACS. In addition, combining electrophysiological and hemodynamic data collected before and after stimulation can give further insight into the effects of tACS on the brain. For these reasons we decided to apply NIRS to the frontal cortices before, during, and after tACS.

During tACS stimulation the participants' eyes were open in a dim room (low endogenous alpha power). The participants were asked to silently subtract from 1000 in steps of 7 during the entire course of stimulation—since the room was dark, this task prevented sleepiness and ensured that the participants were mentally active. The main characteristics of the experiments are summarized In Table 4.1.

In experiment 1, no evidence was found to support the hypothesis that upper alpha tACS affects upper alpha power. On the other hand, upper alpha tACS significantly increased intermediate alpha power (Fig. 4.7A). Further, as can be seen in Fig. 4.7B, deoxy-Hb increased after stimulation (verum compared to

TABLE 4.1 Main Characteristics of the Experiments Analyzing the Influence of Alpha tACS on Neuroelectric and Hemodynamic Brain Activity

Time		Prestimulation (3 min)	tACS sham/ verum (15 min)	Poststimulation (3 min)
Exp. 1	Condition	Eyes open	Eyes open/ IAF + 1 Hz	Eyes open
	Recording	EEG + NIRS	NIRS	EEG + NIRS
	Task	NO	Subtracting 7 from 1000	NO
Exp. 2	Condition	Eyes open	Eyes open/IAF—3 Hz	eyes open
	Recording	EEG + NIRS	NIRS	EEG + NIRS
	Task	NO	Subtracting 7 from 1000	NO

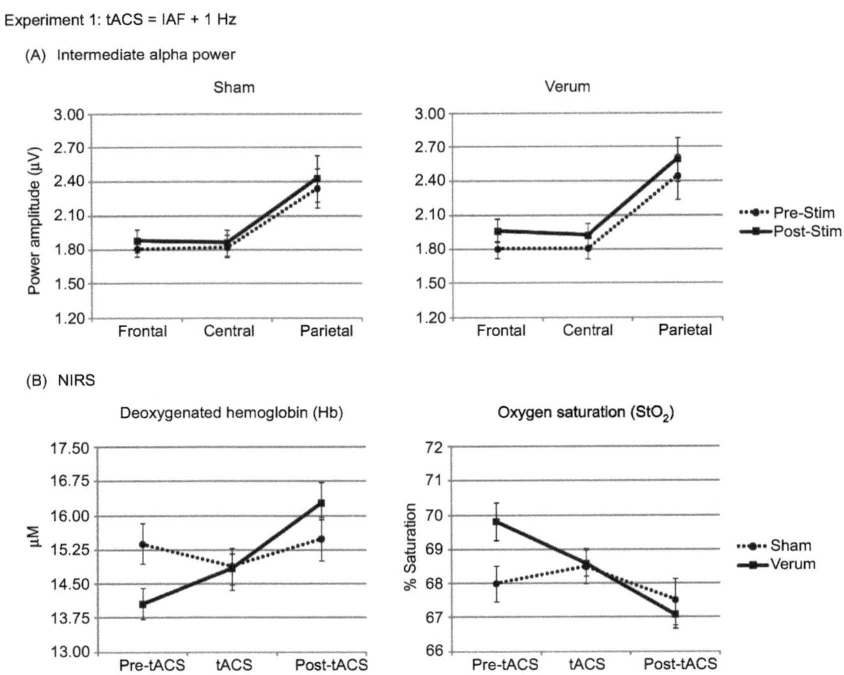

FIGURE 4.7 (A) The effects of sham stimulation and upper alpha tACS on intermediate alpha power. (B) The effects of sham stimulation and upper alpha tACS on the concentration of deoxygenated hemoglobin and on oxygen saturation before, during, and after tACS.

sham), whereas no differences were observed during stimulation itself. There were also some differences in prestimulation levels of deoxy-Hb between sham and verum sessions, and this could be attributed to changes in baseline levels between the 2 days. Deoxy-Hb significantly increased after stimulation (compared to prestimulation) both in the left-frontal and right-frontal areas.

The results of experiment 2 showed that sham tACS somewhat decreased lower alpha power over all locations (Fig. 4.8A). Further, lower alpha tACS somewhat increased upper alpha power in all locations (after stimulation

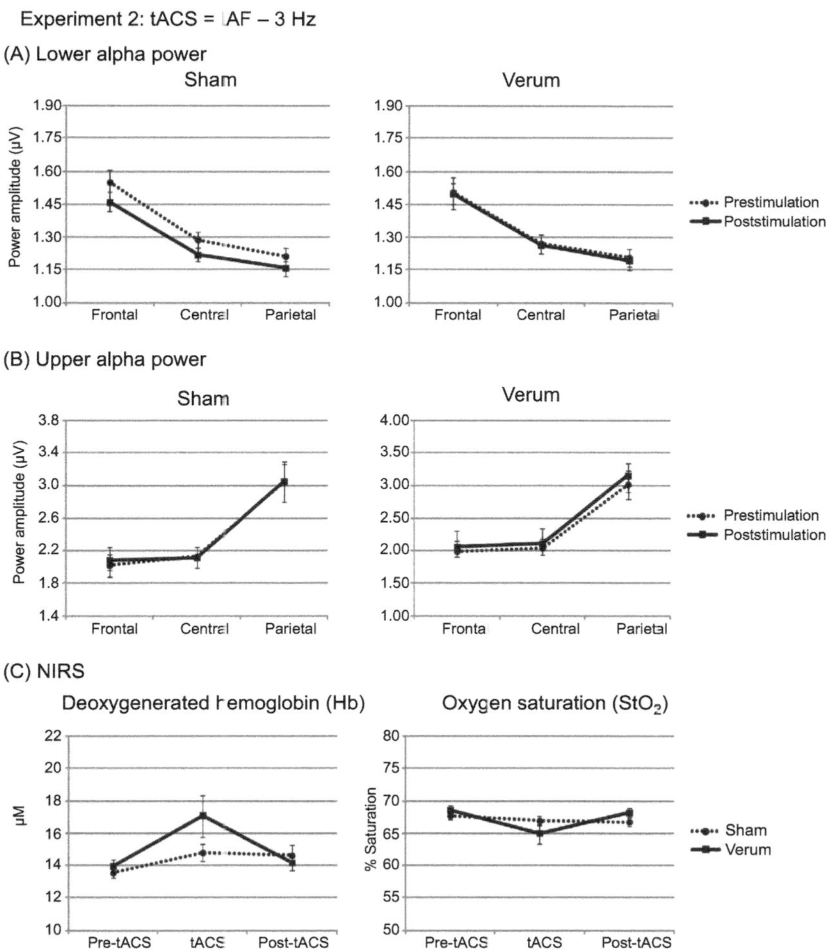

FIGURE 4.8 (A) The effects of sham stimulation and lower alpha tACS on lower alpha power. (B) The effects of sham stimulation and lower alpha tACS on upper alpha power. (C) The effects of sham stimulation and lower alpha tACS on the concentration of deoxygenated hemoglobin and on oxygen saturation before, during, and after tACS.

compared to prestimulation, see Fig. 4.8B). In this experiment, there were no significant main effects or interactions for the concentrations of deoxy-Hb. As can be seen in Fig. 4.8C, verum tACS tended to increase the concentration of deoxy-Hb and decrease the concentration of oxy-Hb during the stimulation period. These effects were not observed during the pre- and poststimulation periods.

Collectively, the findings of our experiments, as well as those included in the review performed by Veniero et al. (2015), indicate that the aftereffects of tACS on neuroelectric brain activity might not be as specific as previously suggested. It seems more likely that complex biological mechanisms of neural plasticity rather than just entrainment[d] and intrinsic resonance[e] contribute to the outlasting neuroelectric tACS aftereffects. Indeed, the results obtained from NIRS supported the hypothesis that verum tACS would result in a deactivation of the stimulated area—however, this effect was only observed for the upper alpha stimulating frequency. Upper alpha tACS was associated with an increase in deoxygenated Hb and a decrease in oxygen saturation in the frontal cortices during the poststimulation period, thereby indicating deactivation (Obrig and Villringer, 2003). The results further showed that the effects of upper alpha tACS were greater after the stimulation period than during the stimulation period, with implications for experimental work in which tACS is paired with a behavioral task. According to our results, the best outcome can be expected when tACS precedes behavioral testing or cognitive training. On the other hand, lower alpha tACS did not have any significant effects on power in the three alpha subbands, nor did it affect the levels of oxygenated and deoxygenated Hb significantly.

In summary, the mechanisms that contribute to the observed tACS effects during and after stimulation are at the moment far from understood. Even though tACS is a promising tool for studying the brain–intelligence relation, it may give conflicting results that will be open to scientific speculation and dispute.

4.1.3 tRNS

Transcranial random noise stimulation (tRNS) is classified as an alternating current approach and is one of the more recent electrical brain stimulation techniques employed (Cohen Kadosh, 2015; Woods et al., 2016). The method was first used by Terney et al. (2008). In this approach alternating current is applied to the scalp with a constantly changing frequency ranging from 0.1 to 640 Hz, the highest physiologically measured human electric brain oscillation in somatosensory-evoked potentials (Gobbelé et al., 2000). The authors reported that tRNS induced excitability increases (e.g., increased the amplitude of motor evoked potentials) lasting 60 min after stimulation. They further concluded that higher frequencies (>100 Hz) were the main generator of this effect that may

d. Inducing a particular oscillation frequency in the brain by means of an external oscillatory force (Thut et al., 2011).

e. The tendency of a system to oscillate with greater amplitude at specific frequencies that are related to the specific structure of the system (Miniussi et al., 2013).

be attributed to the repeated opening of Na^+ channels (Terney et al., 2008). Yet another explanation provided was stochastic resonance (Miniussi et al., 2013). The main principle of this phenomenon is that random-frequency stimulation (tRNS) might induce noise in the system and by that boost the sensitivity of the neurons thereby increasing the signal-to-noise ratio. On the other hand, Fertonani et al. (2011) applied tRNS and tDCS to the visual cortex showing that tRNS exceeded the beneficial influences of tDCS on a visual perceptual learning task. The authors suggested that tRNS might prevent the system to build up a homeostasis by repeated subthreshold stimulations. In a recent study it was further shown that tDCS and tRNS depend on different neural mechanisms (Chaieb et al., 2015). Specifically, tRNS-induced plasticity was dependent on the sodium channel blocker carbamazepine and the $GABA_A$ agonist lorazepam and it was independent of N-methyl-D-aspartate receptor (NMDA). In contrast, pharmacological intervention in combination with tDCS revealed an NMDA receptor dependency (Liebetanz, 2002). It was further suggested that the advantage of tRNS over tDCS is that it has no polarity dependency and causes less perceptible skin sensations (Lavidor, 2016; Pirulli et al., 2016).

The research on the influence of tRNS on performance has been mainly restricted to simple motor or sensory tasks (Fertonani et al., 2011; Romanska et al., 2015; Terney et al., 2008). tRNS was recently also used to enhance effects of cognitive training. Cappelletti et al. (2015) applied three tRNS protocols (sham, motor and parietal), while respondents trained on a numerosity discrimination task. A beneficial tRNS effect on training outcomes was only observed in the parietal random noise group. Similar enhancements induced by tRNS applied to the dorsolateral prefrontal cortex were also reported for learning mathematics (Snowball et al., 2013). Furthermore, both studies showed long lasting training effects for the tRNS groups. Presently, there is insufficient evidence to draw a firm conclusion about the efficacy of tRNS to influence higher cognitive functions. It can be concluded that all three transcranial electric stimulation methods (tDCS, tACS, tRNS) are capable of influencing cognitive processes, although the principles of their impact on brain activity are less clear. At the moment we struggle to share the optimism in the prediction of Kuo and Nitsche (2012) that, in the near future, these methods will serve to explore basic questions in cognition related to brain activity.

4.2 TRANSCRANIAL MAGNETIC STIMULATION (TMS)

Transcranial magnetic stimulation (TMS) was first introduced by Barker et al. (1985). A simple TMS device consists of a few circular turns of copper wire connected to a large electrical discharge system delivering high-voltage (400 V–3 kV), high-current (4–20 kA) pulses that produce a strong time-varying magnetic field (1–2.5 Tesla) that lasts for a short duration ($\geq$1 ms). The idea is to generate a suprathreshold current in the brain and by that depolarize neuronal membranes in order to initiate action potentials (Groppa et al., 2012; Rossini et al., 2015). TMS excites nerves through the same mechanism as transcranial electrical stimulation first used by Merton and Morton (1980). Here the term TES refers to high-voltage electric stimulation delivered through electrodes on the scalp, and should be distinguished from the more general term of transcranial electrical stimulation that includes all types of electrical brain stimulation (e.g., tDCS, tACS, tRNS, otDCS). TMS and TES differ in the direction of the electric fields produced in the brain. The former produces fields that are parallel to the head surface, and are insensitive to skull conductivity, whereas TES produces fields that are parallel and perpendicular to the head surface, and depend on the conductivity of the skull (Groppa et al., 2012). This difference is similar to the one observed in MEG and EEG neuroelectric brain imaging techniques: in relation to head surface, MEG can record only tangentially oriented current flows, whereas EEG can record both the tangentially and radially oriented current flows (Shibasaki, 2008). Another difference between TES and TMS is that TES tends to directly activate cells probably at the proximal axons level (D-waves), whereas TMS mainly generates later discharges via transsynaptic excitation of the corticospinal neurons (I-waves) (Groppa et al., 2012; Rossini et al., 2015).

From a technical perspective the TMS pulse discharge can have a monophasic or biphasic current configuration. The monophasic current pulse consists of a brief (50 ms) initial current with a high amplitude that is balanced by a dampened return current that lasts several hundreds of milliseconds and has no physiological effect in the brain. In contrast, the biphasic pulse has a cosine waveform with an initial current rise followed by a current reversal below zero, and a subsequent rise back (current direction is reversed twice). In a biphasic pulse both phases contribute to electrical cortex stimulation. However, the second reversal phase induces a stronger tissue current, and also causes stimulation of different neural elements in the cortex than the initial rising phase (Groppa et al., 2012; Rossini et al., 2015). Monophasic magnetic pulses are commonly used for single pulse experiments, while a biphasic stimulus waveform is used in repetitive transcranial magnetic stimulation (Sommer et al., 2006).

Another technical problem that transcranial methods of brain stimulation face is that they have to deliver stimulation across the resistance of the scalp, skull, meninges and cerebrospinal fluid. Therefore these methods are not able to activate deep brain regions effectively (Huang et al., 2009). This problem can be partly solved with the shape of the coil used to deliver the magnetic pulses. The first coils had a round shape. Larger round coils (circular coils) with an outer diameter of more than 10 cm are more powerful than figure-of-eight shaped coils (double or butterfly coil), however, the latter allow for more focal stimulation (Groppa et al., 2012). It was established that firing of axons by this type of coils occurs in the cerebral cortex, within 1–2 cm from the coil (Epstein et al., 1990). Further, it was suggested that H-shaped coils (Hesed-coil, HAAD-coil) can stimulate deeper neuronal regions by inducing spatial summation of the induced electric field and reducing the electric field attenuation, at the expense of reduced focality (Huang et al., 2009). However, a more detailed comparison with conventional coils also considering the extent of the stimulated cortical area showed no evidence of deeper stimulation (Fadini et al., 2009).

Yet another problem TMS research faces is the design of adequate sham TMS conditions. It is particularly difficult to replicate the specific skin sensations or muscle twitches and the characteristic sound that is produced by TMS in a sham condition, hence single- and double-blind studies are rare (Lisanby et al., 2001; Lefaucheur et al., 2014).

As stressed by Huang et al. (2009) many new methods and variants of TMS techniques exist. Broadly they can be divided into two paradigms of TMS: single/paired/triple pulse and repetitive transcranial magnetic stimulation (Hampson and Hoffman, 2010). The main characteristics of the most often used TMS methods are summarized in Table 4.2.

As revealed in Table 4.2, although many different types of TMS exist, most of them have researched brain–behavior relationships associated with the primary motor cortex (M1) or to a lesser extent with the visual cortex (V1). There is a growing amount of studies that employ TMS methodology to target brain areas that are not related to M1 or V1 (Luber and Lisanby, 2014; Nevler and Ash, 2015). These studies investigated cortical functions in decision making processes (Jeurissen et al., 2014; Soutschek et al., 2015), language (Devlin and Watkins, 2007), and working memory that will be discussed in more detail in Section 4.4. Furthermore, TMS and especially different types of repetitive transcranial magnetic stimulation have the potential to modulate brain plasticity (Freitas et al., 2013; Nevler and Ash, 2015). Neuroplasticity refers to the ability of neurons to change their functional and structural parameters in relation to environmental demands resulting in either: (1) long-term potentiation—strengthening of synaptic power lasting hours/days, or (2) long-term weakening of synapses, called long-term depression (Nevler and Ash, 2015). TMS has been increasingly used as a therapeutic intervention in abnormal cognitive processes (Lefaucheur et al., 2014; Wassermann and Zimmermann, 2012). A group of European experts commissioned to establish guidelines on the therapeutic use of low-frequency (LF) and high-frequency (HF) repetitive transcranial magnetic stimulation (rTMS) came to the following conclusions (Lefaucheur et al., 2014):

- Definite efficacy of HF rTMS showing:
  - analgesic effects of the primary motor cortex contralateral to the pain,
  - antidepressant effect of the left dorsolateral prefrontal cortex.
- Probable efficacy of rTMS showing:
  - antidepressant effect of LF rTMS of the right DLPFC,
  - reduction of negative symptoms of schizophrenia (HF-rTMS of the left DLPFC),
  - contralesional M1 in chronic motor stroke (LF-rTMS of contralesional M1).
- Possible efficacy of LF rTMS showing:
  - effects in tinnitus and auditory hallucinations (LF-rTMS of left temporoparietal cortex).

TABLE 4.2 Main Types of TMS
Single pulse TMS
In this TMS approach, monophasic current pulses are typically used. It is assumed that they interfere with activity in the stimulated region, thereby acting as a temporary lesion (Hampson and Hoffman, 2010). The method has been mainly used for research on the human corticospinal motor system (Rothwell et al., 1991), and to a lesser extent for the study of the sensory system (Marg and Rudiak, 1994). Research on higher cognitive processes revealed that a single pulse does not always have an inhibitory effect, but can show facilitating effects (Grosbras and Paus, 2002). It was suggested that the effect of the TMS pulse depends not only on the parameters of the coil and pulse, but also on the activity state of the tissue itself (Nevler and Ash, 2015).
Paired pulse TMS
Paired pulse TMS can modulate intracortical inhibition and facilitation. It is usually used to investigate interactions between motor or visual regions of the brain. In this approach two sequential pulses are delivered to the same or different brain areas (dual coil experimentation). The first one is referred to as the conditioning stimulus that is followed by a test stimulus. The latter is applied to a region with an observable output response. In this experimental paradigm the amplitude of the motor evoked response (MEP; compared with the amplitude produced by the test stimulus alone) serves as an indicator for an existing relation between the two probed cortical areas (Hampson and Hoffman, 2010; Rossini et al., 2015). A number of original experimental designs have been used to examine connectivity in the motor system, and its role in behavior. For example, it was shown that an aggressive personality style was positively correlated with functional interhemispheric connectivity. In this TMS experimental design the left-to-right transcallosal inhibition condition consisted of a conditioning TMS pulse delivered to the left hemisphere, followed by a TMS test pulse to the right hemisphere 10 ms later. As a result the MEP of the conditioned response was smaller than the unconditioned MEP of the TMS test pulse alone. It was observed that higher levels of left-to-right transcallosal inhibition significantly correlated with higher aggression scores (Hofman and Schutter, 2009).
Repetitive TMS
In repetitive TMS (rTMS) short trains (e.g., 2 s on, 28 s off) of high-frequency stimulation ($\geq$5 Hz) or long trains (10–20 min) of low-frequency stimulation ($\leq$1 Hz) with extremely brief interstimulus intervals (e.g., 10 or 20 ms) are used (Rossini and Caramia, 1992). rTMS has a modulatory effect on cortical excitability that outlasts the stimulation period. By examining behavioral, clinical, or cognitive performance before and after rTMS this approach can be used to study how perturbations in activity in a certain brain area affect these functions. However, most research was done on the effects of stimulation in the motor cortex (Hampson and Hoffman, 2010). In general, high-frequency stimulation produces an increase in cortical excitability (e.g., MEP amplitude), whereas low-frequency stimulation results in depression (reduction in MEP size) of the target brain area (Pascual-Leone et al., 1994; Speer et al., 2000). A special case of rTMS is the I-wave repetitive paired-pulse TMS intervention also known under the acronym iTMS. In this approach a paired-pulse TMS of equal intensity at an interpulse interval of 1.5 ms (I-waves have a periodicity of $\approx$1.5 ms) is delivered. The idea for rTMS interventions based on I-wave periodicity is that it might be possible to increase the efficacy of transsynaptic events involved in their generation (Huang et al., 2009).

TABLE 4.2 ***(cont.)***

Another example of a patterned stimulation is the quadri-pulse paradigm (QPS). In this approach one train consists of four monophasic pulses at the same intensity separated by 1.5 ms, repeatedly given at 0.2 Hz. Research has shown that QPS induced locally restricted facilitation of motor cortical excitability for up to 75 min without affecting motor thresholds. It was suggested that this approach might be a promising method for inducing plastic changes in the brain (Hamada et al., 2008).
Theta burst TMS
Theta burst stimulation (TBS) is the most established patterned form of TMS. The basic pattern of TBS is a three-pulse burst at 50 Hz given every 200 ms (5 Hz) that can be delivered continuously (cTBS) or intermittently (iTBS). In cTBS a continuous train of bursts for 20 s (300 pulses) or 40 s (600 pulses) is delivered. When applied to the motor cortex, a significant decrease in MEP lasting 20–60 min was observed (Huang et al., 2009). The iTBS paradigm consists of 20 cycles of 50 Hz burst delivered for 2 s every 10 s. One hundred ninety seconds of iTBS over the primary motor area facilitated MEPs for about 20 min (Huang et al., 2007).

Another prospect for the use of TMS is the study of connectivity and neural networks in the human brain. As already stressed in Chapter 2, in the last decade a strong interest in network dynamics can be observed, especially in the opposed activity levels in task related (e.g., attention, fronto–parietal; CEN) and task unrelated (e.g., default mode; DMN) networks and performance on a variety of tasks (Wokke et al., 2015). For example, Eldaief et al. (2011) used rTMS and functional connectivity MRI (fcMRI) to modulate cortical activity in one node of the default network and assessed the effect of this intervention throughout the network. As recently proposed by Emiliano Santarnecchi (2016) on the International seminar on Advances on Intelligence Research held in Madrid in 2016 (https://www.youtube.com/playlist?list=PLQ8yHMSBC8UCwGZBosBXT6msaCvzN8P4S), a combined approach of rTMS and fcMRI could be used for the determination of the causal relationship between brain function and intelligence. The experimental design would include several steps. In a first step network nodes and connections essential for intelligent performance should be identified (Santarnecchi et al., 2015; Pineda-Pardo et al., 2016). The identified nodes and network connections would serve as targets for subsequent rTMS stimulation. This step would also include loading participant's functional connectivity data into a frameless stereotactic optical tracking neuronavigation system that would permit key nodes for intelligent performance to be reliably stimulated within subsequent rTMS sessions (neuronavigated brain stimulation). Neuronavigated TMS enhances the precision of coil placement and by that minimizes inter- and intrasubject variability of the intervention effects. It can be used in real time to monitor/guide the location of the coils in relation to the subject's head position (Freitas et al., 2013).

The next steps would then involve fcMRI to assess possible changes in brain networks targeted during rTMS, and analysis of behavioral measures (IQ tests) to establish possible beneficial/disruptive effects of rTMS. This part could be accompanied by fMRI, or other measures of brain imaging. As stressed by Luber and Lisanby (2014), such stimulation approaches, especially when used in conjunction with brain imaging, are expensive and time consuming—as discussed in the next section, these approaches are often replaced by cheaper although less exact alternatives, such as tACS or tDCS.

4.3 tDCS AND tACS AS A TOOL TO MODULATE INTELLIGENCE

The brain is the source of intelligent behavior as was vividly illustrated by Mountcastle's (1975, p. 131) quote, making the analogy between a prison and the brain that will never allow us to encounter the real world directly. All we will ever know are fabrications made by our brain. Thus, it could be speculated that interventions aimed directly toward the structure and functional connectivity of the brain would have a more pronounced influence on measures of fluid intelligence than different types of cognitive training described in the previous chapter. A direct approach might also have a methodological advantage since it can rule out similarity effects between assessment and training tasks and other environmental variables that are difficult to control or exclude.

Studies employing direct interventions can be divided into three broad categories: neurofeedback training, pharmacological interventions, and brain stimulation. The first two will be described in the next chapter. The third approach that can influence brain activity directly is the recently rediscovered field of transcranial electrical stimulation. To date, there are only few studies that tried to influence fluid intelligence performance by directly stimulating brain activity with tACS or tDCS (for a recent review, see Herrmann et al., 2015; Sarkar and Cohen Kadosh, 2016; Schutter and Wischnewski, 2016). These studies will be discussed in detail, along with research conducted in our lab and in collaboration with our colleagues in Graz (Institute of Psychology, University of Graz).

The first study that used tDCS to influence certain aspects of intelligence was conducted by Cerruti and Schlaug (2009). The test they used to measure the stimulation effects (random association test, RAT) is actually a creativity test (Mednick, 1962) with only assumed relations to verbal intelligence (Mednick and Andrews, 1967). The RAT builds on the fact that in US culture there are associative bonds between words like ham and eggs, pool-hall, whole-wheat, and similar. The test itself provides words from mutually distant associative clusters. The subjects are asked to find a common linguistic associate that forms a compound noun or two-word phrase with each cue word. For example, the words rat, blue, and cottage can be associated with the word

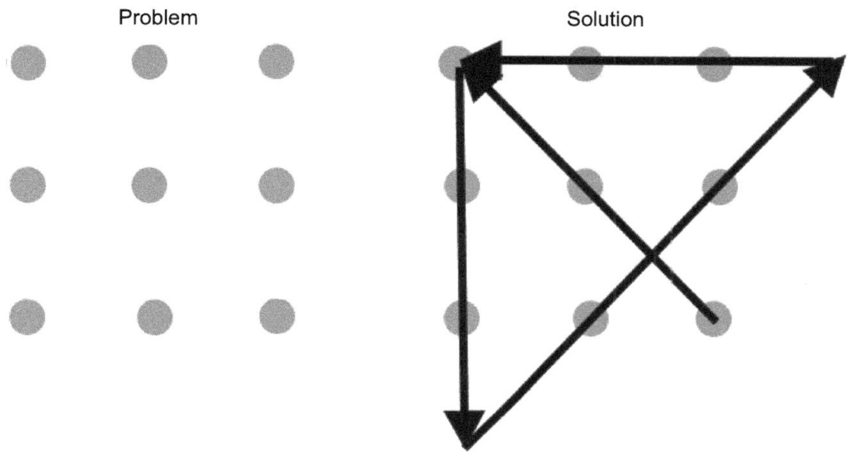

FIGURE 4.9 **The nine-dot problem and its solution.**

cheese that can be combined into word pairs, such as, rat-cheese, blue-cheese, and cottage-cheese.

Cerruti and Schlaug (2009) conducted two experiments in which anodal, cathodal, and sham tDCS was delivered to the left and right dorsolateral prefrontal cortex (F3/F4), with the reference electrode positioned over the contralateral supraorbital region. To reduce the influence of cathodal stimulation on brain activity, they placed a larger electrode over the supraorbital region, reducing the current density to about half the intensity of the anode. The data analysis showed that anodal tDCS (stimulation intensity = 1 mA) delivered to the left dorsolateral prefrontal cortex increased performance on the RAT, but not on a verbal fluency test. No similar effects were observed for cathodal and sham stimulation, or for anodal stimulation of the right dorsolateral prefrontal cortex. The authors concluded that the left dorsolateral prefrontal cortex is involved in solving complex problems with creative and logical components (Cerruti and Schlaug, 2009).

Likewise, Chi and Snyder (2011, 2012) did not employ tests that would measure fluid intelligence directly. In their first study they used matchstick arithmetic problems to measure the influence of tDCS on the mental set effect (Chi and Snyder, 2011). The nine-dot problem (Maier, 1930) that the authors employed in their second study (Chi and Snyder, 2012) could be also classified as an insight problem (Kershaw and Ohlsson, 2004). In this problem, nine dots arranged in a 3 × 3 square must be connected with exactly four straight lines without lifting the pen from the paper and without retracing (Fig. 4.9). The expected solution rate under a time limit of just a few minutes is 0%. Although the problem is regarded as a classic insight problem, research has suggested that higher spatial WM capacity is related to faster solutions. It was further proposed that this finding "...diminishes the likelihood that a third factor that correlates with executive WM, such as general fluid intelligence, might account for the relationship between WM and nine-dot performance" (Chein et al., 2010; p. 890). Because these findings suggest the possibility of a weak relation of the nine-dot problem with spatial WM capacity and intelligence, I will only describe the second study (Chi and Snyder, 2012).

In a parallel group (sham/verum tDCS) single-blind experimental design, 1.6 mA intensity tDCS was applied for 10 min over participants' left (cathode) and right (anode) anterior temporal lobes. It was hypothesized that cathodal stimulation (decreasing excitability) of the left anterior temporal lobe, together with anodal stimulation (increasing excitability) of the right anterior temporal lobe, would facilitate performance on the nine-dot insight problem. This prediction was based on evidence that the right/left anterior temporal brain areas are associated with insight (Kounios and Beeman, 2009). Participants were given 9 min to solve the problem (3 min prior, during, and after tDCS). None of the 11 subjects of the sham condition solved the problem, whereas 5 out of the 11 verum subjects solved the problem during tDCS stimulation. A similar beneficial influence of the same tDCS protocol on the solution of matchstick arithmetic problems was also observed in their first study. In this study it was shown that during tDCS three times more respondents solved the task than in the sham condition (Chi and Snyder, 2011). The authors concluded that the observed effect is due to inhibiting networks associated with top down processes related to prior knowledge that was achieved in particular by inhibiting the areas in the left hemisphere and at the same time increasing the excitability of the right hemisphere.

The first published study that explicitly aimed at influencing fluid intelligence with transcranial electrical stimulation was the one by Santarnecchi et al. (2013). In this study, fluid intelligence was assessed with an extended version of Raven's matrices that included more simple relational tasks based on perceptual relations and more complex tasks that involved high level of abstraction required by logical conditional arguments. tACS with an intensity of 750 μA was delivered in a sham and four verum conditions at theta (5 Hz), alpha (10 Hz), beta (20 Hz), and gamma (40 Hz) frequencies, while participants solved the Raven tasks. The target electrode was placed over the left middle frontal gyrus, whereas the reference (return) electrode was placed on the vertex corresponding to Cz. The only significant difference between sham and verum tACS conditions was observed for gamma stimulation, resulting in significantly shorter latencies (15% faster) on the more difficult Raven tasks requiring logic inference. No beneficial influence was observed for the number of correct responses. The authors concluded that the interactional effect of tACS improvements during gamma band stimulation with task difficulty might suggest a causal involvement of gamma oscillatory activity in the neural processes of higher order cognition, and highlights the importance of the left prefrontal cortex for inferential processes based on conditional arguments (Santarnecchi et al., 2013).

These findings were further elaborated in a recent study by the same authors (Santarnecchi et al., 2016). In this study the authors investigated whether tACS modulates brain activity by changing cortico-spinal excitability or by influencing targeted brain rhythms. In addition, they explored the extent to which improvements in logical reasoning can be explained by indirect enhancement of other cognitive components related to *g*, such as WM. The authors conducted two experiments with a tACS protocol similar to the one described previously. In addition, they used a working memory task measuring scope of attention (Luck and Vogel, 1997), described in Section 3.2.1. In the first experiment tACS was delivered in theta and gamma band frequency bands, whereas in the second experiment theta frequency tACS was replaced with high frequency transcranial random white noise (101–640 Hz) stimulation (tRNS). Like in their first study, the analysis of response time showed a significant stimulation effect for the more difficult logic problems with a marginal effect for the relational problems and no effect for the WM task. On the other hand, no significant effects for task accuracy were observed. The comparison of the different tACS protocols on response time for correctly solved logic and relational problems was only significant for logic problems in the gamma tACS condition and showed no differences between theta tACS, high frequency tRNS, and sham conditions. Furthermore, by collapsing the data of both experiments it was shown that gamma tACS was most beneficial for slower performers at baseline. The authors concluded that the results suggest that higher cognitive functions such as WM or logic inference are dissociable by frequency-specific neuromodulatory effects possibly induced with entrainment of specific brain rhythms (Santarnecchi et al., 2016).

The objective of the study conducted in our lab (Pahor and Jaušovec, 2014) was to explore the influence of offline transcranial alternating current stimulation (tACS) on measures of fluid intelligence. To our knowledge, this was the first study that not only analyzed behavioral tACS effects, but also investigated the influence of tACS on neuroelectric brain activity during rest and during performance on the tasks after sham/verum tACS. Here I present some new data not published in the original paper by Pahor and Jaušovec (2014). In our study, fluid intelligence was assessed with two tests, a modified version of Raven's progressive matrices and the paper folding and cutting (PF&C) subtest of the Stanford-Binet IQ test (Rideout and Laubach, 1996), described in Section 3.6. We used three different tACS protocols using theta frequency (individually determined based on the individual's IAF) stimulation delivered to the left parietal, left frontal, and parieto–frontal brain areas with the return electrode placed over the right orbitofrontal cortex (see Fig. 4.10 for more information about the experimental procedures and tACS parameters). To increase focality of stimulation, the electrode size of the target electrode was smaller than the size of the return electrode placed on the forehead (Moliadze et al., 2010). The magnitude of the stimulating current was based on individually determined thresholds for skin sensations induced by tACS. This resulted in stimulation intensity between 1000 and 2000 μA, delivered in the verum condition for 15 min.

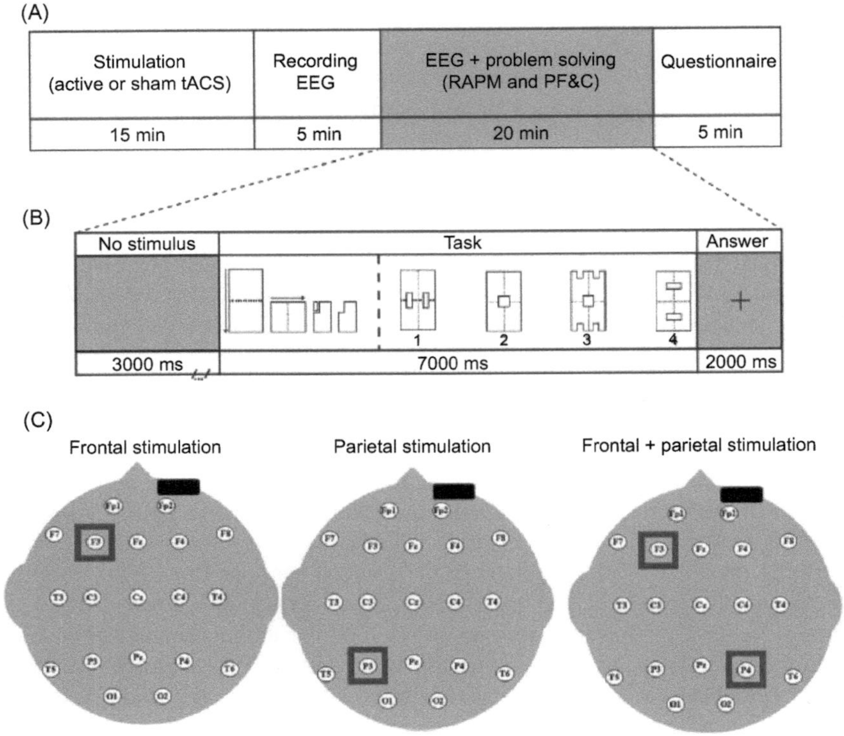

FIGURE 4.10 (A) Sequence of the experiment. (B) Procedure of the PF&C test that also applies to the RAPM test. (C) Experimental setup: tACS target electrodes positioned over left frontal (F3) or left/right parietal sites (P3/P4) and a return electrode positioned above the right eyebrow (Fp2) of the 10/20 system.

The behavioral data analysis for the left parietal tACS protocol in the first experiment showed a significant beneficial influence on the performance of the RAPM and PF&C tasks. These differences were more pronounced for the difficult/complex test items than for the easy ones. In the verum setting, the participants scored notably higher on both tests mainly solving more difficult/complex test items than in the sham setting (Fig. 4.11).

The analysis of neuroelectric brain activity during performance on the tasks (after sham/verum tACS) showed that in the verum setting the participants displayed higher lower-2 alpha ERD (more brain activity) than in the sham setting (Fig. 4.12). A similar trend was also observed in the upper alpha band. Increased upper alpha ERD (more brain activity) in the active setting compared to the sham setting was mainly observed in the left hemisphere during performance on the PF&C test.

These results confirmed the hypothesis that theta tACS applied over the left parietal brain area increases performance on tests of fluid intelligence. This

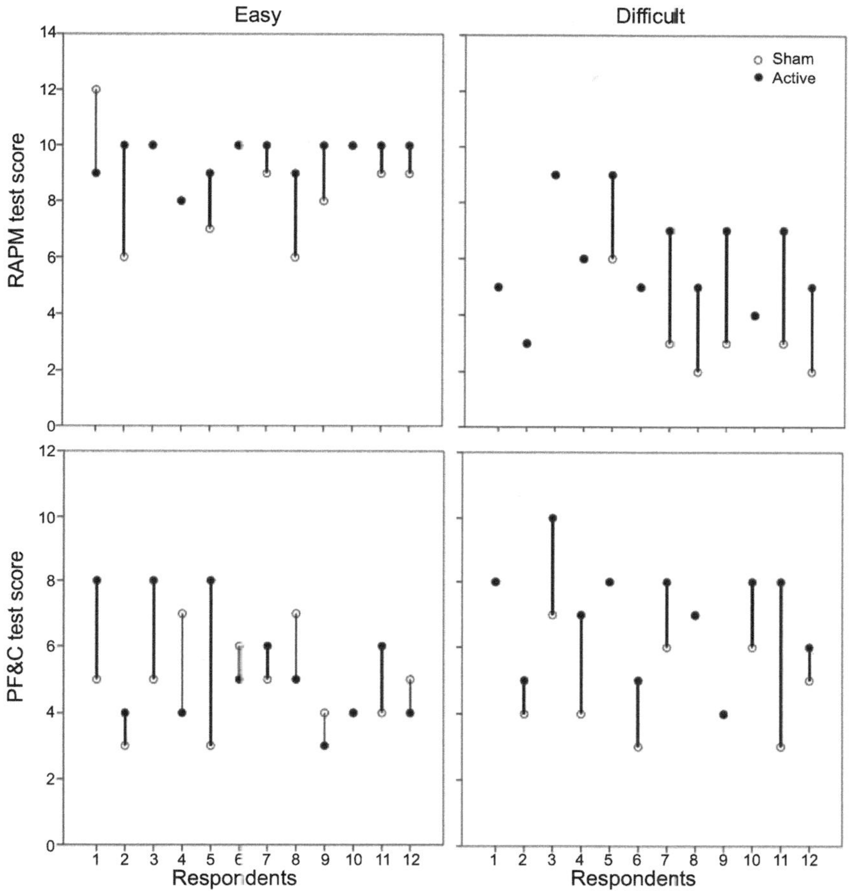

FIGURE 4.11 Drop-line graphs of test performance for easy and difficult test items of RAPM and PF&C tests in sham/active (verum) tACS settings—target electrode positioned over the left parietal site (P3). Bold lines represent increases in test performance after active (verum) tACS.

finding is in line with several studies showing similar structural and functional correlates with IQ tests (Barbey et al., 2012, 2013; Gevins and Smith, 2000; Gläscher et al., 2009; Haier et al., 2003; Jaušovec and Jaušovec, 2004; Jung et al., 2005; Lee et al., 2006) and lends further support to the assumption that short-term storage capacity largely accounts for the relationship between the psychological constructs of working memory and intelligence (Buehner et al., 2005; Colom et al., 2008). The importance of the left parietal lobe for working memory storage capacity has been demonstrated in neuroimaging studies (Cowan et al., 2011; Cowan, 2011; Majerus et al., 2010; Todd and Marois, 2004; Xu and Chun, 2006, see also Chapter 3), studies using rTMS (Postle et al., 2006), and in tDCS studies (Tseng et al., 2012). The results of a

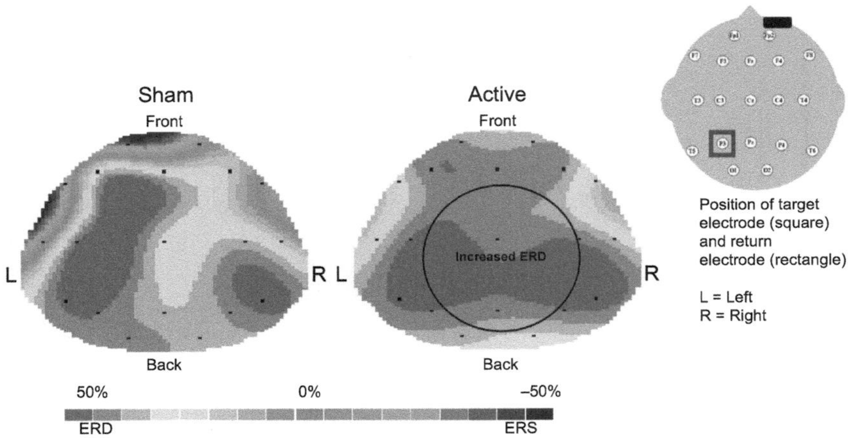

FIGURE 4.12 Topographic power map of neuroelectric ERD/ERS activity in the lower-2 alpha frequency band, while respondents solved the RAPM and PF&C tests after sham and active (verum) theta tACS—target electrode positioned over the left parietal site (P3).

recent study conducted in our lab (Jaušovec and Jaušovec, 2014) also support these findings. For more details see Section 4.4.1 showing that theta tACS applied to the left parietal area significantly increased working memory capacity. Furthermore, the ERP analysis indicated a decrease in P300 latency suggesting an improvement in the participants' ability to rapidly allocate attention resources needed to solve the task. A similar finding was also reported by Tseng et al. (2012) showing that tDCS mainly improved the ability of low performers to suppress irrelevant distractors.

Let's go back to the tACS-fluid intelligence study conducted in our lab. The neuroelectric patterns in the active setting showed increased alpha ERD (lower-2 and upper alpha) in the whole brain and were not, as expected, topographically differentiated between frontal and parietal brain areas. However, as stressed by Neubauer and Fink (2009) this frontal/parietal distinction in neural efficiency is mainly observed in studies using working memory tasks (Gevins and Smith, 2000; Jaušovec and Jaušovec, 2004), or tasks that can be solved using working memory capacities (Rypma et al., 2006). Another aspect that should be considered is that neural efficiency is related to task difficulty—it is mainly observed when participants work on tasks of low to medium difficulty (Neubauer and Fink, 2009). On the other hand, a tendency in the opposite direction has been observed for difficult tasks (Doppelmayr et al., 2005). The RAPM and PF&C tests used in our study were rather difficult to solve; on average, the participants correctly solved only 50% of items (sham setting); this could explain the higher levels of brain activity during the solution process after theta tACS. This explanation makes sense, since the observed increased performance in the active setting was mainly due to increased performance on more difficult/complex test items.

In the second experiment, the association between left frontal brain activity and performance on tests of fluid intelligence was explored. We expected that tACS delivered to the left dorsolateral prefrontal cortex (F3) would increase performance on tests that measure fluid intelligence. This prediction was based on Duncan's multidemand (MD) system theory (Duncan, 2010; see Section 2.3.4) and on Gray's (Gray et al., 2003) conflict monitoring resolution model for the relation between brain activity and performance on IQ tests. In both models, prefrontal brain activity is assumed essential for performance on IQ tests because of two reasons. First, processes such as attention, control, monitoring, and conflict resolution occurrences in cognition are fundamental for problem solving. Second, for most of these processes, there is a similar recruitment of the prefrontal cortex.

The analysis of the differences between sham and verum tACS for behavioral data showed that the only significant difference was for the easy RAPM test items. Namely, in the active setting the participants solved more easy RAPM test items than in the sham setting (Fig. 4.13).

The analysis of ERD/ERS neuroelectric brain activity showed significant differences between sham/verum tACS only in the theta band. Differences in theta ERD/ERS between active and sham settings were more pronounced in the RAPM test. In the active setting, respondents showed a time related (3000–6000 ms) increase in theta ERD (less brain activity), while solving the RAPM mainly in the frontal brain areas (Fig. 4.14). This neuroelectric pattern was accompanied by increased lower-1 alpha ERD (higher brain activity).

The analysis of behavioral data did not confirm our hypothesis that theta tACS delivered to the frontal brain area would have a major positive influence on tests of fluid intelligence. Theta tACS applied to the left DLPFC had a minor influence on test performance, and tended to be more pronounced on the RAPM (easy test items). The sham/verum differences observed in the patterns of neuroelectric brain activity were also more pronounced while respondents solved the RAPM test. In the active setting, a time-related decrease in theta activity was accompanied by increased lower-1 alpha desynchronization. It is assumed that theta frequency is mainly associated with storage and retrieval in episodic working memory (Sauseng et al., 2010). On the other hand, different narrow alpha band frequencies are associated with attention–suppression and selection (Klimesch, 2012). Therefore, it could be speculated that the obtained ERD/ERS pattern revealed a synchronous time related decay in working memory encoding with simultaneous increases in the focus of attention. Further support for this interpretation comes from the fact that theta and lower-1 alpha frequency bands have a harmonic frequency relationship that allows for maximal between-frequency coupling. Indeed, a recent study showed that stimulation of the left parietal areas had a positive effect on WM storage capacity, whereas no such influence was observed for the stimulation of the left frontal area (Jaušovec et al., 2014; see Section 4.4.1). In contrast, stimulation of the left frontal area had a significant influence on the 1-back task (measuring storage and executive processes of WM).

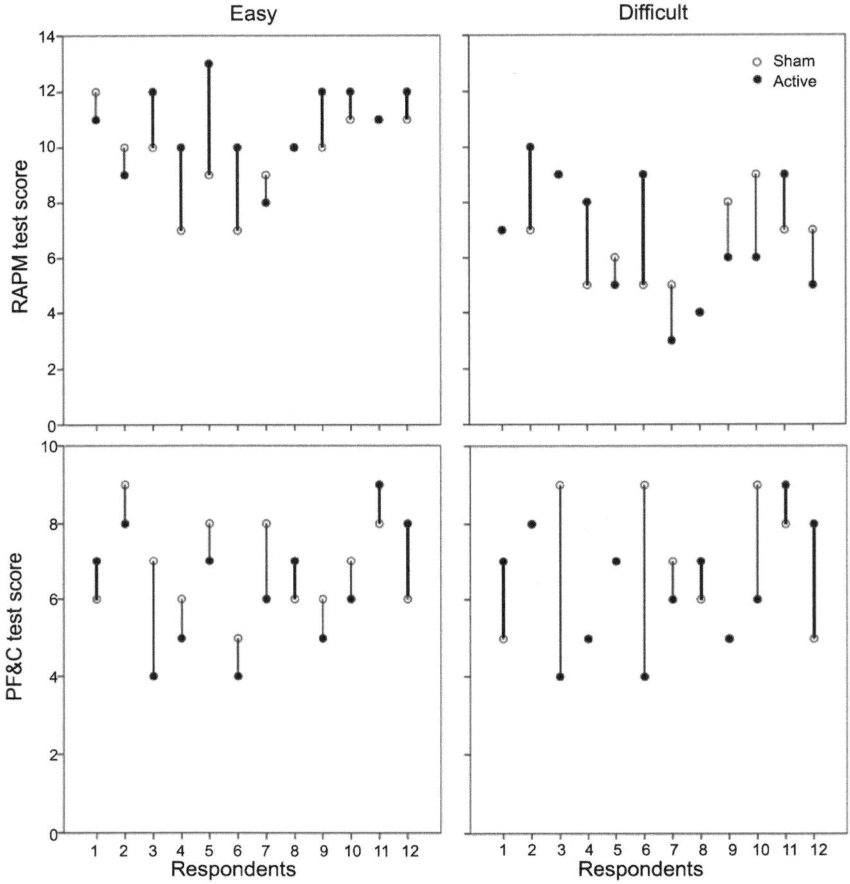

FIGURE 4.13 Drop-line graphs of test performance for easy and difficult test items of RAPM and PF&C tests in sham/active (verum) tACS settings—target electrode positioned over the left frontal site (F3). Bold lines represent increases in test performance after active (verum) tACS.

Overall, the observed behavioral and neuroelectric data obtained after left frontal theta tACS do not lend support to the assumption that *g* is mainly executed within processing networks of the prefrontal cortex (Duncan, 2010), as reported in recent correlational studies (Pineda-Pardo et al., 2016; Santarnecchi et al., 2015) and in tACS intervention studies showing that frontal gamma stimulation had a beneficial influence on respondents' performance on IQ tests (Santarnecchi et al., 2013, 2016). However, several factors might have biased the results. First, even though the left DLPFC is central for attention and control, theta frequency used for stimulation might not have triggered or maintained these processes. There is growing research evidence that attention

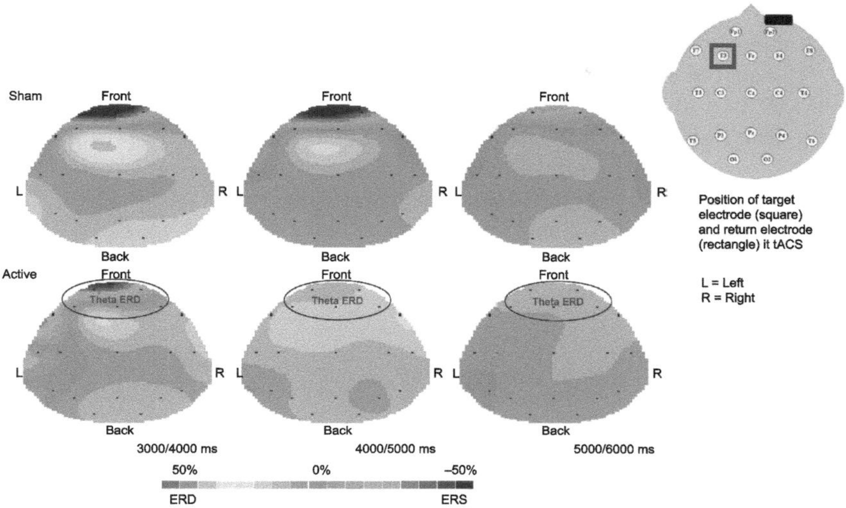

FIGURE 4.14 **Topographic power map of neuroelectric ERD/ERS activity in the theta frequency band, while respondents solved the RAPM test after sham and active (verum) theta tACS—target electrode positioned over the left frontal site (F3).**

processes are related to alpha frequency, assuming that alpha oscillations suppress distracting environmental input and direct the focus of attention on relevant information (Freunberger et al., 2011; Jensen and Mazaheri, 2010; Klimesch, 2012). The neuroelectric patterns of theta and lower-1 alpha coupling discussed earlier also point in this direction, along with the studies that applied tACS in the gamma frequency band (Santarnecchi et al., 2013, 2016). Second, the participants in Experiment 2 had high average baseline performance scores on the RAPM and PF&C tests (compared to those in Experiment 1); therefore, these differences might have additionally influenced the obtained results. A similar finding was also reported by Santarnecchi et al. (2016) showing that the beneficial influence of gamma tACS on processing speed of RAPM tasks was most pronounced for respondents that were slow on baseline testing.

In the third experiment we examined the influence of concurrent parieto–frontal brain stimulation on tests of fluid intelligence. It was expected that such simulation would have the most pronounced influence on test performance. This expectation was based on the P-FIT model of intelligence (Jung and Haier, 2007), which assumes that an efficient interaction of parietal and frontal brain regions, linked by white matter structures, underpins individual differences in performance on intelligence tests. Additional support for the hypothesis was provided by a cognitive training study conducted in our lab (Jaušovec and Jaušovec, 2012; see Section 3.6): we found that improved performance on IQ tests initiated by WM training was associated with fronto–parietal brain activity (right parietal and left frontal). Therefore, a similar pattern of neuroelectric brain activity was expected after active tACS simultaneously delivered to the right parietal and left frontal areas.

The analysis of behavioral data showed a significant influence of sham/active tACS on test performance. Differences between the sham and active settings were significant for the PF&C test, but not for the RAPM. This difference was more pronounced for the difficult/complex test items than for the easy ones. In the active setting, the participants scored notably higher on the PF&C test solving more difficult/complex test items than in the sham setting (Fig. 4.15).

As illustrated in Fig. 4.16, differences in theta ERD/ERS between the active and sham settings were only significant for the PF&C test. In the active setting,

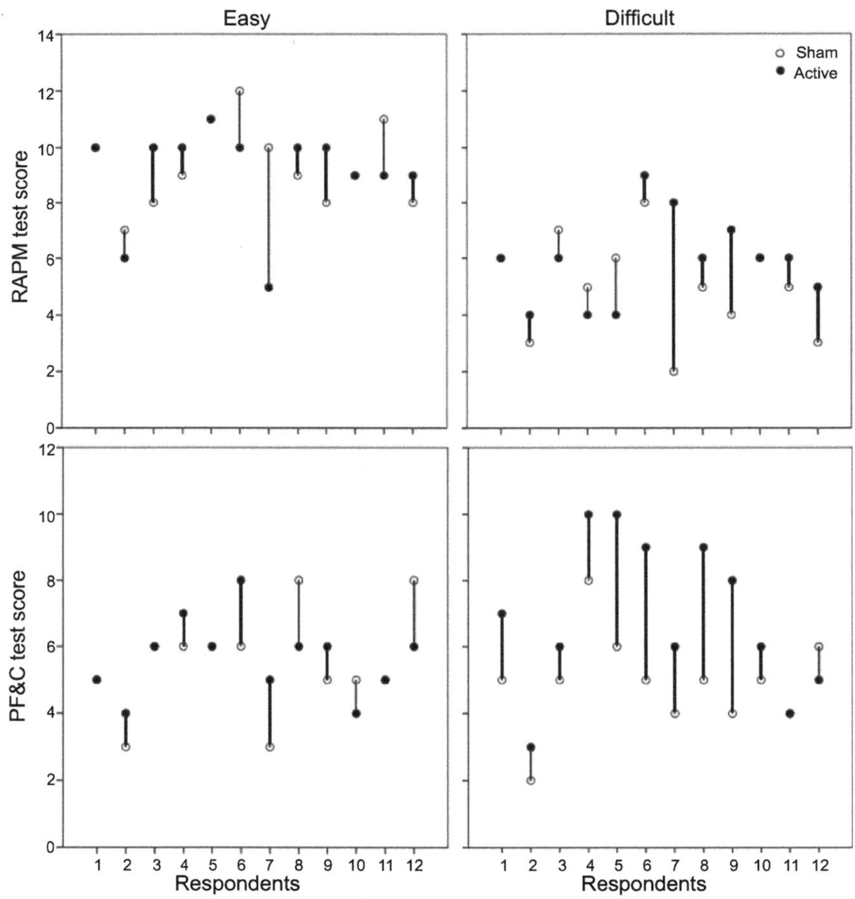

FIGURE 4.15 Drop-line graphs of test performance for easy and difficult test items of RAPM and PF&C tests in sham/active (verum) tACS settings—target electrodes positioned over the left frontal (F3) and right parietal (P4) sites. Bold lines represent increases in test performance after active (verum) tACS.

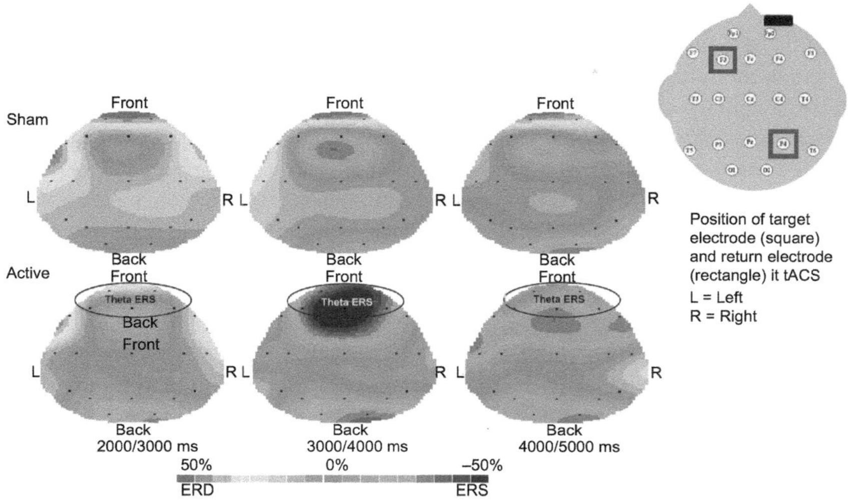

FIGURE 4.16 **Topographic power map of neuroelectric ERD/ERS activity in the theta frequency band, while respondents solved the PF&C test after sham and active (verum) theta tACS—target electrodes positioned over the left frontal (F3) and right parietal (P4) sites.**

respondents showed a time related increase in theta ERS (more brain activity), mainly between 2000 and 5000 ms from stimulus onset.

The behavioral results in experiment 3 did not confirm our expectation that tACS applied to the parieto–frontal brain areas would have the most pronounced influence on tests of fluid intelligence. However, a positive influence on the PF&C spatial reasoning test was observed. Albeit both the left and the right parietal areas have been associated with spatial cognition, lesion studies suggest that mental imaging is associated with the left hemisphere, whereas operations performed on that image (e.g., mental rotation) is a right hemisphere process (Kolb and Whishaw, 2009). This would explain the observed increase in performance for the PF&C spatial reasoning test after right parietal stimulation.

The neuroelectric patterns observed in the third experiment corresponded to the brain areas stimulated by theta tACS, as well as with improvements in test performance. Differences in brain activity between active and sham settings were most pronounced in the theta frequency band during performance on the PF&C test. The characteristic of the impact of tACS on neuroelectric brain activity was a time-related synchronization (ERS, more activity) in the theta frequency bands during performance on the PF&C test, which was most pronounced in frontal and right occipito–parietal brain areas.

In general the results of the three experiments revealed a positive effect of all tACS protocols on performance of fluid intelligence tests. This influence was most pronounced when stimulation was applied to the left parietal area, followed by simultaneous stimulation of fronto–parietal areas, and less pronounced for stimulation of the left frontal area. In the first experiment, verum tACS influenced performance on the RAPM and PF&C tests equally, whereas the influence of protocols used in experiments 2 and 3 were more diverse. Fronto–parietal stimulation had a positive effect on the mental rotation task PF&C whereas frontal stimulation tended to influence performance on the RAPM (easy test items) to a greater extent. The observed verum tACS influences on behavioral variables were also accompanied by changes in neuroelectric activity. Parietal stimulation had a general influence on ERD/ERS patterns, whereas frontal and fronto-parietal stimulation showed a task dependent influence that was related to performance increases on specific tests: frontal stimulation on ERD/ERS during performance on the RAPM test, and fronto–parietal stimulation while solving the PF&C test.

From a broader theoretical perspective the finding that tACS applied to the parietal areas, especially the left parietal brain area, had the most pronounced influence on tests of fluid intelligence is in accordance with the hypothesized short-term storage function of the left parietal brain area (Cowan et al., 2011). This finding is also in accordance with the assumption that short-term storage capacity largely accounts for the relationship between the psychological constructs of WM and intelligence (Buehner et al., 2005; Colom et al., 2008; Cowan et al., 2005). On the other hand, stimulating the left frontal area had a minor effect on test performance. This finding is surprising given the robust evidence of the importance of frontal areas in intelligence and WM (for a detailed discussion see Section 3.2). The most likely explanation is that theta frequency used for stimulation did not generate or maintain these processes. There is growing evidence that inhibition and timing in the alpha frequency band are crucial for the attention processes of suppression and selection of information (Klimesch, 2012; Jensen and Mazaheri, 2010). Therefore, stimulating the frontal areas in a narrow alpha frequency band or in the gamma frequency band as in the studies of Santarnecchi et al. (2013, 2016) might have a more pronounced influence on tests of fluid intelligence. This hypothesis was partly tested in a recent study conducted in our lab (Pahor and Jaušovec, 2016).

The analysis of tACS influences on easy/difficult test items revealed that left parietal tACS had a more pronounced influence on difficult test items (RAPM and PF&C), whereas left frontal tACS had a more pronounced influence on easy test items (RAPM). This suggests that left frontal/parietal tACS differently influenced cognitive processes involved in the fluid intelligence construct. A vague explanation could be that left parietal tACS increased WM storage capacity and in that way increased performance on difficult test items, whereas left frontal tACS influenced attentional components resulting in performance increases on easy test items. Support for this interpretation comes from the relation between working memory processes and neuronal rhythms in frontal and parietal brain areas (Sauseng et al., 2010), and from recent tACS studies (Jaušovec and Jaušovec, 2014; Jaušovec et al., 2014).

Three neurobiological models have been proposed to explain the relation between intelligence and brain activity: the neural efficiency hypothesis (Haier et al., 1988), the parieto–frontal integration (P-FIT) model (Jung and Haier, 2007), and the MD system theory (Duncan, 2010). On the whole, the results of our study lend some support to the neural efficiency hypothesis, as well as to the importance of parietal areas for intelligent reasoning put forward by the P-FIT and to some extent also by the MD system theory.

As discussed in Chapter 2, the neural efficiency theory explains the brain-intelligence relation with a more focused use of task relevant brain areas and the nonuse of brain areas that are irrelevant in solving a task. This relation seems to be corroborated mostly when participants work on tasks of low to medium difficulty or complexity (Neubauer and Fink, 2009). However when tasks are difficult, low-performers display a decrease in activation from easy to difficult tasks, whereas the opposite is true for high-performers (Doppelmayr et al., 2005). A similar tendency after tACS stimulation was also observed in experiments 1 and 3. In line with this explanation was also the finding that gains in test performance observed in experiments 1 and 3 were predominantly observed for the more difficult/complex test items.

The findings described earlier (Pahor and Jaušovec, 2014) were replicated in a recent study in which theta tACS was combined with fMRI (Wammerl et al., 2015). In this double-blind sham-controlled study, the same tasks were used as in our experiment. The tACS protocol was similar to the one we used in the first experiment, except that the return electrode was placed on the vertex location (Cz). The intensity of the stimulating current was held constant at 1500 μA. After sham/verum tACS, the participants solved the two tasks in a 3 T MRI scanner. Significant differences between the sham and verum tACS conditions were only observed for the difficult items in Raven's Advanced Progressive Matrices. Participants solved significantly more (difficult) RAPM items after verum tACS than after sham stimulation (28% vs. 24%). No differences were observed for the PF&C tasks. A comparison of fMRI scans showed that differences were mainly observed for performance on the difficult RAPM items associated with lower activation in the left middle occipital gyrus, as well as in right hemispheric regions of the occipital lobe and frontal lobe. Furthermore, a region of interest (ROI) analysis performed on the clusters showing increased task-specific activation demonstrated that the only brain area that showed a tendency toward a difficulty dependent stimulation effect was the inferior parietal lobe. For difficult items, but not for the easy ones, verum stimulation tended to be associated with lower activation in the left inferior parietal cortex as compared to sham stimulation. The differences in activation patterns displayed by the fMRI scans during performance on the RAPM and PF&C tasks supports the notion that each of the two tasks addresses distinct facets of fluid intelligence performance. In particular, the observed tACS-induced decreases in brain activation were primarily observed in task-negative areas (e.g., precuneus). This finding is in line with research of individual differences in neural efficiency showing that higher intelligence is related to lower brain activation in task-negative brain regions. Two studies reported that the more demanding a task was, the higher was the deactivation in the default mode network (DMN—task-negative brain region), suggesting that less intelligent individuals were confronted with greater cognitive demands, causing a more pronounced deactivation of the DMN and in that way increasing the resources needed to solve the task at hand (Basten et al., 2013; Lipp et al., 2012). The findings reported by Wammerl et al. (2015) could point to tACS-induced neural efficiency of brain activation patterns.

A common characteristic of all of the above reported studies was that TES had a beneficial effect on tasks that are highly saturated with *g* (e.g., Raven's matrices), even though different stimulation protocols with respect to the brain areas stimulated, frequencies employed, and current intensities were used. Another characteristic worth mentioning is that the positive effects were most pronounced for difficult tasks and were less significant (or not significant) for the easy ones. In this direction points also the study by Chi and Snyder (2012) using the extremely difficult nine-dot insight problem. It is puzzling that such a diversity of stimulation parameters resulted in rather similar behavioral effects. In my opinion the only plausible explanation lies in the common characteristic of all these problems. In order to solve a matrix-reasoning test like RAPM, one must switch between different educed relations and correlates (Jaeggi et al., 2010). The more complex the problem becomes, a greater number of alternatives must be maintained and processed in WM (the same holds true for the nine-dot problem). Perhaps the positive effects of theta and gamma tACS on performance on difficult tasks can be best explained by increased WM capacity: scope of attention. In the study by Santarnecchi et al. (2016), an attempt was made to control for WM capacity, however, the difficulty of the WM tasks was probably insufficient to provoke significant differences. The authors used samples with just four elements, and a reduced interstimulus interval between the sample and the probe, making the test easy to solve. In our study (Jaušovec and Jaušovec, 2014) we used samples with six and eight elements that were difficult to solve, given that the scope of attention is estimated to three to four elements (Cowan et al., 2005). Therefore in this experiment a beneficial influence of theta tACS on the scope of attention was established (see Section 4.4.1).

Not all TES studies reported positive intervention effects on fluid intelligence. Sellers et al. (2015) used a double-blind, sham-controlled, pre-, posttest experimental design to test the influence of three different tDCS protocols on intelligence measured with the Wechsler (WAIS-IV edition) IQ test. All subtests measuring verbal comprehension, perceptual reasoning, working memory, and processing speed were included in the test- and posttest sessions, separated by at least 1 week. In the posttest session, the participants were exposed to 20 min tDCS with a current intensity of 2 mA using three different verum electrode placements and a sham condition, followed immediately by WAIS-IV testing. The electrode placements in the verum conditions were: (1) bilateral, anodes over F3 and F4 and cathode over Fz; (2) unilateral, left hemisphere; anode over F3 and cathode over Fz; or (3) unilateral, right hemisphere; anode over F4, cathode over Fz. The results were surprising as all three verum tDCS protocols showed reductions in practice gains on the WAIS-IV. Significant decreases between baseline testing and verum tDCS posttesting were mainly observed for the perceptual reasoning tasks after right frontal (F4) anodal tDCS and at trend level for the other two tDCS protocols—bilateral anodal (F3 and F4) and left anodal (F3). Of the three subtests included in perceptual reasoning, block design, matrix reasoning, and visual puzzles, the latter two exhibited selective impairment based on tDCS: right tDCS decreased performance on matrix reasoning, whereas for visual puzzles, both right and left tDCS reduced practice gains compared to sham stimulation. The authors concluded that some forms of frontal tDCS may produce a lesion effect similar to the one observed in TMS, thereby impairing performance on facets of perceptual reasoning (Sellers et al., 2015).

The second study that showed mixed effects of TES on intelligence was a recent study conducted in our lab (Pahor and Jaušovec, 2016). The objective of this study was to establish a causal relation between individual alpha frequency (IAF) and fluid intelligence (measured with an IQ test). For that purpose we increased IAF of individuals with transcranial alternating current stimulation (tACS). In our study, individual alpha frequency was upregulated by applying tACS in the IAF + 1 Hz frequency band. In the next step we analyzed the differences in behavioral and neuroelectric measures between sham and verum tACS sessions. Based on the results obtained in previous research (for a detailed discussion see Section 2.2) we expected that increased IAF (activation energetic of physiological responding) will have a positive influence on performance on Raven's matrices that will be reflected in neuroelectric brain activation patterns—induced ERD/ERS, and induced event related coherence (ERCoh). Because tACS was delivered in the IAF + 1 Hz frequency band differences in neuroelectric measures were mainly expected in the upper alpha band. The subjects participated in two sessions, sham and verum tACS that took place over two consecutive days, and were counterbalanced. The main characteristics of the experimental design are outlined in Fig. 4.17.

As can be seen in Fig. 4.18, the only significant difference between pre- and post-tACS on IAF was observed in the verum setting. Verum tACS significantly increased IAF (on average for 0.28 Hz), whereas no influence of sham tACS on IAF was observed. Similar influences of verum tACS delivered in the IAF range on power measures have been reported in other studies (Neuling et al., 2013; Vossen et al., 2015; Zaehle et al., 2010).

The analysis of RAPM performance scores showed only a significant interaction effect between tACS and sex. As can be seen in Fig. 4.19, males showed lower RAPM performance in the verum tACS condition compared to the sham tACS condition, whereas females showed increased performance in the verum condition, compared to sham. Subsequent t-tests comparing RAPM

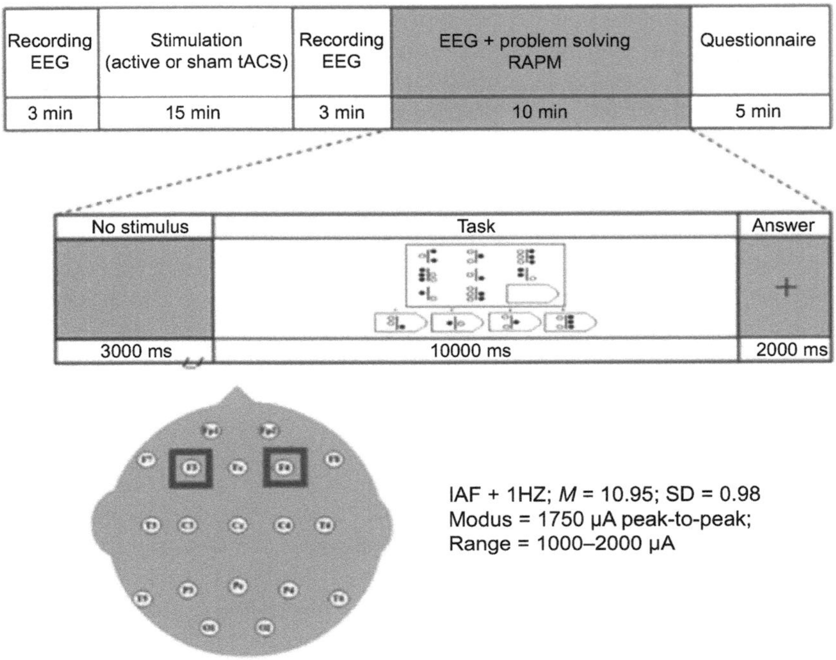

FIGURE 4.17 Experimental setup: tACS target electrodes positioned over left/right frontal sites (F3/F4) of the 10/20 system. On both days, the participant's resting eyes-closed EEG was recorded for 3 min, followed by sham or verum stimulation, after which poststimulation resting eyes-closed EEG was recorded for 3 min, then the participant solved the RAPM tests. At the end of the experiment, they answered a questionnaire about their sensations during the stimulation.

performance in sham and verum tACS settings conducted separately for males and females were not significant. However, the following trend was observed: the disruptive effect of verum tACS on RAPM performance in males was more pronounced ($d = 0.77$) than the beneficial effect for females ($d = 0.38$).

The verum tACS-related disruptive effect on male RAPM performance was unexpected. However, in our opinion it can be explained as a summation of several small effects, such as test characteristics and format of presentation, the sex-specific relation of endophenotypic IAF with factors of fluid intelligence, and preferred solution strategies. The Raven's matrices test is based on Spearman's (1927) principles of education of relations and correlates (Deary, 2000), and is regarded as one of the best measures of *g*. It is based on abstract figures and therefore requires visuo–spatial ability (Abad et al., 2004). Endophenotypic posterior IAF is strongly related to performance on mental rotation and shape manipulation tasks (rotation factor) in males. In contrast, anterior IAF is in males related to orientations involving attentional focus on details one at a time (Pahor and Jaušovec, 2016). In addition, the reported male advantage in mental rotation tests

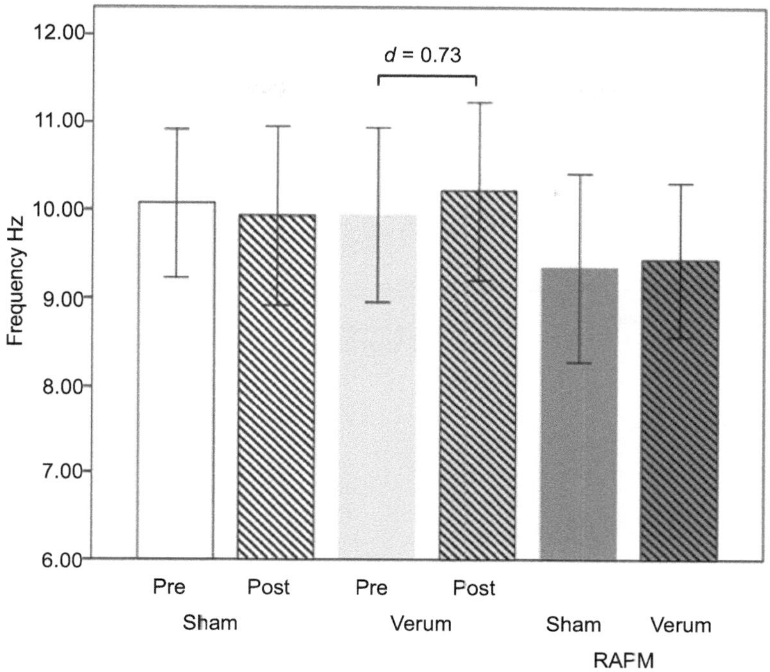

FIGURE 4.18 Histograms of average IAF (Hz) in sham and verum settings recorded prior and after tACS and during performance on RAPM. Shown are Cohen's effect size measures for significant differences determined with a t-test for repeated measures. Error bars = ± 1 SD.

has been most often attributed to sex-specific differences in strategy use (Butler et al., 2006; Gootjes et al., 2008a,b; Hugdahl et al., 2006; Jordan et al., 2002). It was hypothesized that women performed mental rotation tasks in an effortful, "top–down" way, whereas men used an automatic, "bottom–up" strategy. This corresponds also to the reported negative relation of time limits when solving rotation tasks observed in females (Voyer, 2011). Therefore it could be speculated that tACS delivered to the left and right dorsolateral prefrontal cortex enhanced ordered step-wise problem solving processes that correspond to female solution strategies and conflict with male problem solving strategies. Two recent studies reported similar findings: tACS delivered to parietal brain areas had a much more positive effect on RAPM performance than when delivered to frontal brain areas (see earlier). Moreover, the study by Sellers et al. (2015) indicated that the most pronounced disruptive effects of frontal tACS were observed for perceptual reasoning subtests and in particular for matrix reasoning. This subtest is similar to the modified version of Raven's matrices used in our study and its scores are also speed-dependent, lending further support for the provided explanation of verum tACS disrupting effects on male RAPM performance.

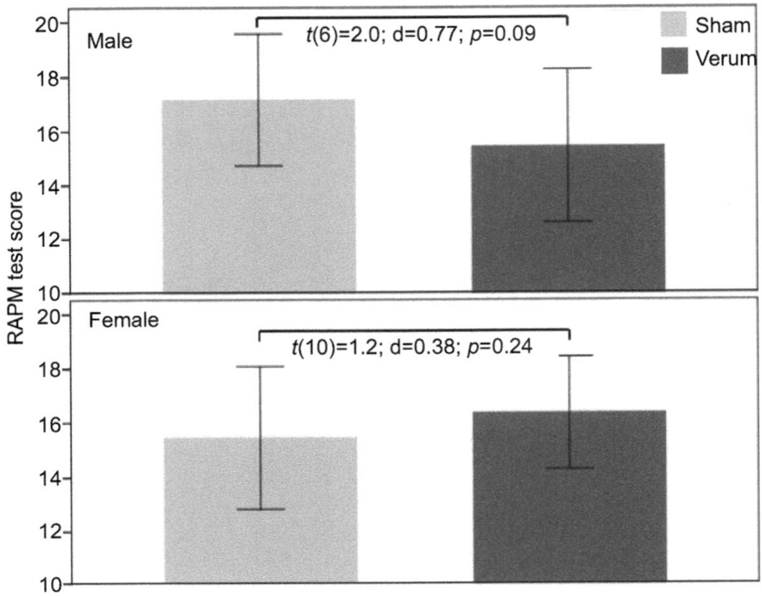

FIGURE 4.19 Histograms of average performance on the RAPM test in sham and verum tACS settings separated for males and females. Shown are *t*-tests with significance levels and Cohen's effect size measures. Error bars = ± 1 SD.

ERD/ERS analysis showed a significant main effect for tACS (sham/verum) only in the upper alpha frequency band. No significant sex-specific interaction effects were observed. As can be inferred from Fig. 4.20, participants in the verum tACS condition as compared to the sham condition showed increased upper alpha desynchronization (ERD) pointing to increased brain activation. A similar pattern of brain activity, although accompanied by substantial gains on several fluid intelligence tests, was also reported in our study where we delivered theta tACS to parietal locations (Pahor and Jaušovec, 2014). Differences in induced ERD/ERS measures were only observed in the upper alpha band, which is in line with our expectation, given that tACS was delivered in the IAF + 1 Hz frequency band, representing the mid frequency of the upper alpha band.

Significant differences in Fisher-Z transformed induced ERCohs were only obtained in upper alpha and theta frequency bands (Fig. 4.20). All values showed increased decoupling mainly over a hub located in the right temporo–parietal area. In the upper alpha band the decoupling was observed between midfrontal and right parietal areas, whereas in the theta band it was mainly between right temporal and central locations. These locations are close to the main hubs of the salience and central executive networks; this may suggest switching of activity from the default mode network to the central executive network (Bressler and Menon, 2010; Raichle et al., 2001; Raichle, 2006; Seeley et al., 2007; Sridharan

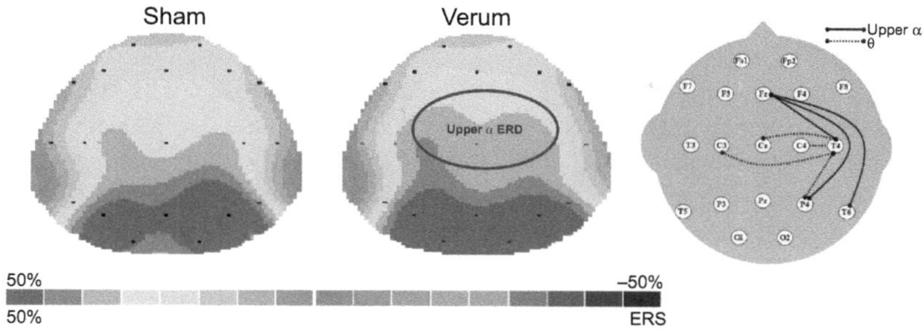

FIGURE 4.20 Topographic maps of ERD/ERS activity in the upper alpha band during performance on RAPM in sham and verum tACS conditions. Connectivity map of significant differences in ERCoh values in the upper alpha and theta frequency band between the sham and verum tACS settings (FDR corrected).

et al., 2008). Furthermore, a similar fronto–parietal dissociation was observed in an fMRI study by Takeuchi et al. (2013) in relation to working memory training-induced changes in resting state brain activation. After 4 weeks of training, decreased resting functional connectivity between the medial prefrontal cortex, the right posterior parietal cortex and the right lateral prefrontal cortex (key nodes of the default mode network and the external attentional system) was observed. This pattern could suggest that higher cognitive performance is related to a dissociation between the two networks. Similar small-world activation patterns in relation to working memory performance obtained with high-density EEG recordings were also reported by Langer et al. (2013). However, in our study only females showed increased cognitive performance, and the same brain activation pattern was in males related to decreased performance. As discussed earlier, this difference could stem from sex-specific solution strategies.

This overview of studies that have tried to influence fluid intelligence with transcranial electrical stimulation has demonstrated the great variety of possibilities the technique provides, in particular with respect to protocol design—which represents the positive part. On the negative side, one could critically remark that the studies are fishing in troubled waters, with no clear theoretical concepts that would justify why specific protocols were employed. Given that only a few studies have been conducted, it is probably premature to make a final assessment on the usefulness of TES to either study or enhance intelligence.

4.4 NONINVASIVE BRAIN STIMULATION IN WORKING MEMORY RESEARCH

In recent years there has been an exponential rise in the interest for noninvasive brain stimulation (NIBS) as a research tool to elucidate neural underpinnings of cognition. A similar conclusion was put forward in Miniussi's (2016) editorial accompanying the special issue of the *European Psychologist,* devoted to the use of NIBS in psychology. However, the previous section showed that cognition and especially fluid intelligence is still not part of mainstream research employing different NIBS techniques, such as rTMS, tACS, tDCS, and tRNS. For example, out of the seven papers published in the special issue of the *European Psychologist,* just one study by Lavidor (2016) addresses the use of NIBS in cognitive research, while the vast majority focuses on the use of NIBS in therapy (poststroke, memory enhancement in older individuals, aphasia treatment, and similar). Among cognitive processes most often modulated with TES was memory. A recent review of TES used to enhance cognition estimated that approximately 30% focused on working memory, followed by language (19%), long-term memory (16%) and short-term memory (14%), with the remaining 21% divided among six cognitive processes: creativity, problem solving, learning, attention, cognitive control and mathematical reasoning (Santarnecchi et al., 2015). A similar distribution of studies was also reported in a metaanalysis of the effects of tACS on cognition (Schutter and Wischnewski, 2016). The authors analyzed 24 studies, including 51 experiments of which 8 studies and 15 experiments investigated memory processes.

In this section I will provide a brief summary of studies employing TES and TMS that investigate neural underpinnings of working memory, mainly based on recent review papers and metaanalyses (Balconi, 2013; Bartrés-Faz and Vidal-Piñeiro, 2016; Brunoni and Vanderhasselt, 2014; Hill et al., 2016; Horvath et al., 2015), summarized in Table 4.3. In addition, studies conducted in our lab investigating tACS influences on WM performance will be presented. The main focus will be on WM storage capacity (scope of attention) and processes involved in solving the *n*-back task. This selection is mainly guided by the fact that as discussed in previous chapters it is assumed that short-term storage capacity largely accounts for the relationship between the psychological constructs of WM and intelligence (Buehner et al., 2005; Colom et al., 2008; Cowan et al., 2005) and that there is tentative evidence that *n*-back training can improve intelligence (Jaeggi et al., 2008).

The metaanalysis summarized in Table 4.3 revealed that the most often used NIBS technique in working memory research was tDCS. On the other hand, there are just a few studies that have used tACS, TMS, or rTMS to explore WM. The area targeted with NIBS in almost all studies was the dorsolateral prefrontal cortex (DLPFC). This is not surprising given that neuropsychological and neuroimaging evidence indicated the central role of frontal brain areas in WM processes, especially when manipulation of information was required. It was further suggested that the role of the DLPFC is in the maintenance of information by directing attention to internal representations of information stored in more posterior regions (Curtis and D'Esposito, 2003; see also Sections 2.4.2 and 3.2.1 for a detailed discussion on WM processes). Therefore posterior parieto–occipital regions have been mainly studied in relation to storage and retrieving/recognition WM functions (Pergolizzi and Chua, 2015; Barbieri et al., 2016).

TABLE 4.3 Overview of Review Papers and Metaanalyses on the Use of Noninvasive Brain Stimulation (NIBS) in Working Memory Research

Balconi (2013)
NIBS technique: TMS The review included 11 TMS and rTMS studies targeting the dorsolateral prefrontal cortex (DLPFC). The studies confirmed the importance of the DLPFC in working memory performance, especially when manipulation of information was required, but being to a lesser extent involved in retention of information (Postle et al., 2006). Single pulse TMS further showed a propagation of information flow from posterior to anterior cortical sites converging in the left prefrontal cortex. Furthermore, a different involvement of the left/right DLPFC was revealed, suggesting a parallel processing of semantic and object features of stimuli (Mottaghy et al., 2003; Mull and Seyal, 2001). Single pulse TMS demonstrated the involvement of the middle frontal gyrus in tactile working memory (Hannula et al., 2010). The authors concluded that all studies reviewed agreed that the DLPFC plays a crucial role in working and episodic memory, suggesting that the function of the DLPFC is the activation of an episodic buffer that mediates between old and new information. The contribution of the left/right DLPFC in this process remains unknown.
Brunoni and Vanderhasselt (2014)
NIBS technique: rTMS and tDCS In the meta-analysis 8 tDCS (19 experiments) and 4 rTMS (14 experiments) studies were included. The main criteria of inclusion were: sham controlled, the use of *n*-back tasks, stimulating the DLPFC, offline design, and using standard procedures for the methods of randomization, blinding and sham stimulation. The main findings can be summarized as follows: • Participants after active compared to sham NIBS had significantly faster response time for correct answers. • Active compared to sham brain stimulation presented a higher percentage of correct responses. • Active compared to sham brain stimulation presented a lower percentage of errors. The authors concluded that noninvasive brain stimulation of the dorsolateral prefrontal cortex (DPFC) induces WM improvement in both healthy and clinical samples. It was further found that participants receiving tDCS, but not rTMS, were faster, but not more accurate. This relation was also related to the type of the experimental design showing that parallel group designs showed improved accuracy rates, whereas in within-group (crossover) designs only faster response times were observed. This might have biased the tDCS/rTMS differences in relation to accuracy, because most tDCS studies used crossover designs. A circumstance that makes the comparison between these two techniques difficult; therefore, it is still unclear whether the effects of tDCS and rTMS on cognition are different.

(Continued)

TABLE 4.3 *(cont.)*

Tremblay et al. (2014)

NIBS technique: tDCS
The main inclusion criteria of the metaanalysis were the application of tDCS to DLPFC and its influence on cognition. Sixty-three papers were included, 14 of which were related to WM. With respect to different types of stimulation (target regions and polarity), the authors identified seven electrode montages that were used in the included papers. Their influence on working memory performance can be summarized as follows:
• Working memory increases were observed for:
  • cathodal tDCS over the left DLPFC (one study),
  • anodal tDCS over the left DLPFC (nine studies),
  • anodal tDCS over the right DLPFC (one study).
• Working memory decreases were observed for:
  • cathodal tDCS over the left DLPFC (one study),
  • anodal tDCS over the left DLPFC (one study),
  • tDCS over bilateral DLPFC (left anodal/right cathodal; one study).
The metaanalysis revealed that similar tDCS protocols can lead to opposite results, and that specific tDCS protocols can induce cognitive effects (facilitation/inhibition) over a wide variety of functions. This makes it difficult to attribute an observed effect on a specific task to a single mechanism.

Horvath et al. (2015)
NIBS technique: tDCS The authors used a rather rigorous inclusion criterion, selecting only studies in which all outcome measures were replicated by at least two different research groups. This ensured that tDCS outcome measures were not attributed to a unique device, protocol, or condition utilized in a single lab. The data were analyzed on two precision levels: • Including only studies that employed the same cognitive task with identical tDCS current densities and reference electrode locations, and • Including all studies which used the same outcome measures (lower level of homogeneity). A third level of analysis was performed for memory tasks, combining all papers that targeted the same neural region and used tasks that were assumed to measure WM. The analysis was based on 24 memory studies. Of the 59 analyses performed on all cognitive tasks (including WM, level 1 and 2 inclusion criteria), tDCS was not found to generate a significant effect on any of these measures. The same null results were reported for the third level of analysis performed only for the effects of tDCS on WM. The explanation provided by the authors was that the differential state-dependent effects between different studies included in the analysis influenced the null results. tDCS influences on study outcome do not only depend on tDCS parameters (intensity of applied current, anodal/cathodal position), but also on the cortical activation state of the targeted neural region.
Fröhlich et al. (2015)
NIBS technique: tACS The authors included 23 studies using tACS to influence diverse cognitive processes. In four of them, tACS was used to influence WM processes. In all of the studies, tACS in the theta frequency band was delivered to left/right frontal and parietal sites, and had a positive influence on different WM measures. The authors concluded that although some methodological shortcomings were observed in these studies, they provided appealing evidence that theta oscillatory activity in parietal and frontal areas might contribute to WM performance.

TABLE 4.3 *(cont.)*

Bartrés-Faz and Vidal-Piñeiro (2016)
NIBS technique: TMS and tDCS Eight (three TMS) papers were included in the analysis with a special focus on the use of NIBS of WM in older adults. The studies reported small memory enhancements that were in three studies (one of which was rTMS) moderate to high (with rather high effect size measures > 0.60). The authors concluded that the use of NIBS to enhance memory in older adults is a promising approach, although dependent on several factors not wholly related to parameters of the stimulation protocols that were administered.
Hill et al. (2016)
NIBS technique: tDCS The authors included 16 sham controlled tDCS studies investigating WM memory in either healthy volunteers (n = 170) or individuals suffering from a neuropsychiatric illness (n = 182). The studies had to be at least single-blind and had to target the DLPFC. The metaanalysis for the healthy cohort showed: • tDCS produced a small but significant reduction in reaction times on the WM tasks, insignificant for online, but significant for offline WM tasks. A similar trend was observed for accuracy results with only a trend for significance for offline WM tasks. The metaanalysis for the clinical cohort showed: • tDCS had no effect on reaction times on the WM tasks, offline or online. tDCS produced a significant and moderate improvement in accuracy for WM tasks performed online, whereas no change was observed for offline WM tasks. The authors concluded that tDCS has some limited possibilities to enhance WM in healthy and neuropsychiatric populations especially when delivered with higher current densities and longer stimulation durations.
Papers are ordered by year of publication

The metaanalysis further revealed that a great variety of tDCS protocols was used. However, the same protocols (anodal/cathodal placement, area stimulated) have led to opposite WM outcomes (Tremblay et al., 2014). Hence, it is not surprising that a more rigorously designed metaanalysis showed that tDCS had no significant effect on cognitive processes, including working memory outcome and language production task (Horvath et al., 2015). The authors suggested that the most likely explanation for the null tDCS influence on cognition are state-dependency effects. State-dependency means that the behavioral and perceptual effects of an external stimulus (e.g., tDCS or TMS) are determined by initial neural activation states (Silvanto et al., 2008). For example, it was shown that two consecutive sessions of anodal tDCS divided by a 10 min interval did not improve WM performance, whereas the same tDCS protocol with cathodal tDCS increased WM performance. It can be therefore assumed that tDCS polarity effects on WM performance depend on the previous modification of the activity in a neuronal network (Carvalho et al., 2015).

Taken together, the reviews highlighted several limitations of NIBS that are not related just to protocol characteristics, such as placement of electrodes, delivered current density, or frequency in the case of tACS and rTMS, but also depend on the initial state of neural circuits targeted with the stimulation. The latter can be more easily controlled in research involving sensory and motor functions, but is difficult to control in cognitive research. In an offline NIBS experimental design, for example, the stimulation is delivered prior to the probe tasks, usually when the participants are asked to relax (the level of their cognitive activity is not controlled for). On the other hand, in an online experimental design, stimulation is delivered during performance on the probe tasks (high cognitive engagement). A recent metaanalysis showed that study outcomes on WM performance (response time and accuracy) significantly differed between offline and online tDCS studies (Hill et al., 2016). Unless these methodological issues are satisfactory solved, NIBS research will remain a promising but highly speculative research tool regarding its efficacy in cognitive research.

4.4.1 Research From Our Lab

The review of metaanalyses of NIBS research on WM has shown that few studies used tACS or tRNS (Table 4.3). Two of them were conducted in our lab.

In our first study we aimed to investigate the influence of tACS on working memory storage capacity (Jaušovec and Jaušovec, 2014). As already mentioned, most studies investigated the function of the DLPFC in working memory processes (e.g., central executive functions, such as inhibition of irrelevant items, monitoring of ongoing performance, and updating representations in memory). In contrast, only a few studies investigated the influence of anodal tDCS of the DLPFC on WM storage capacity (Fregni et al., 2006; Andrews et al., 2011). tDCS significantly increased performance on forward and backward digit-span tasks (Fregni et al., 2006), and on forward digit-span tasks (Andrews et al., 2011). However, it is difficult to generalize this finding with regard to the relationship between the left DLPFC and WM storage capacity for several reasons. First, the respondents who participated in the study by Fregni et al. (2006) were patients with depression, second, the individuals were not exposed to verum and sham conditions. A third issue related to the Andrews et al. (2011) study is that the digit-span task is not a pure measure of WM storage capacity as it enables rehearsal and grouping strategies (Cowan et al., 2005; see also Section 2.4.2). Even less research has been conducted on stimulating parietal brain areas. Jacobson et al. (2012) used anodal tDCS over the left intraparietal sulcus, and this had a positive effect on performance on a verbal recognition memory task. In another study, Polanía et al. (2012) demonstrated that left fronto–parietal coupling in the theta band (0 degree phase tACS stimulation) increased the speed of performance on a delayed letter discrimination task, as compared to a sham condition, as well as to an experimental condition of left fronto–parietal decoupling in the theta band (180 degree phase tACS stimulation).

Therefore the main objectives of the first tACS study conducted in our lab were (1) to examine the causal association between left parietal/frontal brain activity and WM storage capacity and (2) to determine the neurobiological underpinnings of WM storage capacity. It was expected that theta tACS applied to the left parietal brain area would have a more pronounced influence on WM storage capacity, compared to theta tACS applied to the left DLPFC. This assumption was based on recent neuroimaging studies showing that the left intraparietal sulcus (LIS) demonstrated load-dependent activity for visual stimuli or visual and auditory stimuli (Cowan et al., 2011; Cowan, 2011). Additionally, it was reported that LIS activity is closely associated with capacity-limited working memory maintenance (Majerus et al., 2006, 2010; Todd and Marois, 2004; Xu and Chun, 2006) and that disruptive rTMS had a more pronounced influence on WM storage capacity when applied to parietal areas than when applied to DLPFC (Postle et al., 2006). Furthermore, by analyzing differences between sham/verum settings in ERP components (N1–P1, P300 latency and amplitude) and their relation to global stages of information processing, we attempted to determine the neurobiological underpinnings of WM storage capacity. Evidence suggests that the P1 amplitude is associated with suppression of irrelevant information while the N1 is associated with processing of attended, relevant information. Thus, it was suggested that the P1 reflects inhibitory and the N1 excitatory processes (for a review, see Klimesch et al., 2004). The P300 has been related to the closure of the perceptual process (Verleger, 1988), and to an update of the information in the working memory (Donchin and Coles, 1988). P300 represents the last phase in the identification of a relevant stimulus caused by its significance or by the requirements of a task (Hillyard and Münte, 1984). As stressed by Kok (2001), the P300 amplitude reflects the degree of match between the stimulus presented and the internal representation of the stimulus relevant for the task. Because we predicted a beneficial influence of tACS applied to the left parietal area on WM storage capacity, it was expected that this would also be reflected in the P1–N1 and P300 ERP components during the solution of a task that measures the scope of attention. Specifically, we expected to find higher amplitude and shorter latency, as compared to tACS applied to the left DLPFC and sham stimulation.

Sham/verum tACS with individually determined theta frequency (based on IAF) was applied to the left parietal (target electrode = P3) or frontal (target electrode = F3) brain areas (return electrode above the right eyebrow). After sham and verum stimulation, the participants solved a visual-array comparison task (Luck and Vogel, 1997; for a description see Section 3.2.1) that measures the scope of attention, while their EEG was recorded. The results showed that, in general, verum tACS significantly increased performance on the visual-array comparison task, as compared to sham stimulation. This increase was much more pronounced for the parietal position of the target electrode than for the frontal position. As revealed in Fig. 4.21, after verum frontal stimulation, the participants correctly solved an approximately equal number of task items as

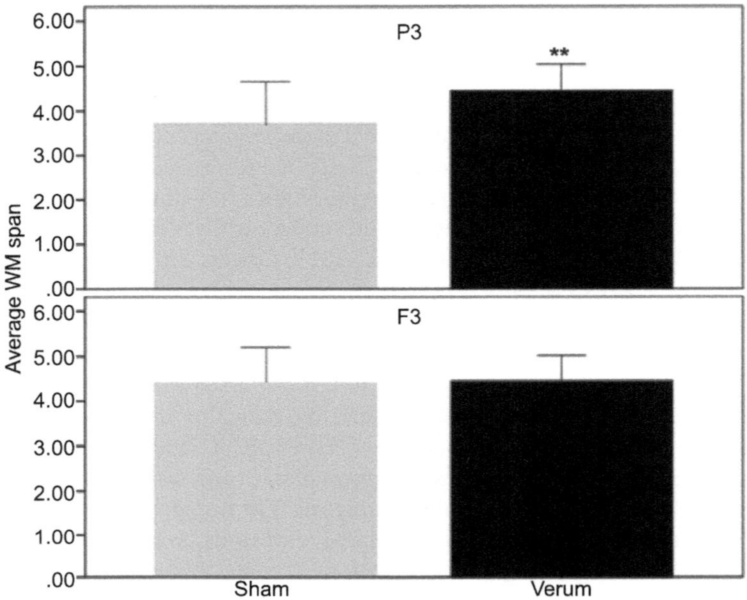

FIGURE 4.21 Average memory capacity score (Pashler, 1988) on the visual-array comparison task in sham and verum tACS settings of respondents of the frontal (target electrode = F3) and parietal group (target electrode = P3). Error bars represent 1 SD.

after sham stimulation. By contrast, after verum parietal tACS, respondents showed a significant increase in WM storage capacity, as compared to sham stimulation.

The only significant difference in ERP values between sham and verum stimulation was observed for P300 latency. As can be seen in Fig. 4.22, shorter P300 latency in the verum condition, as compared to sham, was mainly observed in the left parieto–occipital location.

The central finding of the study was that verum tACS with the target electrode positioned over the left parietal area significantly increased WM storage capacity, as compared to sham tACS. On the other hand, no such influence was observed for verum stimulation when the target electrode was positioned over the left frontal area. This supports the hypothesis that the left parietal brain area plays a central role in WM storage capacity. Additional evidence for the significance of the left parietal area for WM storage capacity was provided by the analysis of ERP components during performance on the visual-array comparison task. This analysis showed that theta tACS decreased P300 latency in left parietal brain areas in the parietal tACS group. Given that P300 latency is an index of classification speed proportional to the time required to detect and evaluate a target stimulus (Polich, 2007), it can be hypothesized that theta tACS increased the respondents' ability to allocate resources needed to solve

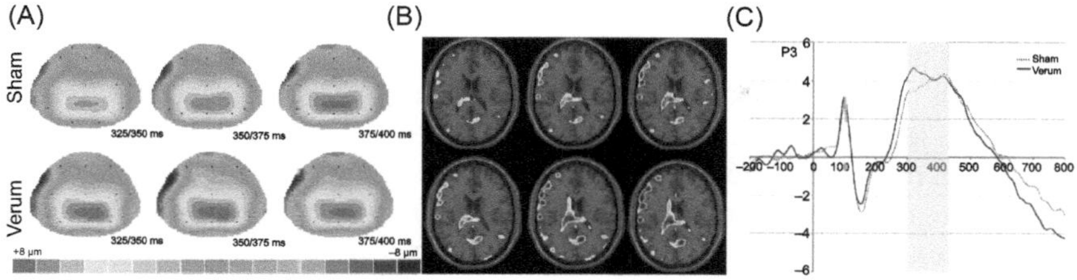

FIGURE 4.22 (A) Topographic ERP maps (325–400 ms) while respondents were solving the visual-array comparison task after sham/verum tACS with the target electrode positioned over the left parietal brain area (P3). (B) Dorsal view of sLORETA (Pascual-Marqui, 2007), source reconstruction for ERP (325–400 ms) while respondents were solving the visual array comparison task after sham/verum tACS. (C) Grand average (digitally filtered at 30 Hz) for sham/verum tACS during the solution of the visual-array comparison task for the P3 electrode (solid line, verum tACS; dotted line, sham tACS).

the WM task more rapidly. The decrease in P300 latency was not reflected in shorter reaction times, but only in the scope of attention. This indicates that theta tACS mainly influenced cognitive processes and, to a lesser extent, motor and executive processes, in line with the general claim that P300 is generated in the service of memory storage (Polich, 2007). The finding that tACS had no influence on the earlier occurring ERP components of P1 and on N1 associated with inhibitory and excitatory processes involved in sensory processing of stimuli (Klimesch et al., 2004), provided further support for the explanation that theta tACS mainly influenced cognitive processes. In comparison, verum stimulation of the left DLPFC had an opposite effect on P300 latency, increasing latency in the left hemisphere and decreasing latency in the right hemisphere; this did not influence WM storage capacity. This result lends further support for the unique importance of the left parietal brain area for the temporary storage of information.

The aim of our second study was similar to the first one, with a more detailed focus on different executive processes (Jaušovec et al., 2014). In a single blind sham controlled experiment respondents solved different working memory tasks (forward and backward digit-span and Corsi block-tapping tasks, as well as a dual *n*-back task) after theta tACS was applied to left frontal, left parietal and right parietal areas. The central finding was that stimulation of parietal areas (target electrodes positioned at P3 or P4) had a positive effect on WM storage capacity, whereas no such influence was observed for the stimulation of the left frontal area. This supports the hypothesis that parietal brain areas play a central role for WM storage capacity. This finding was also observed in our first study.

A second finding was that left parietal theta tACS had a more pronounced influence on backward recall than on forward recall, which was not related to task content (spatial or verbal). This would suggest that different brain areas are involved in forward and backward recall of data. Given the relation of parietal brain areas with spatial orientation and navigation (Kolb and Whishaw, 2009), a more spatial coding strategy for backward recall could be supposed. As stressed by Rudel and Denckla (1974), forward recall involves some form of verbal coding, in contrast, backward recall requires the translation of a given serial order into left–right spatial coordinates. However, this explanation is questionable given that right parietal stimulation did not show a similar backward recall advantage, on the contrary, a tendency for better forward recall was observed. It has been also suggested that successful performance on backward recall tasks requires central executive function due to the additional requirement of manipulation of information within temporary storage (Groeger et al., 1999; Lezak, 2012). Our study does not support this hypothesis. tACS with the target electrode positioned over the left DLPFC had no notable influence on the backward recall tasks. Ambivalent are also the results of neuroimaging studies. In a near infrared spectroscopy study Hoshi et al. (2000) showed that the backward digit task activated the DLPFC of each hemisphere more than the forward digit span task. These results suggest that the backward recall task engaged the working memory system to a greater extent than the forward task, and that the backward recall task implicated visuospatial imagery, as well as verbal representation. In contrast, a positron emission study by Gerton et al. (2004) did not replicate these findings. The digit forward and backward tasks relied upon a largely overlapping functional neural system associated with working memory. The ambiguity of the results obtained does not allow for a firm conclusion on the differences in brain activation related to forward and backward recall. A tentative explanation based on the data obtained in our study would be that increased storage capacity generated by left parietal stimulation more efficiently supported a spatial recall strategy for the backward tasks. In this direction also points an analysis of the Corsi block task based on the multi component model of WM (Baddeley and Hitch, 1974) indicating that in the backward presentation some reliance on the phonological loop was also present for longer sequences. It was suggested that as the task becomes more difficult subjects may have called on verbal support to cope with the situation (Vandierendonck et al., 2004, see also Section 3.4.1).

Another objective of the study was to explore the influence of theta tACS on processing components of WM (e.g., inhibition of irrelevant items, monitoring of ongoing performance, updating representations in memory and binding processes between spatial positions and temporal context). The obtained pattern of differences in performance on the dual *n*-back tasks induced by the three tACS protocols was not in line with the hypothesized expectations that were mainly based on previous neuroimaging evidence obtained in correlational studies showing that executive processes were related to the activity of frontal brain areas (Champod and Petrides, 2010; Collette and Van der Linden, 2002; Curtis and D'Esposito, 2003; Olson and Berryhill, 2009; Sauseng et al., 2010), as well as with experimental evidence employing tDCS (Andrews et al., 2011; Fregni et al., 2005, 2006; Zaehle et al., 2011; see also the metaanalysis by Brunoni and Vanderhasselt, 2014). The most pronounced influence on dual *n*-back performance was observed for the right parietal tACS protocol that had a major influence on the dual 2-back task and a minor on the dual 1-back task. The stimulation of the left parietal area showed a more diverse effect: it was significant for the performance on the dual 2-back task, yet negative for the dual 1-back task. The stimulation of the left frontal area had a minor influence on the dual 1-back task. A vague explanation of the obtained pattern of performance enhancements on the dual *n*-back tasks in relation to the area of stimulation could be that for more complex tasks (higher level of *n*) WM storage capacity is crucial for task performance, whereas for easier tasks (lower level of *n*) attentional control is dominant. This explanation also seems plausible from the requirements of the dual *n*-back task used in our study. The dual *n*-back tasks, beside the executive processes of inhibition, monitoring, and updating, also required binding between spatial positions and temporal context. In Baddeley's multicomponent model of WM this integrative function is assigned to the episodic buffer (Baddeley et al., 2011). Its function is the temporary storage and binding of information streaming from the phonological loop (aurally presented single letters), and the visuo–spatial sketchpad (blue squares in different spatial locations) into unitized episodes or chunks. On the other hand, following the logic of the explanation that with a higher level of *n*, storage capacity is essential for task solution, whereas with a lower level of *n* attentional control is important, it would be expected that parietal stimulation (especially left parietal) would also affect performance on the 3-back tasks. However, just a trend indicating the expected increase could be observed. Left parietal stimulation compared to the other settings had the greatest impact on the solution of the dual 3-back tasks, although it was not statistically significant. A similar finding in relation to the *n*-back difficulty level (*n*) was reported by the study of Hussey et al. (2015), as well as in the metaanalysis of Brunoni and Vanderhasselt (2014) showing that tDCS influences were only present for more difficult *n*-back tasks.

Another variable that might have influenced the obtained tACS enhancement patterns on the dual *n*-back tasks is the content of stimulus material. Left DLPFC stimulation showed the expected positive influence on *n*-back performance only when symbolic/verbal stimulus material was used (Fregni et al., 2005; Zaehle et al., 2011), whereas for figural stimuli less pronounced or no influence was reported (Mulquiney et al., 2011). Yet another possible explanation could be that the attentional component of WM is not reflected just in the theta frequency band, but also in alpha band activity. According to Klimesch (2012), alpha-band oscillations have inhibitional and timing roles that are related to attentional functions of suppression and selection of information. Monitoring new information, for example, elicits an increase in the theta frequency band accompanied by alpha synchronization. On the other hand alpha-band activity can also act as an inhibitory filter that keeps target information activated (Klimesch, 2012). From this view point it could be speculated that stimulating the frontal areas in one of the alpha bands might have a more pronounced influence on attention—the so-called attentional buffer that keeps target information activated that would translate to a more prominent effect on *n*-back performance.

These two studies revealed the potential strengths of tACS for cognitive research, yet also demonstrated its weaknesses that are shared with NIBS methodology. An additional shortcoming of the method, as mentioned in the previous sections, is the fairly unclear biological background of tACS on brain activity that can be biased by the fact that stimulation with two electrodes induces oscillations that are out-of-phase (180 degree phase difference, Polanía et al., 2012). This aspect was not addressed or controlled for in the two reported studies conducted in our lab and might have influenced the behavioral outcomes.

REFERENCES

Abad, F.J., Colom, R., Rebollo, I., Escorial, S., 2004. Sex differential item functioning in the Raven's Advanced Progressive Matrices: evidence for bias. Pers. Individ. Dif. 36 (6), 1459–1470.

Anderson, M.L., Kinnison, J., Pessoa, L., 2013. Describing functional diversity of brain regions and brain networks. NeuroImage 73, 50–58.

Andrews, S.C., Hoy, K.E., Enticott, P.G., Daskalakis, Z.J., Fitzgerald, P.B., 2011. Improving working memory: the effect of combining cognitive activity and anodal transcranial direct current stimulation to the left dorsolateral prefrontal cortex. Brain Stimul. 4 (2), 84–89.

Antal, A., Paulus, W., 2013. Transcranial alternating current stimulation (tACS). Front. Hum. Neurosci. 7, 317.

Antal, A., Nitsche, M.A., Kruse, W., Kincses, T.Z., Hoffmann, K., Paulus, W, 2004. Direct current stimulation over V5 enhances visuomotor coordination by improving motion perception in humans. J. Cogn. Neurosci. 4 (16), 521–527.

Antal, A., Boros, K., Poreisz, C., Chaieb, L., Terney, D., Paulus, W., 2008. Comparatively weak after-effects of transcranial alternating current stimulation (tACS) on cortical excitability in humans. Brain Stimul. 1 (2), 97–105.

Baddeley, A.D., Hitch, G.J., 1974. Working memory. In: Bower, G.A. (Ed.), The psychology of learning and motivation: advances in research and theory. Academic Press, New York, NY, pp. 47–89.

Baddeley, A.D., Allen, R.J., Hitch, G.J., 2011. Binding in visual working memory: the role of the episodic buffer. Neuropsychologia 49 (6), 1393–1400.

Balconi, M., 2013. Dorsolateral prefrontal cortex, working memory and episodic memory processes: insight through transcranial magnetic stimulation techniques. Neurosci. Bull. 29 (3), 381–389.

Barbey, A.K., Colom, R., Solomon, J., Krueger, F., Forbes, C., Grafman, J., 2012. An integrative architecture for general intelligence and executive function revealed by lesion mapping. Brain 135 (4), 1154–1164.

Barbey, A.K., Colom, R., Grafman, J., 2013. Architecture of cognitive flexibility revealed by lesion mapping. NeuroImage 82, 547–554.

Barbieri, M., Negrini, M., Nitsche, M.A., Rivolta, D., 2016. Anodal-tDCS over the human right occipital cortex enhances the perception and memory of both faces and objects. Neuropsychologia 81, 238–244.

Barker, A.T., Jalinous, R., Freeston, I.L., 1985. Non-invasive magnetic stimulation of human motor cortex. Lancet 1, 1106–1107.

Bartrés-Faz, D., Vidal-Piñeiro, D., 2016. Noninvasive brain stimulation for the study of memory enhancement in aging. Eur. Psychol. 21 (1), 41–54.

Basten, U., Stelzel, C., Fiebach, C.J., 2013. Intelligence is differentially related to neural effort in the task-positive and the task-negative brain network. Intelligence 41 (5), 517–528.

Bikson, M., Rahman, A., Datta, A., 2012. Computational models of transcranial direct current stimulation. Clin. EEG Neurosci. 43 (3), 176–183.

Bindman, J.L., Lippold, O.C.J., Redfearm, J.W.T., 1964. The action of brief polarizing currents on the cerebral cortex of the rat (1) during current flow and (2) in the production of long-lasting after-effects. J. Physiol. 172, 369–382.

Boros, K., Poreisz, C., Münchau, A., Paulus, W., Nitsche, M.A., 2008. Premotor transcranial direct current stimulation (tDCS) affects primary motor excitability in humans. Eur. J. Neurosci. 27 (5), 1292–1300.

Bressler, S.L., Menon, V., 2010. Large-scale brain networks in cognition: emerging methods and principles. Trends Cognit. Sci. 14 (6), 277–290.

Brunoni, A.R., Vanderhasselt, M.-A., 2014. Working memory improvement with non-invasive brain stimulation of the dorsolateral prefrontal cortex: a systematic review and meta-analysis. Brain Cognit. 86, 1–9.

Brunoni, A.R., Nitsche, M.A., Bolognini, N., Bikson, M., Wagner, T., Merabet, L., Fregni, F., 2012. Clinical research with transcranial direct current stimulation (tDCS): challenges and future directions. Brain Stimul. 5 (3), 175–195.

Buehner, M., Krumm, S., Pick, M., 2005. Reasoning = working memory≠attention. Intelligence 33 (3), 251–272.

Bunson, M., 2002. Encyclopedia of the Roman Empire. Facts On File, New York.

Butler, T., Imperato-McGinley, J., Pan, H., Voyer, D., Cordero, J., Zhu, Y.-S., Silbersweig, D., 2006. Sex differences in mental rotation: Top–down versus bottom–up processing. NeuroImage 32 (1), 445–456.

Cappelletti, M., Pikkat, H., Upstill, E., Speekenbrink, M., Walsh, V., 2015. Learning to integrate versus inhibiting information is modulated by age. J. Neurosci. 35 (5), 2213–2225.

Carvalho, S., Boggio, P.S., Gonçalves, Ó.F., Vigário, A.R., Faria, M., Silva, S., Leite, J., 2015. Transcranial direct current stimulation based metaplasticity protocols in working memory. Brain Stimul. 8 (2), 289–294.

Cerruti, C., Schlaug, G., 2009. Anodal transcranial direct current stimulation of the prefrontal cortex enhances complex verbal associative thought. J. Cognit. Neurosci. 21 (10), 1980–1987.

Chaieb, L., Antal, A., Paulus, W., 2015. Transcranial random noise stimulation-induced plasticity is NMDA-receptor independent but sodium-channel blocker and benzodiazepines sensitive. Front. Neurosci. 9, 125.

Champod, A.S., Petrides, M., 2010. Dissociation within the frontoparietal network in verbal working memory: a parametric functional magnetic resonance imaging study. J. Neurosci. 30 (10), 3849–3856.

Chein, J.M., Weisberg, R.W., Streeter, N.L., Kwok, S., 2010. Working memory and insight in the nine-dot problem. Mem. Cognit. 38 (7), 883–892.

Chi, R.P., Snyder, A.W., 2011. Facilitate insight by non-invasive brain stimulation. PLoS One 6 (2), e16655.

Chi, R.P., Snyder, A.W., 2012. Brain stimulation enables the solution of an inherently difficult problem. Neurosci. Lett. 515 (2), 121–124.

Cohen Kadosh, R., 2015. Modulating and enhancing cognition using brain stimulation: Science and fiction. J. Cognit. Psychol. 27 (2), 141–163.

Collette, F., Van der Linden, M., 2002. Brain imaging of the central executive component of working memory. Neurosci. Biobehav. Rev. 26 (2), 105–125.

Colom, R., Abad, F.J., Quiroga, M.Á., Shih, P.C., Flores-Mendoza, C., 2008. Working memory and intelligence are highly related constructs, but why? Intelligence 36 (6), 584–606.

Cowan, N., 2011. The focus of attention as observed in visual working memory tasks: making sense of competing claims. Neuropsychologia 49 (6), 1401–1406.

Cowan, N., Elliott, E.M., Scott Saults, J., Morey, C.C., Mattox, S., Hismjatullina, A., Conway, A.R.A., 2005. On the capacity of attention: its estimation and its role in working memory and cognitive aptitudes. Cognit. Psychol. 51 (1), 42–100.

Cowan, N., Li, D., Moffitt, A., Becker, T.M., Martin, E.A., Saults, J.S., et al., 2011. A neural region of abstract working memory. J. Cognit. Neurosci. 23 (10), 2852–2863.

Curtis, C.E., D'Esposito, M., 2003. Persistent activity in the prefrontal cortex during working memory. Trends Cognit. Scie. 7 (9), 415–423.

Datta, A., Elwassif, M., Battaglia, F., Bikson, M., 2008. Transcranial current stimulation focality using disc and ring electrode configurations: FEM analysis. J. Neural. Eng. 2, 163–174.

Deary, I., 2000. Looking Down on Human Intelligence. Oxford University Press, Oxford. Retrieved from http://www.oxfordscholarship.com/view/10.1093/acprof:oso/9780198524175.001.0001/acprof-9780198524175

Devlin, J.T., Watkins, K.E., 2007. Stimulating language: insights from TMS. Brain 130 (3), 610–622. https://doi.org/10.1093/brain/awl331

Dmochowski, J.P., Bikson, M., Parra, L.C., 2012. The point spread function of the human head and its implications for transcranial current stimulation. Phys. Med. Biol. 57 (20), 6459–6477.

Donchin, E., Coles, M.G.H., 1988. Is the P300 component a manifestation of context updating? Behav. Brain Sci. 11 (03), 357–374.

Doppelmayr, M., Klimesch, W., Hödlmoser, K., Sauseng, P., Gruber, W., 2005. Intelligence related upper alpha desynchronization in a semantic memory task. Brain Res. Bull. 66 (2), 171–177.

Duncan, J., 2010. The multiple-demand (MD) system of the primate brain: mental programs for intelligent behaviour. Trends Cognit. Sci. 14 (4), 172–179.

Eldaief, M.C., Halko, M.A., Buckner, R.L., Pascual-Leone, A., 2011. Transcranial magnetic stimulation modulates the brain's intrinsic activity in a frequency-dependent manner. Proc. Natl. Acad. Sci. USA 108 (52), 21229–21234.

Epstein, C.M., Schwartzenberg, D.G., Davey, K.R., Sudderth, D.B., 1990. Localizing the site of magnetic brain stimulation in humans. Neurology 40, 666–670.

Fadini, T., Matthäus, L., Rothkegel, H., Sommer, M., Tergau, F., Schweikard, A., Nitsche, M.A., 2009. H-coil: Induced electric field properties and input/output curves on healthy volunteers, comparison with a standard figure-of-eight coil. Clin. Neurophysiol. 120 (6), 1174–1182.

Fehér, K.D., Morishima, Y., 2016. Concurrent electroencephalography recording during transcranial alternating current stimulation (tACS). J. Vis. Exp. 107, e53527.

Felleman, D.J., Van Essen, D.C., 1991. Distributed hierarchical processing in the primate cerebral cortex. Cereb. Cortex 1 (1), 1–47.

Ferrari, M., Mottola, L., Quaresima, V., 2004. Principles, techniques, and limitations of near infrared spectroscopy. Can. J. Appl. Physiol. 29 (4), 463–487.

Fertonani, A., Pirulli, C., Miniussi, C., 2011. Random noise stimulation improves neuroplasticity in perceptual learning. J. Neurosci. 31 (43), 15416–15423.

Filmer, H.L., Mattingley, J.B., Marois, R., Dux, P.E., 2013. Disrupting prefrontal cortex prevents performance gains from sensory-motor training. J. Neurosci. 33 (47), 18654–18660.

Filmer, H.L., Dux, P.E., Mattingley, J.B., 2014. Applications of transcranial direct current stimulation for understanding brain function. Trends Neurosci. 37 (12), 742–753.

Fregni, F., Boggio, P.S., Nitsche, M., Bermpohl, F., Antal, A., Feredoes, E., Pascual-Leone, A., 2005. Anodal transcranial direct current stimulation of prefrontal cortex enhances working memory. Exp. Brain Res. 166 (1), 23–30.

Fregni, F., Boggio, P.S., Nitsche, M.A., Rigonatti, S.P., Pascual-Leone, A., 2006. Cognitive effects of repeated sessions of transcranial direct current stimulation in patients with depression. Depress. Anxiety 23 (8), 482–484.

Freitas, C., Farzan, F., Pascual-Leone, A., 2013. Assessing brain plasticity across the lifespan with transcranial magnetic stimulation: why, how, and what is the ultimate goal? Front. Neurosci. 7, 42.

Freunberger, R., Werkle-Bergner, M., Griesmayr, B., Lindenberger, U., Klimesch, W., 2011. Brain oscillatory correlates of working memory constraints. Brain Res. 1375, 93–102.

Fröhlich, F., Sellers, K.K., Cordle, A.L., 2015. Targeting the neurophysiology of cognitive systems with transcranial alternating current stimulation. Exp. Rev. Neurother. 15 (2), 145–167.

Fuchs, M., Kastner, J., Wagner, M., Hawes, S., Ebersole, J.S., 2002. A standardized boundary element method volume conductor mode. Electroencephalogr. Clin. Neurophysiol. 113, 702–712.

Gerton, B.K., Brown, T.T., Meyer-Lindenberg, A., Kohn, P., Holt, J.L., Olsen, R.K., et al., 2004. Shared and distinct neurophysiological components of the digits forward and backward tasks as revealed by functional neuroimaging. Neuropsychologia 42 (13), 1781–1787.

Gevins, A., Smith, M.E., 2000. Neurophysiological measures of working memory and individual differences in cognitive ability and cognitive style. Cereb. Cortex 10 (9), 829–839.

Gläscher, J., Tranel, D., Paul, L.K., Rudrauf, D., Rorden, C., Hornaday, A., et al., 2009. Lesion mapping of cognitive abilities linked to intelligence. Neuron 61, 681–691.

Gobbelé, R., Waberski, T.D., Kuelkens, S., Sturm, W., Curio, G., Buchner, H., 2000. Thalamic and cortical high-frequency (600 Hz) somatosensory-evoked potential (SEP) components are modulated by slight arousal changes in awake subjects. Exp. Brain Res. 133, 506–513.

Gootjes, L., Bruggeling, E.C., Magnee, T., Van Strien, J.W., 2008a. Sex differences in functional connectivity during mental rotation: an EEG study. Int. J. Psychophysiol. 69 (3), 228.

Gootjes, L., Bruggeling, E.C., Magnée, T., Van Strien, J.W., 2008b. Sex differences in the latency of the late event-related potential mental rotation effect. NeuroReport 19 (3), 349–353.

Gray, J.R., Chabris, C.F., Braver, T.S., 2003. Neural mechanisms of general fluid intelligence. Nat. Neurosci. 6 (3), 316–322.

Groeger, J.A., Field, D., Hammond, S.M., 1999. Measuring memory span. Int. J. Psychol. 34, 359–363.

Groppa, S., Bergmann, T.C., Siems, C., Mölle, M., Marshall, L., Siebner, H.R., 2010. Slow oscillatory transcranial direct current stimulation can induce bidirectional shifts in motor cortical excitability in awake humans. Neuroscience 166, 1219–1225.

Groppa, S., Oliviero, A., Eisen, A., Quartarone, A., Cohen, L.G., Mall, V., Siebner, H.R., 2012. A practical guide to diagnostic transcranial magnetic stimulation: report of an IFCN committee. Clin. Neurophysiol. 123 (5), 858–882.

Grosbras, M.H., Paus, T., 2002. Transcranial magnetic stimulation of the human frontal eye field: effects on visual perception and attention. J. Cognit. Neurosci. 14, 1109–1120.

Guleyupoglu, B., Schestatsky, P., Edwards, D., Fregni, F., Bikson, M., 2013. Classification of methods in transcranial electrical stimulation (tES) and evolving strategy from historical approaches to contemporary innovations. J. Neurosci. Methods 219 (2), 297–311.

Haier, R.J., Siegel, B.V., Nuechterlein, K.H., Hazlett, E., Wu, J.C., Paek, J., Browning, H.L., Buchsbaum, M.S., 1988. Cortical glucose metabolic rate correlates of abstract reasoning and attention studied with positron emission tomography. Intelligence 12, 199–217.

Haier, R.J., White, N.S., Alkire, M.T., 2003. Individual differences in general intelligence correlate with brain function during nonreasoning tasks. Intelligence 31 (5), 429–441.

Hamada, M., Terao, Y., Hanajima, R., Shirota, Y., Nakatani-Enomoto, S., Furubayashi, T., Matsumoto, H., Ugawa, Y., 2008. Bidirectional long-term motor cortical plasticity and metaplasticity induced by quadropulse transcranial magnetic stimulation. J. Physiol. 586, 3927–3947.

Hampson, M., Hoffman, R.E., 2010. Transcranial magnetic stimulation and connectivity mapping: tools for studying the neural bases of brain disorders. Front. Syst. Neurosci. 4, pii: 40.

Hannula, H., Neuvonen, T., Savolainen, P., Hiltunen, J., Ma, Y.-Y., Antila, H., Pertovaara, A., 2010. Increasing top-down suppression from prefrontal cortex facilitates tactile working memory. NeuroImage 49 (1), 1091–1098.

Helfrich, R.F., Schneider, T.R., Rach, S., Trautmann-Lengsfeld, S.A., Engel, A.K., Herrmann, C.S., 2014. Entrainment of brain oscillations by transcranial alternating current stimulation. Curr. Biol. 24 (3), 333–339.

Herrmann, C.S., Rach, S., Neuling, T., Struber, D., 2013. Transcranial alternating current stimulation: a review of the underlying mechanisms and modulation of cognitive processes. Front. Hum. Neurosci. 7, 279.

Herrmann, C.S., Strüber, D., Helfrich, R.F., Engel, A.K., 2015. EEG oscillations: from correlation to causality. Int. J. Psychophysiol., https://doi.org/10.1016/j.ijpsycho.2015.02.003.

Hill, A.T., Fitzgerald, P.B., Hoy, K.E., 2016. Effects of anodal transcranial direct current stimulation on working memory: a systematic review and meta-analysis of findings from healthy and neuropsychiatric populations. Brain Stimul. 9 (2), 197–208.

Hillyard, S., Münte, T., 1984. Selective attention to color and location: an analysis with event-related brain potentials. Percept. Psychophys. 36, 185–198.

Hofman, D., Schutter, D.J.L.G., 2009. Inside the wire: aggression and functional interhemispheric connectivity in the human brain. Psychophysiology 46 (5), 1054–1058.

Horvath, J.C., Forte, J.D., Carter, O., 2015. Quantitative review finds no evidence of cognitive effects in healthy populations from single-session transcranial direct current stimulation (tDCS). Brain Stimul. 8 (3), 535–550.

Hoshi, Y., Oda, I., Wada, Y., Ito, Y., Yamashita, Y., Oda, M., Tamura, M., 2000. Visuospatial imagery is a fruitful strategy for the digit span backward task: a study with near-infrared optical tomography. Cognit. Brain Res. 9 (3), 339–342.

Huang, Y.Z., Chen, R.S., Rothwell, J.C., Wen, H.Y., 2007. The after-effect of human theta burst stimulation is NMDA receptor dependent. Clin. Neurophysiol. 118, 1028–1032.

Huang, Y.-Z., Sommer, M., Thickbroom, G., Hamada, M., Pascual-Leonne, A., Paulus, W., Ugawa, Y., 2009. Consensus: new methodologies for brain stimulation. Brain Stimul. 2 (1), 2–13.

Hugdahl, K., Thomsen, T., Ersland, L., 2006. Sex differences in visuo-spatial processing: an fMRI study of mental rotation. Neuropsychologia 44 (9), 1575–1583.

Hussey, E.K., Ward, N., Christianson, K., Kramer, A.F., 2015. Language and memory improvements following tDCS of left lateral prefrontal cortex. PLoS One 10 (11), e0141417.

Jacobson, L., Koslowsky, M., Lavidor, M., 2012. tDCS polarity effects in motor and cognitive domains: a meta-analytical review. Exp. Brain Res. 216 (1), 1–10.

Jaeggi, S.M., Buschkuehl, M., Jonides, J., Perrig, W.J., 2008. Improving fluid intelligence with training on working memory. Proc. Natl. Acad. Sci. USA 105 (19), 6829–6833.

Jaeggi, S.M., Studer-Luethi, B., Buschkuehl, M., Su, Y.-F., Jonides, J., Perrig, W.J., 2010. The relationship between n-back performance and matrix reasoning: implications for training and transfer. Intelligence 38 (6), 625–635.

Jaušovec, N., Jaušovec, K., 2004. Intelligence related differences in induced brain activity during the performance of memory tasks. Person. Individ. Dif. 36 (3), 597–612.

Jaušovec, N., Jaušovec, K., 2012. Working memory training: Improving intelligence—changing brain activity. Brain Cognit. 79 (2), 96–106.

Jaušovec, N., Jaušovec, K., 2014. Increasing working memory capacity with theta transcranial alternating current stimulation (tACS). Biol. Psychol. 96, 42–47.

Jaušovec, N., Jaušovec, K., Pahor, A., 2014. The influence of theta transcranial alternating current stimulation (tACS) on working memory storage and processing functions Acta Psychol. 146, 1–6.

Jensen, O., Mazaheri, A., 2010. Shaping functional architecture by oscillatory alpha activity: gating by inhibition. Front. Hum. Neurosci. 4, 1–8.

Jeurissen, D., Sack, A.T., Roebroeck, A., Russ, B.E., Pascual-Leone, A., 2014. TMS affects moral judgment, showing the role of DLPFC and TPJ in cognitive and emotional processing. Front. Neurosci. 8, 18.

Jordan, K., Wüstenberg, T., Heinze, H.J., Peters, M., Jäncke, L., 2002. Women and men exhibit different cortical activation patterns during mental rotation tasks. Neuropsychologia 40 (13), 2397–2408.

Jung, R.E., Haier, R.J., 2007. The Parieto–Frontal Integration Theory (P-FIT) of intelligence: converging neuroimaging evidence. Behav. Brain Sci. 30, 135–187.

Jung, R.E., Haier, R.J., Yeo, R.A., Rowland, L.M., Petropoulos, H., Levine, A.S., Brooks, W.M., 2005. Sex differences in N-acetylaspartate correlates of general intelligence: an 1H-MRS study of normal human brain. NeuroImage 26 (3), 965–972.

Jurcak, V., Tsuzuki, D., Dan, I., 2007. 10/20, 10/10, and 10/5 systems revisited: their validity as relative head-surface-based positioning systems. Neuroimage 34, 1600–1611.

Kershaw, T.C., Ohlsson, S., 2004. Multiple causes of difficulty in insight: the case of the nine-dot problem. J. Exp. Psychol. Learn. Mem. Cognit. 30 (1), 3–13.

Klimesch, W., 2012. Alpha-band oscillations, attention, and controlled access to stored information. Trends Cognit. Sci. 16 (12), 606–617.

Klimesch, W., Schack, B., Schabus, M., Doppelmayr, M., Gruber, W., Sauseng, P., 2004. Phase-locked alpha and theta oscillations generate the P1-N1 com-plex and are related to memory performance. Cognit. Brain Res. 19 (3), 302–316.

Kok, A., 2001. On the utility of P3 amplitude as a measure of processing capacity. Psychophysiology 38 (3), 557–577.

Kolb, B., Whishaw, I.Q., 2009. Fundamentals of Human Neuropsychology, 6th ed. Worth Publishers, New York, NY.

Kounios, J., Beeman, M, 2009. The Aha! moment: the cognitive neuroscience of insight. Curr. Dir. Psychol. Sci. 18, 210–216.

Krause, V., Wach, C., Südmeyer, M., Ferrea, S., Schnitzler, A., Pollok, B., 2014. Cortico-muscular coupling and motor performance are modulated by 20 Hz transcranial alternating current stimulation (tACS) in Parkinson's disease. Front. Hum. Neurosci. 7, 928.

Kuo, M.-F., Nitsche, M.A., 2012. Effects of transcranial electrical stimulation on cognition. Clin. EEG Neurosci. 43 (3), 192–199.

Langer, N., von Bastian, C C., Wirz, H., Oberauer, K., Jäncke, L., 2013. The effects of working memory training on functional brain network efficiency. Cortex 49 (9), 2424–2438.

Lavidor, M., 2016. tES stimulation as a tool to investigate cognitive processes in healthy individuals. Eur. Psychol. 21 (1), 15–29.

Lee, K.H., Choi, Y.Y., Gray, J.R., Cho, S.H., Chae, J.H., Lee, S., Kim, K., 2006. Neural correlates of superior intelligence: stronger recruitment of posterior parietal cortex. NeuroImage 29, 578–586.

Lefaucheur, J.-P., André-Obadia, N., Antal, A., Ayache, S.S., Baeken, C., Benninger, D.H., Garcia-Larrea, L., 2014. Evidence-based guidelines on the therapeutic use of repetitive transcranial magnetic stimulation (rTMS). Clin. Neurophysiol. 125 (11), 2150–2206.

Lezak, M.D. (Ed.), 2012. Neuropsychological assessment. 5th ed. Oxford University Press, Oxford; New York.

Liebetanz, D., 2002. Pharmacological approach to the mechanisms of transcranial DC-stimulation-induced after-effects of human motor cortex excitability. Brain 125 (10), 2238–2247.

Lipp, I., Benedek, M., Fink, A., Koschutnig, K., Reishofer, G., Bergner, S., Neubauer, A., 2012. Investigating neural efficiency in the visuo-spatial domain: an fmri study. PLoS One 7 (12), e51316.

Lisanby, S.H., Gutman, D., Luber, B., Schroeder, C., Sackeim, H.A., 2001. Sham TMS: intracerebral measurement of the induced electrical field and the induction of motor-evoked potentials. Biol. Psychiatry 49, 460–463.

Luber, B., Lisanby, S.H., 2014. Enhancement of human cognitive performance using transcranial magnetic stimulation (TMS). NeuroImage 85, 961–970.

Luck, S.J., Vogel, E.K., 1997. The capacity of visual working memory for features and conjunctions. Nature 390 (6657), 279–281.

Luria, A.R., 1973. The working brain. Penguin, Harmondsworth, England.

Mahmoudi, H., Haghighi, A.B., Petramfar, P., Jahanshahi, S., Salehi, Z., Fregni, F., 2011. Transcranial direct current stimulation: electrode montage in stroke. Disabil. Rehabil. 33 (15–16), 1383–1388.

Maier, N.R.F., 1930. Reasoning in humans. I. On direction. J. Comp. Psychol. 10 (2), 115–143. https://doi.org/10.1037/h0073232

Majerus, S., Poncelet, M., Van der Linden, M., Albouy, G., Salmon, E., Sterpenich, V., Maquet, P., 2006. The left intraparietal sulcus and verbal short-term memory: focus of attention or serial order? NeuroImage 32 (2), 880–891.

Majerus, S., D'Argembeau, A., Martinez Perez, T., Belayachi, S., Van der Linden, M., Collette, F., Maquet, P., 2010. The commonality of neural networks for verbal and visual short-term memory. J. Cognit. Neurosci. 22 (11), 2570–2593.

Marg, E., Rudiak, D., 1994. Phosphenes induced by magnetic stimulation over the occipital brain: description and probable site of stimulation. Optom. Vis. Sci. 71, 301–311.

Mednick, S., 1962. The associative basis of the creative process. Psychol. Rev. 69 (3), 220.

Mednick, M.T., Andrews, F.M., 1967. Creative thinking and level of intelligence. J. Creative Behav. 1 (4), 428–431.

Merlet, I., Birot, G., Salvador, R., Molaee-Ardekani, B., Mekonnen, A., Soria-Frish, A., Wendling, F., 2013. From oscillatory transcranial current stimulation to scalp EEG changes: a biophysical and physiological modeling study. PLoS One 8 (2), e57330.

Merton, P.A., Morton, H.B., 1980. Stimulation of the cerebral cortex in the intact human subject. Nature 285, 227.

Miniussi, C., 2016. A foreword on the use of noninvasive brain stimulation in pychology. Eur. Psychol. 21 (1), 1–3.

Miniussi, C., Harris, J.A., Ruzzoli, M., 2013. Modelling non-invasive brain stimulation in cognitive neuroscience. Neurosci. Biobehav. Rev. 37 (8), 1702–1712.

Moliadze, V., Antal, A., Paulus, W., 2010. Boosting brain excitability by transcranial high frequency stimulation in the ripple range. J. Physiol. 588 (24), 4891–4904.

Morales-Quezada, L., Castillo-Saavedra, L., Cosmo, C., Doruk, D., Sharaf, I., Malavera, A., Fregni, F., 2015. Optimal random frequency range in transcranial pulsed current stimulation indexed by quantitative electroencephalography. NeuroReport 1, 747–752.

Mottaghy, F.M., Gangitano, M., Krause, B.J., Pascual-Leone, A., 2003. Chronometry of parietal and prefrontal activations in verbal working memory revealed by transcranial magnetic stimulation. NeuroImage 18 (3), 565–575.

Mountcastle, V.B., 1975. The view from within: pathways to the study of perception. Johns Hopkins Med. J. 136, 109–131.

Mull, B.R., Seyal, M., 2001. Transcranial magnetic stimulation of left prefrontal cortex impairs working memory. Clin. Neurophysiol. 112 (9), 1672–1675.

Mulquiney, P.G., Hoy, K.E., Daskalakis, Z.J., Fitzgerald, P.B., 2011. Improving working memory: exploring the effect of transcranial random noise stimulation and transcranial direct current stimulation on the dorsolateral prefrontal cortex. Clin. Neurophysiol. 122 (12), 2384–2389.

Neubauer, A.C., Fink, A., 2009. Intelligence and neural efficiency: measures of brain activation versus measures of functional connectivity in the brain. Intelligence 37 (2), 223–229. https://doi.org/10.1016/j.intell.2008.10.008.

Neuling, T., Rach, S., Herrmann, C.S., 2013. Orchestrating neuronal networks: sustained after-effects of transcranial alternating current stimulation depend upon brain states. Front. Hum. Neurosci. 7, 161.

Neuling, T., Ruhnau, P., Fuscà, M., Demarchi, G., Herrmann, C.S., Weisz, N., 2015. Friends, not foes: magnetoencephalography as a tool to uncover brain dynamics during transcranial alternating current stimulation NeuroImage 118, 406–413.

Nevler, N., Ash, E.L., 2015. TMS as a tool for examining cognitive processing. Curr. Neurol. Neurosci. Rep. 15 (8), 52.

Nitsche, M.A., Paulus, W., 2000. Excitability changes induced in the human motor cortexes by weak transcranial direct current stimulation. J. Physiol. 527, 633–639.

Nitsche, M.A., Fricke, K., Henschke, U., Schlitterlau, A., Liebetanz, D., Lang, N., Paulus, W., 2003. Pharmacological modulation of cortical excitability shifts induced by transcranial direct current stimulation in humans. J. Physiol. 553 (Pt 1), 293–301.

Obrig, H., Villringer, A., 2003. Beyond the visible: imaging the human brain with light. J. Cereb. Blood Flow Metabol. 23 (1), 1–18.

Olson, I.R., Berryhill, M., 2009. Some surprising findings on the involvement of the parietal lobe in human memory. Neurobiol. Learn. Mem. 91 (2), 155–165.

Pahor, A., Jaušovec, N., 2014. Theta–gamma cross-frequency coupling relates to the level of human intelligence. Intelligence 46, 283–290.

Pahor, A., Jaušovec, N., 2016. Making Brains run Faster: are they Becoming Smarter? The Spanish journal of psychology, 19.

Pascual-Leone, A., Valls-Sole, J., Wassermann, E.M., Hallett, M., 1994. Responses to rapid-rate transcranial magnetic stimulation of the human motor cortex. Brain 117 (Pt 4), 847–858.

Pascual-Marqui, R.D., 2007. Discrete, 3D distributed, linear imaging methods of electric neuronal activity. Part 1 exact, zero error localization. arXiv:0710.3341.[math-ph]. http://arxiv.org/pdf/0710.3341.

Pashler, H., 1988. Familiarity and visual change detection. Percept. Psychophys. 44 (4), 369–378.

Paulus, W., 2011. Transcranial electrical stimulation (tES—tDCS; tRNS, tACS) methods. Neuropsychol. Rehabil. 21 (5), 602–617.

Pergolizzi, D., Chua, E.F., 2015. Transcranial direct current stimulation (tDCS) of the parietal cortex leads to increased false recognition. Neuropsychologia 66, 88–98.

Pineda-Pardo, J.A., Martínez, K., Román, F.J., Colom, R., 2016. Structural efficiency within a parieto–frontal network and cognitive differences. Intelligence 54, 105–116.

Pirulli, C., Fertonani, A., Miniussi, C., 2016. On the functional equivalence of electrodes in transcranial random noise stimulation. Brain Stimul. 9 (4), 621–622.

Polanía, R., Nitsche, M.A., Korman, C., Batsikadze, G., Paulus, W., 2012. The importance of timing in segregated theta phase-coupling for cognitive performance. Curr. Biol. 22 (14), 1314–1318.

Poldrack, R.A., Halchenko, Y.O., Hanson, S.J., 2009. Decoding the large-scale structure of brain function by classifying mental states across individuals. Psychol. Sci. 20 (11), 1364–1372.

Polich, J., 2007. Updating P300: an integrative theory of P3a and P3b. Clin. Neurophysiol. 118 (10), 2128–2148.

Postle, B.R., Ferrarelli, F., Hamidi, M., Feredoes, E., Massimini, M., Peterson, M., Tononi, G., 2006. Repetitive transcranial magnetic stimulation dissociates working memory manipulation from retention functions in the prefrontal, but not posterior parietal, cortex. J. Cognit. Neurosci. 18 (10), 1712–1722.

Priori, A., Berardelli, A., Rona, S., Accornero, N., Manfredi, M., 1998. Polarization of the human-motor cortex through the scalp. NeuroReport 9, 2257–2260.

Raichle, M.E., 2006. The brain's dark energy. Science 314 (5803), 1249.

Raichle, M.E., MacLeod, A.M., et al., 2001. A default mode of brain function. Proc. Natl. Acad. Sci. USA 98, 676–682.

Rango, M., Cogiamanian, F., Marceglia, S., Barberis, B., Arighi, A., Biondetti, P., Priori, A., 2008. Myoinositol content in the human brain is modified by transcranial direct current stimulation in a matter of minutes: A ^{1} H-MRS study. Magn. Reson. Med. 60 (4), 782–789.

Rideout, B.E., Laubach, C.M., 1996. EEG correlates of enhanced spatial performance following exposure to music. Percept. Motor Skills 82 (2), 427–432.

Robinovitch, L.G., 1914. Electrical analgesia, sleep and resuscitation. In: Gwathmey, J.T. (Ed.), Anesthesia. Appleton, New York, NY.

Romanska, A., Rezlescu, C., Susilo, T., Duchaine, B., Banissy, M.J., 2015. High-frequency transcranial random noise stimulation enhances perception of facial identity. Cereb. Cortex 25 (11), 4334–4340.

Rossini, P.M., Caramia, M.D., 1992. Central conduction studies and magnetic stimulation. Curr. Opin. Neurol. Neurosurg. 5, 697–703.

Rossini, P.M., Burke, D., Chen, R., Cohen, L.G., Daskalakis, Z., Di Iorio, R., Ziemann, U., 2015. Non-invasive electrical and magnetic stimulation of the brain, spinal cord, roots and peripheral nerves: Basic principles and procedures for routine clinical and research application. An updated report from an I. F. C. N. Committee. Clin. Neurophysiol. 126 (6), 1071–1107.

Rothwell, J.C., Thompson, P.D., Day, B.L., Boyd, S., Marsden, C.D., 1991. Stimulation of the human motor cortex through the scalp. Exp. Physiol. 76, 159–200.

Rudel, R.G., Denckla, M.B., 1974. Relation of forward and backward digit repetition to neurological impairment in children with learning disabilities. Neuropsychologia 12, 109–118.

Rypma, B., Berger, J.S., Prabhakaran, V., Bly, B.M., Kimberg, D.Y., Biswal, B.B., D'Esposito, M., 2006. Neural correlates of cognitive efficiency. NeuroImage 33, 969–979.

Sale, M.V., Mattingley, J.B., Zalesky, A., Cocchi, L., 2015. Imaging human brain networks to improve the clinical efficacy of non-invasive brain stimulation. Neurosci. Biobehav. Rev. 57, 187–198.

Salvador, R., Mekonnen, A., Ruffini, G., Miranda, P.C., 2010. Modeling the electric field induced in a high resolution head model during transcranial current stimulation. Conf. Proc. IEEE Eng. Med. Biol. Soc., 2073–2076, 2010.

Santarnecchi, E., Rossi, S., 2016. Advances in the Neuroscience of Intelligence: from Brain Connectivity to Brain Perturbation. The Spanish journal of psychology, 19.

Santarnecchi, E., Polizzotto, N.R., Godone, M., Giovannelli, F., Feurra, M., Matzen, L., Rossi, S., 2013. Frequency-dependent enhancement of fluid intelligence induced by transcranial oscillatory potentials. Curr. Biol. 23, 1449–1453.

Santarnecchi, E., Brem, A.-K., Levenbaum, E., Thompson, T., Kadosh, R.C., Pascual-Leone, A., 2015. Enhancing cognition using transcranial electrical stimulation. Curr. Opin. Behav. Sci. 4, 171–178.

Santarnecchi, E., Muller, T., Rossi, S., Sarkar, A., Polizzotto, N.R., Rossi, A., Kadosh, R.C., 2016. Individual differences and specificity of prefrontal gamma frequency-tACS on fluid intelligence capabilities. Cortex 75, 33–43.

Sarkar, A., Cohen Kadosh, R., 2016. Transcranial electrical stimulation and numerical cognition. Can. J. Exp. Psychol. 70 (1), 41–58.

Sauseng, P., Griesmayr, B., Freunberger, R., Klimesch, W., 2010. Control mechanisms in working memory: a possible function of EEG theta oscillations. Neurosci. Biobehav. Rev. 34 (7), 1015–1022.

Schutter, D.J.L.G., Wischnewski, M., 2016. A meta-analytic study of exogenous oscillatory electric potentials in neuroenhancement. Neuropsychologia 86, 110–118.

Seeley, W.W., Menon, V., Schatzberg, A.F., Keller, J., Glover, G.H., Kenna, H., Greicius, M.D., 2007. Dissociable intrinsic connectivity networks for salience processing and executive control. J. Neurosci. 27 (9), 2349–2356.

Sellers, K.K., Mellin, J.M., Lustenberger, C.M., Boyle, M.R., Lee, W.H., Peterchev, A.V., Fröhlich, F., 2015. Transcranial direct current stimulation (tDCS) of frontal cortex decreases performance on the WAIS-IV intelligence test. Behav. Brain Res. 290, 32–44.

Shibasaki, H., 2008. Human brain mapping: hemodynamic response and electrophysiology. Clin. Neurophysiol. 119 (4), 731–743.

Shin, Y.-I., Foerster, Á., Nitsche, M.A., 2015. Transcranial direct current stimulation (tDCS)—application in neuropsychology. Neuropsychologia 69, 154–175.

Silvanto, J., Muggleton, N., Walsh, V., 2008. State-dependency in brain stimulation studies of perception and cognition. Trends Cognit. Sci. 12 (12), 447–454.

Snowball, A., Tachtsidis, I. Popescu, T., Thompson, J., Delazer, M., Zamarian, L., Cohen Kadosh, R., 2013. Long-term enhancement of brain function and cognition using cognitive training and brain stimulation. Curr. Biol. 23 (11), 987–992.

Sommer, M., Alfaro, A., Rummel, M., Speck, S., Lang, N., Tings, T., Paulus, W., 2006. Half sine, monophasic and biphasic transcranial magnetic stimulation of the human motor cortex. Clin. Neurophysiol. 117 (4), 838–844.

Soutschek, A., Sauter, M., Schubert, T., 2015. The importance of the lateral prefrontal cortex for strategic decision making in the prisoner's dilemma. Cognit. Affect. Behav. Neurosci. 15 (4), 854–860.

Spearman, C., 1927. The abilities of man. Macmillan, London, England.

Speer, A.M., Kimbrell, T.A., Wassermann, E.M., J, D.R., Willis, M.W., Herscovitch, P., Post, R.M., 2000. Opposite effects of high and low frequency rTMS on regional brain activity in depressed patients. Biol. Psychiatry 48, 1133–1141.

Sridharan, D., Levitin, D.J , Menon, V., 2008. A critical role for the right fronto–insular cortex in switching between central-executive and default-mode networks. Proc. Natl. Acad. Sci. USA 105 (34), 12569–12574.

Stagg, C.J., Best, J.G., Stephenson, M.C., O'Shea, J., Wylezinska, M., Kincses, Z.T., Johansen-Berg, H., 2009. Polarity-sensitive modulation of cortical neurotransmitters by transcranial stimulation. J. Neurosci. 29 (16), 5202–5206.

Strüber, D., Rach, S., Neuling, T., Herrmann, C.S., 2015. On the possible role of stimulation duration for after-effects of transcranial alternating current stimulation. Front. Cell. Neurosci. 9, 148.

Takeuchi, H., Taki, Y., Nouchi, R., Hashizume, H., Sekiguchi, A., Kotozaki, Y., Kawashima, R., 2013. Effects of working memory training on functional connectivity and cerebral blood flow during rest. Cortex 49 (8), 2106–2125.

Terney, D., Chaieb, L., Moliadze, V., Antal, A., Paulus, W., 2008. Increasing human brain excitability by transcranial high-frequency random noise stimulation. J. Neurosci. 28 (52), 14147–14155.

Thut, G., Miniussi, C., 2009. New insights into rhythmic brain activity from TMS-EEG studies. Trends Cognit. Sci. 13, 182–189.

Thut, G., Veniero, D., Romei, V., Miniussi, C., Schyns, P., Gross, J., 2011. Rhythmic TMS causes local entrainment of natural oscillatory signatures. Curr. Biol. 21, 1176–1185.

Todd, J.J., Marois, R., 2004. Capacity limit of visual short-term memory in human posterior parietal cortex. Nature 428 (6984), 751–754.

Toro, R., Fox, P.T., Paus, T., 2008. Functional coactivation map of the human brain. Cereb. Cortex 18 (11), 2553–2559.

Tremblay, S., Lepage, J.-F., Latulipe-Loiselle, A., Fregni, F., Pascual-Leone, A., Théoret, H., 2014. The uncertain outcome of prefrontal tDCS. Brain Stimul. 7 (6), 773–783.

Tseng, P., Hsu, T.-Y., Chang, C.-F., Tzeng, O.J.L., Hung, D.L., Muggleton, N.G., Juan, C.-H., 2012. Unleashing potential: transcranial direct current stimulation over the right posterior parietal cortex improves change detection in low-performing individuals. J. Neurosci. 32 (31), 10554–10561.

Utz, K.S., Dimova, V., Oppenlander, K., Kerkhoff, G., 2010. Electrified minds: transcranial direct current stimulation (tDCS) and galvanic vestibular stimulation (GVS) as methods of non-invasive brain stimulation in neuropsychology: a review of current data and future implications. Neuropsychologia 48 (10), 2789–2810.

van den Heuvel, M.P., Sporns, O., 2013. Network hubs in the human brain. Trends Cognit. Sci. 17 (12), 683–696.

Vandierendonck, A., Kemps, E., Fastame, M.C., Szmalec, A., 2004. Working memory components of the Corsi blocks task. Br. J. Psychol. 95 (1), 57–79.

Veniero, D., Vossen, A., Gross, J., Thut, G., 2015. Lasting EEG/MEG aftereffects of rhythmic transcranial brain stimulation: level of control over oscillatory network activity. Front. Cell. Neurosci. 9, 477.

Verleger, R., 1988. Event-related potentials and cognition: a critique of the context updating hypothesis and an alternative interpretation of P3. Behav. Brain Sci. 11, 343–356.

Vossen, A., Gross, J., Thut, G., 2015. Alpha power increase after transcranial alternating current stimulation at alpha frequency (α-tACS) reflects plastic changes rather than entrainment. Brain Stimul. 8 (3), 499–508.

Voyer, D., 2011. Time limits and gender differences on paper-and-pencil tests of mental rotation: a meta-analysis. Psychon. Bull. Rev. 18 (2), 267–277.

Wagner, T., Valero-Cabre, A., Pascual-Leone, A., 2007. Noninvasive human brain stimulation. Annu. Rev. Biomed. Eng. 9 (1), 527–565.

Wammerl, M., Benedek, M., Jauk, E., Jaušovec, N., 2015. The influence of transcranial alternating current on fluid intelligence. A fMRI study. Brain Stimul. 8, 311.

Wassermann, E.M., Zimmermann, T., 2012. Transcranial magnetic brain stimulation: therapeutic promises and scientific gaps. Pharmacol. Ther. 133 (1), 98–107.

Wokke, M.E., Talsma, L.J., Vissers, M.E., 2015. Biasing neural network dynamics using non-invasive brain stimulation. Front. Syst. Neurosci. 8.

Woods, A.J., Antal, A., Bikson, M., Boggio, P.S., Brunoni, A.R., Celnik, P., Nitsche, M.A., 2016. A technical guide to tDCS, and related non-invasive brain stimulation tools. Clin. Neurophysiol. 127 (2), 1031–1048.

Xu, Y., Chun, M.M., 2006. Dissociable neural mechanisms supporting visual short-term memory for objects. Nature 440 (7080), 91–95.

Zaehle, T., Rach, S., Herrmann, C.S., 2010. Transcranial alternating current stimulation enhances individual alpha activity in human EEG. PLoS One 5 (11), e13766.

Zaehle, T., Sandmann, P., Thorne, J., Jäncke, L., Herrmann, C., 2011. Transcranial direct current stimulation of the prefrontal cortex modulates working memory performance: combined behavioural and electrophysiological evidence. BMC Neurosci. 12 (1), 2.

Chapter 5

Other Approaches: From Neurofeedback to Cognitive-Enhancing Drugs

Cognitive training has been approached by two lines of research: behavioral training and neurostimulation/neurofeedback techniques. Research involving behavioral training is discussed in Chapter 3—in short, it assumes that repeated performance on a certain cognitive task not only improves task-specific skills but also broad cognitive abilities, and that this process is associated with neuroplastic changes. On the other hand, the goal of neurostimulation techniques, such as transcranial electric stimulation (TES) and transcranial magnetic stimulation (TMS), and neurofeedback is to upregulate neural mechanisms that are thought to underlie specific cognitive functions, thereby inducing performance increases. Both transcranial alternating current stimulation (tACS) and electroencephalography (EEG)-based neurofeedback can be used to manipulate brain oscillatory activity in specific frequency bands that are associated with the function of interest. All of these methods rely on the concept of neural plasticity, which can occur at various levels, ranging from molecular and cellular levels to systemic alterations (Enriquez-Geppert et al., 2013a).

Low-cost alternatives to neurostimulation/neurofeedback techniques, such as physical activity, diet and meditation, can also positively affect cognitive abilities. There is accumulating evidence that exercise and a healthy diet provide substantial benefits for brain function. Animal and human research suggests that physical activity improves learning and memory, which is thought to be partially mediated by hippocampal neurogenesis (van Praag, 2009). Likewise, dietary factors can affect specific molecular systems and mechanisms that maintain mental function. Diets rich in omega-3 fatty acids support cognitive processes whereas diets that are high in saturated fat have the opposite effect, and may even increase the risk of neurological dysfunction. The combination of exercise and certain diets may have additive effects on synaptic plasticity and cognitive function (Gómez-Pinilla, 2008). Furthermore, lifestyle modifications, such as exercise, dietary changes, and cognitive training might improve cognitive aging and prevent the onset of dementia (Gill and Seitz, 2015). This was supported in a large-scale controlled study that investigated the effects of a

Increasing Intelligence. http://dx.doi.org/10.1016/B978-0-12-809407-5.00005-1

2-year multidomain intervention on cognition in at-risk elderly people (Ngandu et al., 2015). The multidomain intervention (diet, exercise, cognitive training, and vascular risk monitoring) significantly improved overall cognition, executive functioning, and processing, compared to general health advice (the control group).

It is well established that the brain undergoes functional and structural changes in response to stress; however, these changes are reversible in healthy people and can be influenced by pharmacological agents, and the aforementioned lifestyle factors (McEwen and Morrison, 2013). One of the methods that can be used to reduce psychological stress is meditation (Goyal et al., 2014a), the practice of which may also benefit certain cognitive functions, although the latter is debatable. A far more risky approach in terms of side effects is to enhance cognitive functions via prescription drugs, which will be discussed in the last section of this chapter. As we are not experts in these fields of research, we will merely provide a discussion of key findings.

5.1 NEUROFEEDBACK

Neurofeedback training is an operant conditioning method that teaches the participant to modify his or her neural activity. It requires active engagement of participants thus it differs from neurostimulation techniques in which the participant assumes a passive role. Neurofeedback was popularized by Dr. Joe Kamiya in the 1960s, who observed that some of his research subjects could learn to control the frequency and amplitude of their EEG alpha activity, if provided with EEG feedback (Evans and Abarbanel, 1999, preface). During neurofeedback training, the EEG of an individual is recorded, components of interest are extracted and presented to the individual in the form of audio, visual, or combined audio-visual information. For example, visual feedback can be given via a simple bar that increases or decreases in size corresponding to changes in amplitude or frequency (Vernon, 2005). Unless stated otherwise, neurofeedback is provided during periods of rest, that is, no other cognitive tasks are presented concurrently. The participant is typically not given specific instructions *how* to achieve a certain brain state, but they are encouraged to take note of successful strategies, and to apply them to achieve the best possible result. The ability to self-regulate brain activity appeals not only to researchers who wish to uncover the neurobiological mechanisms subserving cognition but also to those who explore the therapeutic potential of this method (Thibault et al., 2016). Clinical applications of neurofeedback are beyond the scope of this section thus we will focus on neurofeedback-induced changes in cognitive performance in healthy subjects.

Neurofeedback training involves multiple sessions that take place over the time course of days or weeks. Given the time and financial costs associated with neurofeedback, rigorous testing is needed to determine whether training individuals to amplify a particular frequency and inhibit another specifically influences cognitive performance (Vernon et al., 2003). Unfortunately, the validity of neurofeedback treatment paradigms has been questioned mainly due to lack of control groups or nonspecific training effects. In a review of EEG-based neurofeedback studies that involved healthy participants, only 28 out of 86 studies survived the selection process in which the quality of control paradigms was evaluated (Rogala et al., 2016). The authors stressed that this presents a problem because it suggests that this field of research is characterized by poorly controlled experiments. The following recommendations have been made to encourage high-quality neurofeedback research: (1) random assignment of participants to groups, (2) use of active control groups (the control group is subject to the same training procedure as the experimental group except that it receives sham feedback based on the EEG of another participant or based on simulated EEG signal), (3) use of double blind trials, (4) pre- and post-EEG baselines should be measured to monitor changes resulting from neurofeedback training, (5) clear measures of behavior should be obtained pre- and posttraining (6) changes in EEG should be correlated with changes in behavior, (7) if possible, long-term stability of training effects should be evaluated (Enriquez-Geppert et al., 2013a; Vernon, 2005). Zoefel et al. (2011)[a] provided even more detailed criteria for the validation of a neurofeedback parameter:

> *There should be spectral effects within the trained frequency band caused by the training (trainability). These spectral changes should not affect other frequency bands (independence). Finally, it is reasonable to choose a frequency band that is associated with certain cognitive functions to increase the probability of reliable behavioral effects as well as applicability (interpretability). (p. 1427)*

While most neurofeedback research is based on EEG, there is a growing number of studies that employ magnetoencephalography (MEG), functional near-infrared spectroscopy (fNIRS), and functional magnetic resonance imaging (fMRI) for feedback regarding the current brain state: fNIRS and fMRI neurofeedback protocols will be discussed in Section 5.1.2. The advent of real-time imaging in the 2000s revived the field and allowed more focal investigations of higher brain functions and neuroplasticity. The underlying signal differs based on the technique used to provide feedback: electrical activity from pyramidal cells perpendicular to the scalp (EEG), magnetic fields produced by pyramidal cells perpendicular and tangential to cortical surface (MEG), blood oxygenation level (fMRI), or volume of oxygenated and deoxygenated blood (fNIRS). EEG and MEG have a feedback delay of less than 50 ms, whereas for fMRI and fNIRS it is approximately 1.5 s (plus 4–6 s hemodynamic delay). Neurofeedback thus refers to diverse imaging methods that help guide volitional control over electromagnetic and hemodynamic changes in brain activity (Thibault et al., 2016).

a. Reprinted from Zoefel et al. (2011), with permission from Elsevier.

5.1.1 EEG and MEG-Based Neurofeedback

EEG-based neurofeedback has been used as a therapy for a wide range of psychological and behavioral disorders, particularly attention deficit hyperactivity disorder (ADHD) (for reviews, see Arns et al., 2009, 2014), and for the optimization of cognitive performance in healthy volunteers (Rogala et al., 2016). The focus of this section will be on the latter: teaching the participant to reach brain states that are thought to correspond to optimum levels of task performance. This process can be achieved by controlling levels of arousal, attention, and motivation and by optimizing autonomic control and the ability to consciously alternate between states (Vernon, 2005). It has been suggested that neurofeedback produces behavioral effects by enhancing synaptic strength through repeated firing (Niv, 2013).

The theoretical rationale for the use of EEG and MEG-based neurofeedback to enhance cognitive performance is based on research findings linking EEG frequency components with performance on specific cognitive tasks. For example, theta activity (4–7 Hz) has been associated with working memory (WM) (Lisman, 2010), and frontal midline theta is thought to reflect cognitive control (Cavanagh and Frank, 2014). Alpha activity (8–12 Hz) has been linked to memory, attention, and information processing (Klimesch, 2012) thus it is not surprising that these frequency bands are often used for neurofeedback training aimed at improving performance on a wide range of cognitive tasks. Theta and alpha training can be applied separately (i.e., participants are encouraged to modulate either their level of theta or alpha activity) or concurrently, which aims typically at increasing theta activity over that of alpha (Gruzelier, 2014). Further examples of EEG-based neurofeedback research include beta band (13–30 Hz) training, which has been used to boost attentional processes (Egner and Gruzelier, 2004), and gamma band (30–100 Hz) training that was investigated in relation to episodic retrieval and feature binding (Keizer et al., 2010a, b). It has been recently shown that MEG can also be used as a real-time neurofeedback device. It presents an advantage over EEG since it can detect high-frequency activity and has a higher spatial resolution. Successful modulation of brain activity driven by MEG neurofeedback has been reported for sensorimotor rhythms (Mellinger et al., 2007), posterior alpha activity (Okazaki et al., 2015), and specific neural oscillatory components within brain regions of interest (Florin et al., 2014). This field of research is still in its infancy, therefore the rest of this section will be devoted to EEG-based neurofeedback.

Even though most neurofeedback protocols are successful, studies often reported that some participants do not show any modulation of brain activity in response to feedback—these participants have been termed "nonresponders" (Lubar et al., 1995). Few studies have attempted to explain why some people fail to respond to neurofeedback training. Lack of motivation and commitment do not appear to be the reason for nonresponsiveness (Enriquez-Geppert et al., 2014b). According to Halder et al. (2013), structural integrity and myelination quality of deep white matter structures may contribute to individual performance on brain–computer interfaces (*n.b.*, neurofeedback is a form of brain–computer interface). Further evidence for this was provided by Enriquez-Geppert et al. (2013b), who reported that the morphology of midcingulate cortex predicts frontal-midline (fm) theta neurofeedback success. In contrast, inferior, superior, and middle frontal cortices were not related to training success. The results also demonstrated that fm-theta enhancement observed during the second session indicated the amount of fm-theta enhancement at the end of training, suggesting that early training responsiveness serves as predictor of overall training success.

Enriquez-Geppert conducted another study in order to investigate the behavioral and neural effects of fm-theta neurofeedback on executive functions (Enriquez-Geppert et al., 2014a). Both the experimental group and the active control group took part in an individualized eight-session neurofeedback training program and completed pre- and postbehavioral testing (3-back task, Stroop task, better-number task-switching, and stop-signal task). Fm-theta neurofeedback facilitated memory updating and mental set shifting, but did not affect conflict monitoring or motor inhibition. These findings are in line with research showing increases in frontal midline theta during tasks that require working memory (Jensen and Tesche, 2002). Moreover, there is evidence to suggest that fm-theta neurofeedback improves attention and working memory performance of older adults (Wang and Hsieh, 2013).

According to the neural efficiency hypothesis, effective performance on cognitive tasks depends on more efficient brain activation (Haier et al., 1992). One of the ideas that propelled neurofeedback research is that alpha activity may facilitate task performance by inhibiting nonessential processing (Vernon et al., 2009). At least two independent alpha subbands exist—lower and upper—that have been examined separately in neurofeedback research. For example, Hanslmayr et al. (2005) applied neurofeedback in order to increase upper alpha power, and decrease theta power. The frequency bands were adjusted according to individual alpha frequency (upper alpha: IAF to IAF + 2 Hz, theta: IAF − 6 to IAF − 4 Hz). Neurofeedback increased upper alpha power and led to improved performance on a mental rotation task. Moreover, training success positively correlated with improvement on the task. It is surprising that theta neurofeedback did not significantly affect theta power or task performance; in fact, it led to slightly increased upper alpha power. These results highlight the main problems of neurofeedback research: trainability and independence of spectral changes. Nevertheless, it is encouraging that the findings reported by Hanslmayr et al. (2005) were replicated by Zoefel et al. (2011), who showed that neurofeedback training of the upper alpha frequency band improved performance on a mental rotation task. In addition, the latter study fulfilled the criteria of trainability, independence, and interpretability. Alpha neurofeedback has also been examined in relation to memory performance. For example, Nan et al. (2012) reported that the participants that received neurofeedback training of individual alpha frequency were able to increase the relative amplitude in alpha band during training sessions, which in turn drove significant improvements in short-term memory performance (a significant positive relation was observed between these variables). However, the control group also showed a significant improvement in short-term memory performance, which was attributed to test practice.

Studies that demonstrate that different neurofeedback protocols lead to distinct behavioral effects provide further support for the use of neurofeedback for the enhancement of cognition and task performance. For example, Keizer et al. (2010a) conducted a double-blind study in which one group of participants learned to increase gamma band activity, and the other group learned to increase beta band activity. The neurofeedback training consisted of one 30-min training session per day over a period of 10 or 11 days. The EEG signal that was used for feedback was recorded from two electrodes: Fz and Oz. In the gamma training group, a tone was generated when the gamma power of the occipital electrode exceeded the upper thresholds. In the beta training group, a tone was generated when beta power of both electrodes exceeded the upper threshold. Before and after training, the participants solved behavioral tests measuring short-term and long-term binding. The results showed that the participants learned to increase occipital and frontal gamma band activity via gamma training, and to increase coherence between frontal and occipital areas in the beta band through beta training. Gamma band training resulted in greater flexibility in retrieving information in short- and long-term memory. In addition, increased coherence between frontal and occipital areas in the beta band led to improved familiarity memory.

Many of the studies described earlier were included in a review of neurofeedback studies by Rogala et al. (2016) that involved healthy participants and appropriate control measures ($N = 28$). The efficacy of EEG based neurofeedback training is typically determined by comparing the amplitudes of targeted EEG bands recorded before and after the training. If this change is significantly different from that observed in the control group, training is deemed successful. The authors of the review evaluated whether the training reported in the studies was successful or not and expressed this in terms of *success ratio* (SR)—the percentage of studies that were successful within the total number of the studies. Theta and alpha band training successfully induced changes in EEG spectra (80 and 89%, respectively), whereas beta training was less successful ($SR = 40\%$). Only 20% of the studies that used multiband protocols were successful, and this confirmed the authors' observation that the number of frequency bands used for training negatively correlates with EEG changes. The results further showed that only theta training was highly effective at producing behavioral changes ($SR = 80\%$). In the next step, studies that had specific behavioral goals were identified and allocated to two groups: memory ($N = 9$) and attention ($N = 6$). Different frequency bands were used for training within each group. The SR of the desired behavioral effects of the memory group was 67%—this group also successfully induced EEG changes ($SR = 78\%$). Similar results were reported for the attention group: an SR of 50% for behavioral changes and an SR of 67% for EEG changes. There was no frequency band that was exceptionally effective in improving performance on attention or memory tests. One would expect a positive relation between EEG and behavioral success, however, no significant correlations were observed. While these results indicate that theta and alpha neurofeedback can modulate the amplitude of the EEG signal and induce behavioral changes, the effects were not consistent, suggesting that the researchers did not place the electrodes optimally (Rogala et al., 2016).

5.1.1.1 The Effects of Neurofeedback on Measures of Intelligence

There is little research on the effects of neurofeedback on performance on tests of intelligence. Most studies to date are proof of principle, and differ substantially in neurofeedback protocol and participant demographics. While most of them report neurofeedback-related changes in EEG spectra, these changes do not translate to improved performance on intelligence tests. For example, a recent sham controlled EEG neurofeedback study demonstrated that frontal beta training selectively modified EEG beta activity a week after neurofeedback training and at a 3-year follow-up (Engelbregt et al., 2016). However, beta training did not result in improved performance on a Dutch intelligence test.

Keizer et al. (2010b) used neurofeedback to explore the relation between gamma band activity and feature binding in a double-blind randomized controlled trial. Since both feature binding and gamma band activity have been linked to intelligence, the authors decided to add Raven's standard progressive matrices to pre- and postneurofeedback testing. The experimental group aimed to increase power in the gamma band (at the cost of beta activation) whereas the control group were rewarded for increasing power in the beta band (at the cost of gamma activation). While the former succeeded to enhance gamma power, the latter did not produce significant changes in beta or gamma power. Neurofeedback training resulted in a decrease in binding costs in the experimental group, and an increase of binding costs in the control group. On the other hand, there was no significant difference between the performance of the two groups on Raven's matrices at pretest or posttest. The percent change of intelligence score from baseline to posttest correlated significantly with the percent change in gamma power in the experimental group ($r = 0.82$, $p < 0.05$), suggesting that gamma neurofeedback targets a mechanism common to fluid intelligence and binding. A similar neurofeedback protocol was conducted in a double-blind randomized controlled trial (RCT) in order to investigate whether increasing gamma or beta band activity can help improve or maintain cognitive functioning in the elderly (Staufenbiel et al., 2014). In addition to Raven's matrices and the binding task, the participants solved an episodic memory task before and after neurofeedback training. The elderly participants were able to increase brain activity within training sessions (particularly the gamma training group), however, no increases were observed between sessions nor did training translate to behavioral improvements. Both the gamma and the beta training groups showed increases in beta power within training sessions, suggesting that these frequencies cannot be trained independently. These findings indicate that older adults are able to modulate their own brain activity, yet it remains unclear whether such training can produce meaningful behavioral effects.

A more effective approach might be to use neurofeedback to target EEG characteristics that deviate from the normative mean. For example, normal aging has been associated with changes in the occipital alpha rhythm (e.g., Babiloni et al., 2014) and posterior theta power (e.g., Stomrud et al., 2010). Becerra et al. (2012) investigated whether neurofeedback could be used to reduce theta activity in 14 normal elderly individuals who showed abnormally high theta absolute power. The subjects were randomly allocated either to an experimental group or a control group that received placebo treatment. Behavioral testing consisted of the Wechsler adult intelligence scale (WAIS-III) and NEUROPSI. The second application of WAIS-III was conducted 6 months later in order to minimize practice effects. The participants received 30 neurofeedback training sessions over a period of 10–12 weeks. In the experimental group, a reward in the form of a tone was given for suppression of theta activity. The neurofeedback protocol was successful: reduced theta absolute power and increased alpha absolute power was only observed in the experimental group. After neurofeedback training, the experimental group but not the control group showed significant increases in Verbal Comprehension Index and Verbal IQ. However, the control group significantly improved on Executive IQ, indicating the presence of placebo effects. The NEUROPSI memory score improved significantly in both groups, although an improvement in the total score (including attention, memory, and executive functions) was only observed in the experimental group. It is unlikely that neurofeedback could be used to improve *g*, however, the former study suggests that it can be used to improve attention and executive functions, thereby potentially ameliorating the decline of certain cognitive processes. Individually tailored neurofeedback protocols designed to counteract abnormal EEG characteristics might increase the likelihood of desired behavioral effects.

Neurofeedback has also been investigated as a potential tool for cognitive improvement in children. Surmeli and Ertem (2010) reported that 19 out of 23 children with mental retardation showed significant improvement on the Wechsler intelligence scale for children–revised, and on the test of variables of attention (TOVA) after neurofeedback training. Similarly, children with learning problems who had received neurofeedback training significantly improved scores on reading tests, verbal IQ, and full-scale IQ after the intervention compared to a control group (Orlando and Rivera, 2004). Since the former study did not use a control group, and the latter study did not deliver neurofeedback to the participants in the control group, it cannot be verified whether the differences in pre- and posttests resulted from neurofeedback training. This can be determined with the help of double-blind placebo-controlled designs, such as the one employed by de Ruiter et al. (2016). The aim of this study was to investigate the effectiveness of neurofeedback training as a potential intervention for neurocognitive deficits in pediatric patients. The authors recruited participants aged 8–18 years that were treated for a brain tumor more than 2 years prior to the study. Participants' EEG was compared to a normative database in order to determine which neurofeedback protocol would be most appropriate for them (based on deviations from the normative mean): reinforce low beta frequency, reinforce sensory motor rhythm, or inhibit low beta frequency. The primary outcome measures were attention, processing speed, memory, executive functioning, visuomotor integration, and intelligence. Neurofeedback training consisted of 30 sessions of 30 min; participants were assessed at baseline, posttraining, and at 6-months follow-up. The results showed no positive effects of neurofeedback over placebo feedback on any of the primary or secondary measures. Both groups showed significant improvements over time (baseline vs. follow-up) for processing speed, visual short term memory, working memory, and intelligence, which might stem from nonspecific treatment effects.

Given that the studies described above mainly focused on gamma and beta frequency bands, and that these bands do not produce reliable behavioral and EEG changes (Rogala et al., 2016), more research is needed to examine the effects of theta and alpha neurofeedback training on measures of intelligence. However, it seems unlikely that up- or down-regulating the activity of just one frequency band could result in improved performance on tests of intelligence, and that these effects would be stable over time. Neurofeedback is often based on one or two electrodes, yet targeted frequencies are generated by different neural structures and/or by distinct cortical networks. This problem can be avoided by using multiple electrodes located over brain areas that are functionally correlated with the targeted task performance (Rogala et al., 2016). Frontal midline theta represents an exception since it is generally maximal at, or just anterior to, the Fz electrode (Mitchell et al., 2008). Indeed, there is evidence that fm-theta neurofeedback can be used to modulate theta activity, which in turn improves memory updating and mental set shifting (Enriquez-Geppert et al., 2014a), attention and working memory performance (Wang and Hsieh, 2013), and memory consolidation (Reiner et al., 2014). In order to validate neurofeedback protocols, researchers should be able to justify electrode locations and fulfill the criteria of trainability, independence, and interpretability (Zoefel et al., 2011).

5.1.2 fNIRS- and fMRI-Based Neurofeedback

Functional nearinfrared spectroscopy (fNIRS) takes advantage of the fact that brain tissues are relatively transparent to near-infrared range light (650–1000 nm) that is either absorbed by chromophores[b] or scattered in tissues. By illuminating

b. Colored chemical compounds; hemoglobin is one of the main chromophores in blood perfused tissue.

a tissue at two (or more) wavelengths, it is possible to detect changes in the concentration of deoxygenated (deoxy-Hb) and oxygenated hemoglobin (oxy-Hb) and hemoglobin volume (Ferrari and Quaresima, 2012), thereby indirectly measuring neural activity. fNIRS-based neurofeedback is a relatively new method that presents a valuable alternative to fMRI neurofeedback protocols which are expensive and immobile. Given that NIRS instrumentation is smaller, more portable, and less costly, it can be applied in clinical settings more readily (Abdelnour and Huppert, 2009). However, this method also has certain limitations, such as a slow information transfer rate, and high error rates. fNIRS-based neurofeedback research has mainly focused on neurorehabilitation. The most common brain areas that are targeted are the primary motor cortex and the prefrontal cortex. fNIRS-based brain–computer interfaces have been used to restore some of the lost motor or cognitive functions that result from stroke or spinal cord injury, and can serve as a means of communication for individuals with motor disorders, such as amyotrophic lateral sclerosis (ALS) (Naseer and Hong, 2015). For example, Naseer et al. (2014) demonstrated that online classifiers can successfully decode "yes" answers corresponding to increased mental workload from "no" answers, during which the subjects were asked to relax.

Recently, novel cognitive training programs have been developed in which individuals are not only presented with behavioral feedback regarding their performance, but also with their brain activity in targeted neural networks during performance on that task. According to Hosseini et al. (2016), task-based neurofeedback has many advantages: (1) the individuals can evaluate which strategy works best for the task in terms of efficient use of neural resources, (2) cognitive training is more complex and perhaps more engaging due to the added neurofeedback as it places more demands on working memory and attention, and (3) the training can be customized according to individual differences in brain networks. In the study by Hosseini et al. (2016), a verbal working memory task (the Sternberg task) was used for fNIRS-based neurofeedback training. Healthy adults were allocated to an experimental group ($N = 10$) or a control group ($N = 10$) that received sham feedback—note that both groups received real behavioral feedback. The NIRS probe set covered the prefrontal regions thus the goal was to improve executive functioning. In order to prevent task-interference due to regulation of brain activity, neurofeedback was presented intermittently and updated at the end of each trial. The participants completed four training sessions over the course of 2 weeks, and were asked to solve a set of cognitive tests before and after training. The primary outcome measure was performance on a verbal n-back test. There was no significant difference in the practice effect of Sternberg tasks between the experimental and control groups. The experimental group showed significantly improved working memory (WM) performance (n-back accuracy increased by 1.8%) compared to the control group. Furthermore, the experimental group showed reduced activity in the right middle and inferior frontal regions, suggesting that training led to greater efficiency of storing and updating of information in these areas. These preliminary findings need to be tested more rigorously with respect to the transfer effect and with a larger sample size. It remains to be determined whether adding neurofeedback to cognitive training produces behavioral effects that are greater than those elicited by either of these methods alone.

Recent advances in fMRI acquisition and reconstruction times have enabled real-time analysis of fMRI data thereby providing online neurofeedback of the blood oxygen dependent signal (BOLD) (Scharnowski and Weiskopf, 2015). Not only does neurofeedback based on real time fMRI (rtfMRI) change specific localized brain activity, there is evidence to suggest that learned control of brain activity leads to behavioral effects that are specific to the functional role of the target brain area (e.g., Bray et al., 2007). Like neurostimulation techniques, rtfMRI enables researches to draw causal relations between the activity of a certain brain area and behavior. An advantage of rtfMRI is that individuals can learn to modify the BOLD signal within 30 min of training, much faster than the time needed to modulate brain electrical fluctuations (Thibault et al., 2016). Research on the neural correlates of neurofeedback learning is growing but the mechanisms are not well understood, moreover, there are still open questions regarding optimization of study design. Due to high scanning costs, rtfMRI neurofeedback typically involves a limited amount of training (up to 3 h), much less than EEG-based neurofeedback. To date, little is known about the clinical applications of rtfMRI neurofeedback, however, a small number of pilot studies have reported training success in patient groups (for a review of scientific and clinical applications of rtfMRI, see Sulzer et al., 2013). For example, depressed participants were able to self-regulate their amygdala response via rtfMRI neurofeedback that resulted in pre–post decreases in anxiety ratings and increases in happiness ratings compared to the control group. The control group received neurofeedback from the intraparietal sulcus, a region that is presumably not involved in emotional processing (Young et al., 2014).

The premise for cognitive enhancement through rtfMRI is that individual differences in performance on cognitive tasks are related to differences in brain activity. As in other neurofeedback techniques, it is assumed that optimizing brain activity leads to improved cognitive performance. To our knowledge, there are no studies that investigate the effects of rtfMRI neurofeedback on measures of intelligence, but there is evidence that complex cognitive functions, such as working memory, attention, and linguistic processing can be enhanced with the help of this technique.

A small number of studies suggest that self-regulation of cortical activity via rtfMRI improves performance on WM (Shen et al., 2015; Sherwood et al., 2016; Zhang et al., 2013). Zhang et al. (2013) chose the left dorsolateral prefrontal cortex (DLPFC), a region regarded as a key node in WM, as the target ROI for rtfMRI training aimed at enhancing verbal WM performance. The participants were healthy adults that were given two neurofeedback sessions, and the pre- and posttraining tasks. For each individual, the ROI for rtfMRI training was selected based on activation during performance on a digital 3-back task.

Half of the participants ($N = 15$) received sham neurofeedback. The participants were given visual feedback in the form of a thermometer and were instructed to up-regulate activation in the target area with the help of a cognitive strategy. Three types of cognitive tasks constituted pre/post behavioral testing: digit span task (criterion), letter memory task (near transfer), and spatial 3-back and Stroop color word (far transfer). The results demonstrated that the experimental group but not the control group significantly up-regulated left DLPFC activation. Moreover, improvement in the criterion task was significantly greater in the experimental group. A significant within-group enhancement was observed for the near transfer task (letter memory) in the experimental but not in the control group. For the far transfer tasks, significant improvements from pretest to posttest were reported for both groups, and no significant between-group differences were found. Shen et al. (2015) reanalyzed this dataset aiming to determine the role of interactions within the WM network that underlies rtfMRI training. A structural equation model that was fit to the training data revealed that the path from the left prefrontal to left parietal cortices increased significantly as a result of neurofeedback. The alteration of this connection correlated significantly with improved performance in verbal WM. Strengthening the frontoparietal connection via neurofeedback could present another option for enhancing WM performance.

The behavioral effect of self-regulating the BOLD signal in the DLPFC was also investigated by Sherwood et al. (2016). The authors predicted that this would lead to improved synaptic efficiency of WM and attention control networks. A single blind design with an active control group was adopted in order to examine pre- and posttraining performance on *n*-back and vigilance tasks. The first run of the *n*-back task was used to identify the left DLPFC for the experimental group. This group received five training sessions (a total of 32 min of training) that consisted of performance on the *n*-back task followed by self-regulation training via rtfMRI. The vigilance tasks were only presented in sessions one and five. In order to account for practice effects of *n*-back training, a behavioral control group was used that performed the same procedures but were not given neurofeedback. Since this took place outside the scanner, the environment in which training took place differed substantially between the two groups. While the individuals in the experimental group successfully learned to modulate activity in the left DLPFC, training with neurofeedback did not lead to greater improvements in WM performance than behavioral training alone. Nevertheless, differences were observed between the first and the second testing day across the groups: the highest increase was observed in the experimental group. In addition, the rtfMRI group showed greater improvement in the vigilance tasks than the control group. It is perhaps too early to claim that combining rtfMRI with cognitive training enhances task performance to a greater extent than behavioral training without neurofeedback. Still, the results of this study present further evidence for rapid self-regulation of the BOLD signal in specific brain areas through rtfMRI neurofeedback.

New procedures and data analysis methods propel the rapidly evolving field of rtfMRI research. For example, deBettencourt et al. (2015) combined rtfMRI with multivariate pattern analysis (MVPA) thereby creating a continuous feedback signal that reflected moment-to-moment variations in sustained attention during task performance. When participants attended well to the target stimuli (indicated by the amount of task-relevant information detected in their brains), they were rewarded with a stronger stimulus. In contrast, lapses in attention (more task-irrelevant information detected) were punished with a degraded stimulus. The approach in which a task is continually updated based on real-time behavioral performance (or in this case, based on MVPA of whole-brain neuroimaging data) has been termed closed-loop. Closed-loop neurofeedback from MVPA increased the participants' ability to detect neural signals that indicated upcoming attentional lapses, which resulted in improved behavioral performance after one training session, relative to control participants who received sham neurofeedback. Moreover, training was predicted by the extent to which feedback relied on information in a frontoparietal attentional network, suggesting that attention can be trained if relevant neurofeedback is provided. As stated by Awh and Vogel (2015), instead of attempting to boost the participant's existing ability to engage in goal-driven selection of information, the future of rtfMRI might lie in optimizing their attentional capacity by improving detection and correction of attentional lapses. Helping participants to stay focused on the task and make mistakes less often presents a potential for behavioral improvement on a wide range of cognitive tasks.

The studies discussed earlier trained participants to regulate localized brain activity within a target area. Given our knowledge about the role of interregional functional connectivity in higher cognitive functions, it is worth investigating whether participants can be trained to voluntarily control the connectivity between two regions of interest. Koush et al. (2013) were the first to demonstrate successful connectivity feedback based on rtfMRI. In their study, near real time dynamic causal modeling was used for neurofeedback; the signal represented connectivity between left visual cortex and left parietal cortex, or between right visual cortex and right parietal cortex. During neurofeedback trials, the participants solved a visual-spatial attention task in which they were asked to shift visual attention either to the left or right visual field. The results showed that the participants were able to control the feedback signal by covertly shifting their visual attention to the target location. Performance improved slightly but did not reach significance. The results of this proof-of-concept experiment pave the path for studies investigating the behavioral effects of self-regulation of functional brain networks.

At this stage, fNIRS and rtfMRI research shows that modulating one's own brain activity in specific brain areas is possible, and that this can translate to improved performance on WM and attention tests. Further research is needed that will investigate near- and far-transfer effects along with long-term effectiveness of training. New methods such as task-based neurofeedback, closed-loop systems, and connectivity-based neurofeedback provide an exciting prospect for the study of the neural basis of cognitive functions and for clinically relevant investigations, which aim at behavioral modulations in pathology (Weiskopf, 2012). Self-regulation of brain activity in specific areas can improve task performance—however, can the effects of neurofeedback training transfer to situations in which neurofeedback is not presented (Scharnowski and Weiskopf, 2015)? Besides the technical and design challenges that lay ahead, this question will need to be addressed by future real time fNIRS and fMRI studies.

Can neurofeedback make you "smarter," that is, increase your IQ score? Some neurofeedback providers would state that this is the case. However, this claim has not been supported by double-blind sham-controlled studies involving healthy participants. Nevertheless, there is preliminary evidence that neurofeedback presents a promising tool for enhancing specific cognitive processes such as attention and memory. The challenge for future research is to move beyond the proof-of-principle stage and to determine whether the behavioral effects of neurofeedback outweigh the time and financial costs associated with training. Methods that enable self-regulation of functional connectivity between regions of interest may provide the best option for enhancing performance on tests of intelligence.

5.2 EXERCISE

The long-term benefits of physical activity on health are well documented. Not only does it build muscle mass and cardiorespiratory fitness, it helps reduce the risk of cardiovascular disease, stroke, type 2 diabetes, weight gain, colon and breast cancer, osteoporosis, and depression (Physical Activity Guidelines for Americans, 2008), whereas a sedentary lifestyle presents a risk factor for all-cause mortality (van der Ploeg et al., 2012). Moreover, animal and human research suggests that chronic physical activity positively affects brain health and cognition. In this chapter, I will briefly review the molecular and cellular responses to physical activity derived from the animal literature, followed by a more comprehensive review of observational and interventional research on the effects of different types of physical exercise on human cognition and brain plasticity, primarily in healthy adults.

5.2.1 Animal Research

Animal studies have provided the basis of our understanding of the effects of exercise on metabolic, structural, and functional properties of the brain. While a comprehensive review of animal research is beyond the scope of this chapter, I will outline the mechanisms that are thought to exert benefits on cognitive function and brain plasticity in animal and human studies alike.

5.2.1.1 Neurogenesis

Exercise increases the production of new neurons in the dentate gyrus of adult mice, a finding that is highly reproducible (Fabel et al., 2003; Holmes et al., 2004; Marlatt et al., 2012; Trejo et al., 2001; van Praag et al., 1999b; to name a few). Physical exercise, such as wheel running promotes all stages of adult neurogenesis: cell proliferation, neuronal differentiation, and cell survival. In rodent studies, long-term exercise (3–12 weeks) that is voluntary in nature rather than forced produces the most reliable effects on learning and plasticity (Cotman et al., 2007). Tasks sensitive to hippocampal function such as the Morris water maze, radial arm maze, or Y-maze are typically used to assess behavioral performance. These studies suggest that exercise-induced hippocampal neurogenesis improves spatial learning and memory performance (Li et al., 2013; Mustroph et al., 2012; Van der Borght et al., 2007; van Praag et al., 1999a). Given that new neurons are also produced in the human hippocampus (Eriksson et al., 1998), exercise-induced neurogenesis presents one to the potential mechanisms mediating improvements in plasticity and cognition in humans. However, one should keep in mind that rodent studies predominantly focus on hippocampus-dependent learning whereas human research often targets tasks that depend on frontal lobe function.

5.2.1.2 Angiogenesis and Blood Flow

Changes in neuronal circuitry and function require a sufficient supply of nutrients and energy, hence it was hypothesized that long-term exercise promotes the development of new blood vessels (angiogenesis) in order to support neurogenesis. Indeed, it has been shown that exercise increases angiogenesis in the rodent hippocampus (Clark et al., 2009; van Praag et al., 2005), motor cortex (Swain et al., 2003), striatum (Ding et al., 2006), and cerebellum (Black et al., 1990). In addition, there is evidence to suggest that physical activity improves cerebral blood flow (Gertz et al., 2006; Swain et al., 2003), a finding that has also been reported in human studies (Ainslie et al., 2008; Pereira et al., 2007).

5.2.1.3 Growth Factors

Exercise is thought to regulate neurogenesis, angiogenesis, and learning by increasing brain availability of growth factors. Voluntary exercise can increase levels of brain-derived neurotrophic factor (BDNF), a protein that regulates synaptic efficacy, neuronal connectivity and use-dependent plasticity (Cotman and Berchtold, 2002). For example, Adlard et al. (2004) investigated BDNF induction in the rat hippocampus following voluntary exercise. The results demonstrated significant increases in BDNF mRNA during the time course of voluntary exercise (1–28 days) and a significant increase in BDNF protein after 28 days of exercise. Moreover, rats that exercised prior to testing required less time to escape a Morris water maze and showed an enhanced rate of learning compared to sedentary rats. These findings appear to be transferable to humans: the results of two metaanalyses suggested that both acute and chronic aerobic exercise increase peripheral BDNF levels in adults (Huang et al., 2014; Szuhany et al., 2015).

Research indicates that exercise-induced increases in hippocampal BDNF are mediated by neurotransmitters and interactions with peripheral factors such as estrogen, corticosterone, and insulin-like growth factor 1 (IGF-1) (Cotman and Berchtold, 2002). IGF-1 in particular has received a great deal of attention, and is now considered as one of the key factors mediating the broad benefits of exercise on the brain. For example, a study on adult rats reported that after 1 h of treadmill running, serum IGF-1 entered the brain through the blood-CSF pathway and activated target neurons. Neurons accumulating IGF-1 showed enhanced sensitivity to afferent stimulation and increased spontaneous firing rate. This pattern was also observed after exogenous intracarotid injection of IGF-1 (Carro et al., 2000). Furthermore, it has been suggested that age-related downregulation of IGF-1 signaling is related to cognitive decline in aged rats (Markowska et al., 1998) and humans (Dik et al., 2003), however, the role of this signaling pathway in cognition is not completely understood. Another growth factor that mediates the role of exercise on the brain is vascular endothelial growth factor (VEGF) (Fabel et al., 2003). Together with IGF-1, it regulates angiogenesis and hippocampal neurogenesis. Overall, exercise induces growth factor cascades (BDNF, IGF-1, and VEGF) that work together to modulate exercise-related benefits in plasticity and cognitive function (Cotman et al., 2007).

5.2.1.4 Synaptic Plasticity

Based on these findings, it is not surprising that exercise facilitates long-term potentiation (LTP) in the dentate gyrus (Eadie et al., 2005; Farmer et al., 2004; van Praag et al., 1999a). LTP presents one of the leading models of synaptic plastic changes that occur during learning. Similar results have been reported in studies in which mice were raised in enriched environments that included running wheels (Duffy et al., 2001). It is important to note that these events do not occur in isolation. In the study by Farmer et al. (2004), voluntary exercise stimulated plasticity in the dentate gyrus of adult rats on several levels. Exercising rats compared to controls showed increased: (1) neurogenesis, (2) induction and expression of LTP and (3) expression of BDNF, along with changes in a subtype of glutamate receptor.

5.2.1.5 Antiinflammatory Effects

Brain health is affected negatively by the presence of peripheral risk factors for cognitive decline such as metabolic syndrome.[c] A characteristic feature of metabolic syndrome is systemic inflammation, which in turn affects inflammation in

c. A cluster of interrelated risk factors of metabolic origin: atherogenic dyslipidemia, insulin resistance, hypertension, and elevated plasma glucose (Grundy et al., 2005)

the brain, and may even contribute to progression of chronic neurodegenerative disease (Perry, 2004). Regular exercise induces an antiinflammatory response (Pedersen, 2006). Animal studies suggest that antiinflammatory changes in the hippocampus correlate with the proneurogenic and procognitive effects of exercise (for a review, see Ryan and Nolan, 2016).

Collectively, animal research supports the benefits of exercise on brain and cognition. In rodents, voluntary running on wheels improves learning and memory on hippocampus-dependent tasks, which is accompanied by growth factor cascades, neurogenesis, and angiogenesis. While animal studies provide far greater control of environmental and genetic factors than human studies, caution must be exerted in translating research findings from animal to humans. Nevertheless, animal models have provided compelling evidence for the role of exercise in neuroprotection and neuroplasticity.

5.2.2 Human Observational Studies

While human exercise research cannot explain changes at a molecular and cellular level, it can provide an understanding of system-level changes in the central nervous system related to higher-level cognition. There is a lack of research on the relationship between physical activity and cognition in young adults, perhaps due to the conception that there is little room for improvement in cognitive function (Hillman et al., 2008). Among the few studies that tackle this subject, the majority of evidence points toward a positive relationship between physical activity and cognitive function—or, at the very least, suggests that time allocated toward exercise does not hinder academic performance. For example, a cross-sectional study of Spanish adolescents aged between 13 and 18.5 years showed that those who engaged in physical activity during leisure time had significantly better cognitive performance than those who did not (Cohen $d = 0.32$), as measured by the Spanish version of the test of educational ability and self-reported measures of physical activity (Ruiz et al., 2010). Socioeconomic status, family status, cardiorespiratory fitness, BMI, and television viewing did not significantly modify the results when entered in the model. However, these results were not supported by Daley and Ryan (2000): adolescents' self-reported levels of exercise did not significantly correlate with academic performance.

The results of a systematic review of observational studies that included high-school adolescents aged between 13 and 18 years suggested that physical activity is positively related to academic and cognitive performance, but mainly in girls (Esteban-Cornejo et al., 2015). With respect to the relationship between physical activity and cognitive performance, three studies reported positive associations and one longitudinal study determined a negative association. For the physical activity–academic performance relation, 10 studies determined a positive association, 4 studies showed no association, and 1 study showed a negative association. It should be noted that the studies were not weighted thus differences in sample size and design could have affected the outcome. Furthermore, all but one study utilized questionnaires to assess physical activity, which can be prone to social-desirability bias. For this reason, some researchers prefer to use objective measures of physical activity, such as pedometers or accelerometers, or to measure cardiorespiratory fitness (maximal oxygen uptake during exercise tests). For example, a cohort study of 1,221,727 Swedish male subjects (Aberg et al., 2009) examined the relationship between cardiovascular fitness (measured by ergometer cycling) and general cognitive ability. Linear regression models, which were adjusted for parental education level, conscription year, and test center, revealed a positive association between cardiovascular fitness and global intelligence at 18 years of age (regression coefficient $b = 0.172$). An analysis of MZ twin pair data demonstrated similar results. Moreover, the results showed that an improvement in cardiovascular fitness between ages 15 and 18 was associated with higher global intelligence scores, higher educational attainment, and higher socioeconomic status at age 18. A cross-sectional (age 11) and longitudinal study of UK adolescents (Booth et al., 2014) reported similar results. Physical activity was assessed via an accelerometer that was worn by the participants during waking hours for 7 consecutive days. Academic attainment was determined at ages 11, 13, and 16 years. After controlling for a number of confounders, such as birth weight, pubertal status, body mass index, SES, etc., percentage of time spent in moderate-vigorous intensity physical activity at age 11 positively predicted subsequent academic attainment in English, Mathematics, and Science (*n.b.,* gender differences were reported). The results even pointed toward a dose response effect, providing further support for the importance of physical activity in adolescence.

Population aging has become an increasing concern in many countries, as it corresponds with a growing proportion of individuals with cognitive impairment or dementia (Prince et al., 2013). The aging population has substantial economic and social consequences, including high medical expenses associated with cognitive deterioration. It is only appropriate that this has been paralleled by research aimed at discovering methods that could delay the onset and/or ameliorate cognitive decline. Regular physical activity has been identified as one of the most promising methods to counteract age-related declines in brain structure and function. A number of observational studies demonstrated that fitness is positively associated with cognitive function in older adults. For example, a large prospective cohort study of middle-aged British civil servants demonstrated a significant positive association between self-reported levels of physical activity, and fluid intelligence and phonemic fluency. Likewise, low levels of physical activity at baseline were associated with a decline in fluid intelligence 11 years later (Singh-Manoux et al., 2005). Since it is well established that fluid intelligence declines with age (Bugg et al., 2006), low levels of physical activity may present an additional risk factor for middle age.

Barnes et al. (2003) conducted a 6-year study of 349 healthy adults over the age of 55. At baseline, cardiorespiratory fitness measures were determined via treadmill exercise tests and cognitive ability was assessed with a modified version of the mini-mental state examination (MMSE). After 6 years of follow up, the full version of the MMSE was administered again. As predicted, baseline measures of cardiovascular fitness were positively related to the preservation of cognitive function 6 years later. The association was strongest for measures of global cognitive function and attention/executive function, thus the authors concluded that high levels of cardiorespiratory fitness may play a protective role against cognitive dysfunction in older people. Interestingly, self-reported levels of physical activity (transformed into metabolic expenditure scores) did not correlate significantly with MMSE scores 6 years later, suggesting that cardiorespiratory fitness may provide a more accurate measure of the relationship between physical activity and cognitive function.

A negative association between aging and cognition, and a positive association between fitness and cognitive function was also reported in a cross-sectional study of women above the age of 50 (Brown et al., 2010). The sample was divided into active and sedentary groups based on their maximal oxygen uptake (VO_2 max) of age-predicted values. The main outcome measure was overall cognitive function, which was assessed with a series of tests. Both cardiovascular conductance and mean arterial pressure predicted cognitive functioning. Furthermore, a large scale prospective cohort study of women and men above the age 65 demonstrated that at a 5-year follow-up, self-reported exercise (compared to no exercise) was associated with lower risks of dementia, cognitive impairment, and Alzheimer disease, particularly in women (Laurin et al., 2001). Similar findings were reported by a number of prospective cohort studies (Abbott et al., 2004; Chang et al., 2010; Larson, 2006; Rovio et al., 2005; Weuve et al., 2004). Nevertheless, some studies have failed to support the role of physical activity in cognitive decline. For example, Sturman et al. (2005) reported that more hours of physical activity was associated with a small beneficial effect on the rate of cognitive decline over a period of 6 years in individuals aged over 65, however, the coefficients were no longer statistically significant when the model was adjusted for participation in cognitive activities, depression, and vascular illness. The authors noted that the individuals engaged in lower levels of physical activity than those reported in previous studies, which may explain the unsubstantial association with the rate of cognitive decline. A different approach was adopted by a study that included adults aged 75–85 with a 5.1 year follow up: self-reported leisure activities were analyzed in order to determine whether they might lower the risk of dementia (Verghese et al., 2003). The responses to the interview about leisure activities were transformed into physical-activity scores (sports, group exercises, walking, housework, etc.) and cognitive–activity scores (reading, writing, doing cross-word puzzles, group discussions, playing musical instruments). Multivariate models included the following covariates: age at enrollment, sex, educational level, presence or absence of chronic medical illness, and base-line cognitive status. The results showed that a one-point increase in the cognitive–activity score (one activity for 1 day/week) was related to a 7% decrease in the risk of dementia; individuals who scored in the highest third of the cognitive–activity scale had 63% lower risk of dementia compared to individuals who scored in the lowest third. In contrast, the physical-activity score was not associated with a decreased risk of dementia—the only exception was dancing (so put on your dancing shoes!).

Despite some inconsistencies in the literature, metaanalytic studies suggest that regular physical activity can alleviate or help prevent cognitive decline and dementia. An analysis of 15 prospective studies involving 33,816 subjects demonstrated that even low-to-moderate levels of exercise were associated with reduced risk of cognitive impairment (–35%) and subjects with high levels of exercise showed a 38% reduced risk compared to sedentary subjects (Sofi et al., 2011). A metaanalysis of 42 prospective studies published since 2007 showed that individuals with higher levels of physical activity, when compared to those with lower levels, are protected against cognitive decline (relative risk = 0.65) and dementia (relative risk = 0.86) (Blondell et al., 2014). The studies included in the metaanalysis were conducted world-wide on population-based samples, used a prospective design and collected baseline physical activity.

While the majority of observational studies suggest that regular physical activity can mitigate or, at least, delay the onset of cognitive decline and dementia, there are a number of factors that need to be considered (Kramer and Erickson, 2007). First, self-reported measures of physical activity may not accurately represent actual levels of physical activity. A systematic review that compared self-reported versus direct measures of physical activity (e.g., calorimetry, accelometry, pedometry, hear rate monitoring, direct observation) showed patterns of both overestimation and underestimation; overall, low-to-moderate correlations were reported that ranged between –0.71 and 0.98 (Prince et al., 2008). Given that the studies discussed in this chapter focus on different types of exercise (aerobic vs. nonaerobic), which also vary in frequency and intensity, caution should be used when comparing studies across methods. Second, Karp et al. (2006) demonstrated that older adults who participate in activities that contain mental, physical, and social components show a greater reduction in risk of dementia compared to individuals who take part in activities that load heavily on only one of these components. Hence, mental and social components may moderate the effect of physical activity on cognition.

Third, genetic factors must be taken into account. Apolipoprotein E (APOE) e4 allele has been identified as a risk factor for Alzheimer's disease (Farrer, 1997), dementia (Henderson et al., 1995), and coronary heart disease (Wilson et al., 1996), and is related to faster cognitive decline in nonpathological aging (Schiepers et al., 2012). Research on gene–lifestyle interactions in relation to physical activity has provided conflicting results. While Podewils et al. (2005) reported that the reduced risk of dementia and Alzheimer's disease due to regular physical activity may be limited to people who are not APOE e4 carriers, the opposite has also been reported: (1) physical activity may reduce the risk of cognitive decline in elderly men, particularly in APOE e4 carriers (Schuit et al., 2001), (2) aerobic fitness is associated positively with performance on memory tests in female APOE carriers (Etnier et al., 2006), and (3) physical activity at least twice a week in midlife is related to 50% lower odds of dementia and 60% lower odds of Alzheimer's disease compared to sedentary individuals, even after controlling for a number of possible confounders that included APOE genotype (Rovio et al., 2005). It should be noted that these studies were based on epidemiological or cross-sectional data, leading to relatively low sample sizes of APOE carriers. Given that cognitively normal adults with an APOE e4 allele show greater cortical amyloid deposition, and that regular exercise is associated with lower amyloid deposition, it is not surprising that Head et al. (2012) demonstrated that physical exercise can moderate the effects of APOE genotype on amyloid deposition in adults aged 45–88 years. Specifically, the authors reported that sedentary APOE e4 carriers had the highest levels of amyloid deposition whereas active APOE e4 carriers had reduced levels of amyloid—comparable to those of noncarriers of the e4 allele.

The gene encoding brain-derived neurotrophic factor has also been examined in terms of cognitive function. A single nucleotide polymorphism in the BDNF gene leads to a valine (val) to methionine (Met) substitution at codon 66, which has been associated with lower levels of BDNF-protein secretion and, in some studies, with poorer memory performance (Egan et al., 2003). Erickson et al. (2013) proposed that individual variability in the extent to which physical activity improves cognitive function could be explained by polymorphisms in the BDNF gene. An analysis of 1032 adults aged between 20 and 54 years revealed that BDNF genotype interacted with the effect of physical activity on WM (measured via letter and spatial *n*-back tasks), but not with episodic memory, visual-spatial memory, or switching. Among those who were less physically active, Met carriers performed worse on the WM tests than Val homozygotes. This effect was not present among physically active individuals. Exercise intervention studies aimed at increasing BDNF serum levels will be presented in the next section. Collectively, these results suggest that an active lifestyle may be particularly beneficial for individuals with the greatest genetic susceptibility to cognitive decline and Alzheimer's disease (Prakash et al., 2015). Nevertheless, cross-sectional and prospective observational studies offer a limited picture of the physical activity–cognition relationship, and very little data related to measures of intelligence, thus we must examine whether exercise interventions can improve cognitive function. Gaining a better understanding of this interaction may provide the groundwork for personalized exercise interventions based on genotype.

5.2.3 Human Intervention Studies

5.2.3.1 Young and Middle Aged Adults

Intervention studies that have been designed to improve cardiorespiratory fitness in healthy adults provide further insight into the effects of physical activity on cognition. This section will focus on predominantly RCTs with aerobic and nonaerobic exercise interventions. It will become clear that the results of intervention studies are far less consistent than those reported by observational studies. Compared to research on cognitive decline, there are scarcely any studies that investigate the effects of exercise on cognition in healthy young adults. Research on the acute effects of exercise on cognition is more prominent in this age group, but will not be discussed here (for a review, see Chang et al., 2012). To our knowledge, there are no RCTs that would directly tests the effects of an exercise intervention on measures of intelligence in healthy young or middle-aged adults. However, there are some studies that focus on measures of cognitive control, executive functions, and certain components of intelligence. Stroth et al. (2009) conducted an aerobic exercise intervention study on 28 students aged between 17 and 29. The students were matched according to gender and habitual physical activity, and assigned randomly to experimental and control groups. The control group was asked to maintain their habits or join a workout program. The training sessions consisted of three running sessions of 30 min/week, and were individually adapted based on concentration of lactate in blood samples. Fitness, cognitive, and affect testing took place at the beginning and the end of the intervention. Cognitive testing was assessed by measures of visuospatial and verbal memory (the visual and verbal memory test) and the d2 test of attention. After 6 weeks of intervention, the control group participated in the same aerobic program; regrettably, cognitive function and affect were not assessed at the end. The results for 23 students demonstrated a significant post-intervention increase in the running group in positive affect, and visuospatial memory compared to the control group. In contrast, the aerobic intervention did not significantly affect verbal memory and concentration performance. It remains unclear why certain cognitive processes benefited from exercise and others did not. Given the small sample size, drawing conclusions about the efficacy of aerobic training on cognitive functions in young adults would be unwarranted. The next study that will be discussed did not solely focus on young adults, but included adults aged 18–70 who were aerobically active fewer than three times a week (Masley et al., 2009). After additional recruitment of subjects, 71 completed the exercise intervention and 20 were allocated to the control group. It should be noted that only the initial group of participants ($N_{completed} = 46$) was randomly assigned to the two groups. The treatment group undertook 30–45 min long aerobic exercises (at 70%–85% anticipated maximum heart rate), 3–7 days/week for the duration of 10 weeks. Once a week, the participants exercised under the supervision of a personal trainer to ensure adequate technique; the rest of the sessions the participants generated exercise logs. The participants in the control group continued their current level of activity and intake of food. Before and at the end of the intervention, the participants solved a computerized cognitive battery that was used to generate scores for memory, psychomotor speed and information processing time, attention, and cognitive flexibility. In comparison to the control group, the treatment group showed a significant increase in cognitive flexibility (i.e., scores generated from a shifting attention test and a Stroop test) after the intervention, which was accompanied by improved cardiorespiratory fitness, as indicated by VO_2. Moreover, the results demonstrated a dose-responsive relationship between the frequency of aerobic exercise and the extent of improvement in cognitive flexibility.

The last exercise-related study that will be presented in this section was conducted on mice and human adults (21–45 years, mean age = 33) (Pereira et al., 2007). Rodent research has shown that exercise can induce neurogenesis in a subregion of the hippocampus called the dentate gyrus (van Praag et al., 1999b), however, this can only be assessed in postmortem tissue. The goal of the study was to establish an in vivo correlate of exercise-induced neurogenesis in humans. Given that neurogenesis has been associated with angiogenesis, which is in turn related to cerebral blood volume (CBV), it was assumed that MRI measurements of CBV may provide an indirect measure of neurogenesis. In the first step, nonexercising mice were compared to mice that ran voluntarily on wheels for 2 weeks. MRI images were obtained at baseline and 2, 4, and 6 weeks later. In order to test their hypothesis regarding the CBV-neurogenesis relationship, in vivo images were compared to postmortem measurements of neurogenesis in the hippocampus. The group of mice that exercised showed significantly greater dentate gyrus CBV and postmortem measurements of neurogenesis than the nonexercising group. Further analyses confirmed that neurogenesis accounted for the exercise effect on CBV. In the next step, a human study was conducted on 11 participants with below-average levels of aerobic fitness (*n.b.,* there was no control group). The participants took part in a 12-week exercise training protocol, which mostly consisted of aerobic activities. The participants exercised four times a week (session length = 1 h). Hippocampal CBV maps and VO_2 measurements were obtained before and after exercise. Declarative memory was assessed pre- and postexercise with a modified version of the Rey auditory verbal learning test in eight participants. After exercise, there was a significant increase in dentate gyrus CBV and VO_2max, specifically, individual differences in dentate gyrus CBV correlated with individual changes in VO_2max. This correlation was not observed in any other hippocampal subregion. Moreover, the participants performed significantly better on trial 1 learning after exercise, which correlated with VO_2max changes and with dentate gyrus CBV. While it cannot be directly confirmed that changes in dentate gyrus CBV reflected exercised-induced neurogenesis, there is a growing amount of research suggesting that physical activity may promote efficient regulation of cerebral blood flow (Murrell et al., 2013) that may present one of the mechanisms through which regular exercise improves cognitive function (Brown et al., 2010; Guiney et al., 2015).

5.2.3.2 Older Adults

Due to the extensive amount of RCTs of the effect of physical activity on cognitive function in older adults, the results of metaanalyses conducted in the recent years will be discussed. A Cochrane intervention review (Angevaren et al., 2008) examined 11 RCTs that compared aerobic physical activity programs with other types of intervention (or no intervention) in adults over the age of 55, who did not suffer from cognitive dysfunction. Eight out of 11 studies reported that the interventions result in enhanced cardiorespiratory fitness which concurred with improvements in motor function, auditory and visual attention, cognitive speed, and delayed memory functions. However, most of the comparisons did not yield significant results, therefore, the authors concluded that there is insufficient evidence that exercise-induced improvements in cognitive function stem from improved cardiorespiratory fitness. A number of RCTs have been published since, some of which were included in a meta-analytic review conducted in 2010 (Smith et al., 2010). In order to be selected for the review, the mean age of participants had to be above 18 years. Nevertheless, 21 out of 29 studies only included older participants. The selected studies included either healthy adults or adults with mild cognitive impairment, depression, fibromyalgia, multiple sclerosis, hypertension, and cystic fibrosis, giving a total of 2049 participants. The trials lasted at least 6 weeks, and primarily consisted of jogging and brisk walking, contrasted against no intervention or stretching, toning, and relaxation exercises in the control groups. Only 13 studies used blinded assessments, highlighting the need for higher quality RCTs in this field. Effect size analyses revealed that aerobic exercise was associated with moderate improvements in attention and processing speed ($g = 0.158$, $p = 0.003$), executive function ($g = 0.123$, $p = 0.018$), and memory ($g = 0.128$, $p = 0.026$), but not with WM ($g = 0.032$, $p = 0.642$). These results are in line with the findings reported by a metaanalysis that was published in 2003 (Colcombe and Kramer, 2003). In contrast to Colcombe and Kramer's study, neither intensity nor duration of the intervention affected the aforementioned cognitive functions. Interestingly, combined strength and aerobic training improved attention and processing speed, and WM to a greater extent than only aerobic interventions. The results further suggested that exercise may produce larger improvements in memory for individuals with mild cognitive impairment compared to adults with no known cognitive impairment (Smith et al., 2010). A systematic review of 30 RCTs and quasi experiments in adults over the age of 50 years further highlighted the dearth of good quality studies, mainly due to insufficient sample sizes and nonrepresentative sampling, along with lack of intention-to-treat analyses and blinding (Snowden et al., 2011). Moreover, dose-responsive effects often cannot be established due to variation across studies and insufficient information regarding the intensity of exercise—this was also the case in the present review. The 30 studies were assigned to 20 intervention-outcome pairings; the interventions were grouped as cardiorespiratory, muscle strength, and multicomponent exercises, and the outcomes were general cognition, executive function, memory, reaction time, attention, cognitive processing, visuospatial, and language. An expert panel assessed the quality and effectiveness of each pairing. Thirteen intervention-outcome pairings had sufficient numbers of studies, and showed some positive effects on cognition, however, it could not be confirmed that physical exercise improves cognition in older adults.

A metaanalysis that was published a year later compared the effects of aerobic training ($N_{studies} = 17$) and extended cognitive training ($N_{studies} = 25$) on untrained cognitive tasks in healthy older adults (Hindin and Zelinski, 2012). For this purpose, between-group pre–post effect sizes (comparisons of experimental and control group) and within-group pre–post effect sizes were computed. Both aerobic training and extended cognitive training significantly improved untrained cognitive outcomes (i.e., choice response time, memory, and executive functions). The between-group effect sizes for aerobic training (estimated ES = 0.33, 95% CI = 0.10–0.55), and extended cognitive training (estimated ES = 0.33, 95% CI = 0.13–0.52) did not differ in magnitude. However, the between group effect sizes were significantly smaller than the within group effect sizes, suggesting that practise effects from baseline testing magnified the ostensible training benefit.

In summary, two metaanalyses suggested that aerobic training interventions positively affect cognitive functions in older adults (Hindin and Zelinski, 2012; Smith et al., 2010), whereas two metaanalyses concluded that there is insufficient evidence to support this claim (Angevaren et al., 2008; Snowden et al., 2011), which may in part stem from the fact that the RCTs analyzed in these systematic reviews employed substantially different training protocols, and that some of them used small sample sizes and short intervention periods.

What are the effects of other forms of exercise interventions such as balance, flexibility, and resistance training on cognition in older adults? A metaanalysis conducted in 2014 focused solely on tai chi studies (Wayne et al., 2014), a form of exercise that consists of physical, cognitive, social and meditative components. The physical component provides moderate aerobic, mobility, and agility training, whereas the cognitive component involves learning and memorization of patterns of movement. In addition, the meditation component may help reduce anxiety, and has been associated with enhanced attention and executive functions. Twenty studies were incorporated in the review, 11 of which were RCTs. Four RCT studies included older adults without known cognitive impairment (total $N = 421$) and the rest included cognitively impaired older adults (total $N = 843$). Among the former, one study was excluded due to a low quality score. The results for the remaining three RCTs showed a small but significant effect for Tai Chi (compared to nonintervention controls) on measures of executive function (Hedges' $g = 0.394$, $p = 0.004$). In one of these studies, tai chi was contrasted against Western exercise (endurance, resistance, and flexibility exercises) and a nonintervention control group. Participants assigned to tai chi showed improvements in backward digit span after 6 months of intervention, which was maintained at a 12 months follow up (Taylor-Piliae and Froelicher, 2004). Metaanalyses for the high quality RCTs that compared cognitively impaired adults to nonintervention controls also indicated a small, but significant effect for tai chi on global cognition (Hedges' $g = 0.304$, $p = 0.033$), as measured by the MMSE. Similar results were reported for RCTs that compared the effects of tai chi to active interventions in cognitively impaired adults. Further support for these findings comes from a more recent systematic review of prospective controlled trials of the efficacy of tai chi on cognitive ability (Zheng et al., 2015), however, the authors also stated that the methodological quality of the papers was low, and could have been affected by publication bias.

A systematic review and metaanalysis on the effectiveness of yoga compared to other exercise interventions for older adults demonstrated that yoga interventions improved flexibility, aerobic fitness, and self-rated mental-health status, but did not affect cognitive function (Patel et al., 2012). In contrast, a metaanalysis published 3 years later reported that yoga was associated with moderate improvements in cognitive function ($g = 0.33$, $p < 0.001$), particularly attention and processing speed, followed by executive function and memory (Gothe and McAuley, 2015). Most of the RCTs used Hatha yoga, a structured practice that involves physical components, regulated breathing, and meditation. It should be noted that this study analyzed RCTs that involved adults, and a minority of studies even involved children, thus the results cannot be compared to the Patel et al. (2012) metaanalysis.

Resistance training, also known as strength training, is a popular form of exercise that increases body mass, strength, power, and energy expenditure and has been associated with a number of health benefits in the older population. One of the mechanisms through which resistance training may prevent cognitive decline is by increasing IGF-1, a growth hormone that promotes neuronal growth and differentiation in the brain. Animal research suggests that IGF-1 mediates exercise-induced increases in the number of new neurons in the hippocampus (Trejo et al., 2001). There is a limited number of RCTs that examine the effects of resistance training on cognition in older adults. Kimura et al. (2010) reported that a 3-month strength training did not result in significant improvement in executive function measured by task-switching reaction time in older adults. On the other hand, Brown et al. (2009) reported that a group-based exercise composed of resistance and balance training (two classes a week for 6 months) positively affected fluid intelligence scores measured by WAIS-R in seniors, compared to control groups (no-exercise and flexibility exercise). Cassilhas et al. (2007) and Liu-Ambrose et al. (2010) also reported resistance training-related improvements in cognition in the elderly. In the former study, both moderate and high load resistance training groups demonstrated significant improvements compared to controls in two out of three WAIS III subtests: similarities and forward digit span (but not backward digit span), along with significant improvements in the backward version of the Corsi block tapping task, and the Rey-Osterrieth complex figure immediate recall. In addition, the experimental groups had higher IGF-1 serum levels than the control group, which could help explain why resistance exercise was associated with improved cognitive functions in older adults (Cassilhas et al., 2007). The latter study (Liu-Ambrose et al., 2010) was based on a 12-month RCT in which women aged 65 to 75 participated in resistance training once ($N = 54$) or twice a week ($N = 52$) or were assigned to a balance and tone training control group ($N = 49$). Both resistance groups significantly improved performance on the Stroop test compared to the control group by 12.6% and 10.9%, respectively. The Stroop test is a widely used measure of selective attention that requires interference resolution, response inhibition, and response selection. Paradoxically, resistance training was also associated with reduced whole-brain volume that the authors attempted to explain in terms of reduced levels of β-amyloid and other proteins, which could have caused shifts in cerebral fluid and, therefore, a reduction in brain volume. The same study was also analyzed in terms of functional plasticity (Liu-Ambrose et al., 2012): before and after the intervention, the participants performed a modified Eriksen flanker task (a test of selective attention and conflict resolution) during which fMRI data was obtained. The results demonstrated that women who participated in resistance training twice a week significantly improved their performance on the flanker task compared to the control group. This was accompanied by functional changes in the anterior portion of the left middle temporal gyrus and the left anterior insula extending into the lateral orbital frontal cortex, regions that have previously been associated with response inhibition processing.

Perhaps the benefits of resistance interventions do not stem from enhanced endurance and resistance alone, but from improved neuromuscular coordination or nonphysical components of exercise. This was investigated in a study that compared two types of intervention in older adults: resistance training versus multicomponent training that was designed to target neuromuscular coordination, balance, agility, and cognitive executive control (Forte et al., 2013). After a 3-month long intervention, both groups demonstrated improved inhibitory capacity, as measured by a number of executive function tests. Mediation analyses revealed that the resistance training affected executive function indirectly via enhanced muscular strength, whereas multicomponent training appeared to have a direct effect on this measure. Even though the sample size was relatively small ($N = 42$) and there was no control group, this study highlights an important issue: we are still far from understanding the mechanisms through which different types of exercise (and parameters such as intensity, duration, and frequency) affect cognitive functions.

Why do studies that use the same type of exercise intervention report discrepant benefits ranging from executive function, processing speed, to memory and attention, yet a clear pattern is not evident? Genetic factors that determine the extent to which regular physical activity affects cognition must also be considered. Not to mention the issue whether potential benefits in cognition persist after training ends and whether one can claim that regular physical exercises affects global intelligence. A metaanalysis of 25 RCTs that examined the effects of aerobic exercise, resistance training, and tai chi on cognitive performance in healthy older adults was able to provide some explanations for the inconsistent results observed across trials (Kelly et al., 2014). The results showed that (1) resistance training was associated with significant improvements on measures of reasoning, compared to stretching/toning, and (2) tai chi training groups outperformed no-exercise control groups on measures of attention and processing speed. However, these results are based on only two tai chi RCT studies. There was no evidence to suggest that aerobic training, contrasted against no-exercise control groups, significantly improved any of the cognitive measures. This finding is not surprising considering the lack of consistent evidence for aerobic training. The authors proposed that exercise interventions should consist of at least 150 min of moderate intensity aerobic activity per week (Nelson et al., 2007) and that longer intervention and follow-up periods should be employed. Indeed, several cross-sectional and longitudinal studies examined in Section 5.1.2 involved participants that engaged in more intense physical activity for longer periods of time. It is possible that interventions ranging from as little as 12 weeks may not be able to affect cognitive functions comparable to the results of longitudinal studies.

5.2.4 Neuroimaging Research

5.2.4.1 Structural Imaging

In Chapter 2, we discussed the relationship between total brain volume and intelligence—the former has been estimated to account for ~12% of variance in global intelligence (Ritchie et al., 2015). Colom et al. (2013) conducted a morphometry study on healthy young adults in which cortical measures were regressed against psychological constructs of interest. Cortical gray matter volume and cortical surface areas in the middle frontal gyrus showed overlap with fluid intelligence and WM; an overlapping cluster in this area was reported also for crystallized intelligence. In addition, the inferior frontal gyrus showed overlap for crystallized intelligence, spatial intelligence, processing speed, and attention (Colom et al., 2013). Hence one might predict that exercise-induced changes in brain volume, particularly in the frontal lobes, may provide the basis for potential changes in core cognitive functions. Before we consider evidence for this, let's look at the results of a longitudinal MRI twin study, which demonstrated significant phenotypic associations between intelligence and brain volume change (Brouwer et al., 2014). At baseline assessment, the average age of 176 participants (monozygotic twins, dizygotic twins, and siblings) was 29 years (range = 19–56 years). DNA testing, brain imaging, and intelligence testing (estimated with the Dutch version of WAIS III) were conducted at baseline and after 5 years. The following changes in brain volume were found to be heritable: total brain volume (mean = −6.4 mL; 0.5% loss); cerebellum (1.4 mL, 1.0% increase), cerebral white matter (4.4 mL, 0.9% increase), lateral ventricles (0.6 mL; 4.8% increase), and surface area (−19.7 cm^2, 1.1% contraction). No significant sex effects on brain volume change were reported. Higher full scale IQ was associated with higher total brain volume, which could also be explained as a reduction in brain volume loss. Specifically, higher verbal IQ was related to an increase (or attenuated loss) in whole brain volume, cerebellar volume, and white matter volume. In contrast, performance IQ was not significantly associated with brain volume change or surface area change. Genes that affect both IQ and brain volume drove most of the correlations. While measuring dynamic changes in brain volume may present a viable method for examining the effect of exercise on components of intelligence, one must bear in mind that the relationship between intelligence and brain volume is not exclusively determined by phenotype, but is also related to the neuroanatomical course of human brain development.

Research on exercise-related structural plasticity in young and middle aged adults will be presented first, followed by studies involving older adults. Correlational studies have suggested a positive relationship between hippocampal size and aerobic fitness in young and middle-aged adults. For example, one study showed that higher aerobic fitness predicted better learning on a virtual Morris Water task, and larger hippocampal volumes in adolescents (Herting and Nagel, 2012). Similar findings were reported for healthy adults aged between 18 and 45 years: self-reported levels of physical activity correlated with gray matter volume of the right hippocampus (Killgore et al., 2013). The mechanisms by which exercise increases hippocampal volume remain unclear, but presumably include increased blood flow and oxygenation and/or increased production of neurotrophic factors such as BDNF and their receptors. Likewise, Whiteman et al. (2016) reported a strong association between cardiorespiratory fitness and right entorhinal volume in healthy young adults. Moreover, average gray matter volume in bilateral entorhinal cortex was positively related to memory performance.

Understanding whether (and how) regular exercise affects structural plasticity and cognitive function is important since longitudinal research shows that higher cardiorespiratory fitness is positively associated with better performance on verbal memory and psychomotor speed tests 25 years later (Zhu et al., 2014). Hence, midlife physical activity may have neuroprotective effects in older age. On the other hand, a four-decade longitudinal study suggested that the positive relationship between midlife cognitive function and objectively measured cardiorespiratory fitness may be explained by childhood performance on the same cognitive tests (Belsky et al., 2015). Most of the association was explained by the finding that children with higher IQ scores grew up in households with higher socioeconomic status, experienced better healthcare, and chose healthier lifestyles in young adulthood than children with lower IQ scores. In particular, lifetime obesity accounted for a large proportion of the association between childhood IQ and adult cardiorespiratory fitness. These findings suggest that midlife cardiorespiratory fitness may reflect neuroselective rather than neuroprotective effects. However, these findings do not exclude the possibility that fitness interventions improve cognitive function via changes in structural plasticity in young adulthood. Yet again we are faced with a lack of such studies. One study allocated 15 students (average age = 21.2) to 30 sessions of aerobic training (dance) and 15 students (average age = 20.9) to a no-intervention control group (Gondoh et al., 2009). VBM analyses before and after the intervention revealed no changes in the training group, whereas the control group showed a decrease in gray matter volume of the left insula. While the authors interpreted this finding as a neuroprotective effect of aerobic exercise in young adulthood, it is also possible that these results are confounded by neuroselection. A cross-over intervention study reported increased anterior hippocampal volume after 6 weeks of aerobic training in healthy sedentary adults (Thomas et al., 2016). These results were best explained by an increase in myelination. However, the increase in hippocampal volume was not permanent, as it returned to baseline after 6 weeks without aerobic exercise. The aerobic training improved cardiovascular fitness, albeit there was no evidence to suggest that it induced changes in cognitive measures or vasculature. Perhaps the participants were not amenable to exercise-induced changes in cognitive function, or the intervention was too short to induce these changes.

The aging process in healthy adults involves deterioration of specific cognitive abilities, such as executive function, processing speed and episodic memory, which in part results from a decline in total brain volume and expansion of the ventricular system. Different brain regions show different rates of shrinkage (Fjell and Walhovd, 2010). For example, a 5-year longitudinal study of healthy adults reported that the greatest shrinkage was observed in the caudate, cerebellum, and hippocampus, followed by tertiary association cortices, and that substantial individual differences in brain volume change existed in most measured regions (Raz et al., 2005). In Chapter 3 we discussed that even short lasting training can induce changes in brain structure in areas related to the training. Research suggests that structural neuroplasticity is not limited to younger adults, as demonstrated by a 3-month juggling study that reported training-related increased gray matter in the hippocampus on the left side and in nucleus accumbens bilaterally in older adults (Boyke et al., 2008). The question that arises is whether exercise could be used to counteract age-related decline in brain volume?

A cross-sectional study of healthy older adults suggested that this may be the case: higher self-reported levels of exercise over the past 10 years were related to larger superior frontal volumes, and appeared to moderate age-related atrophy of the medial temporal lobe (Bugg and Head, 2011). Greater physical activity in old age has also been related to 2%–2.5% greater average tissue volumes in the white matter of corona radiata extending toward the parieto-occipital junction (Ho et al., 2011). Other factors that explained some of the variance in brain structure were educational level and body mass index. More conclusive evidence can be derived from RCTs involving exercise interventions for older adults. For example, a study published in 2006 reported that aerobic exercise increases brain volume in older adults (Colcombe et al., 2006). Specifically, 59 healthy but predominantly sedentary adults aged between 60 and 79 years were randomly allocated either to an aerobic training group or a whole-body toning and stretching control group. During a 6-month period, both groups attended three 1-h-long sessions per week. Twenty younger adults served as methodological controls, and were not assigned to any type of intervention. High resolution structural MRI images were obtained for all participants at baseline and 6 months later. At the start of the study, the two groups of aging adults did not differ with respect to demographic factors, maximal oxygen uptake (VO_2) scores, or MMSE scores. The aerobic training group showed a significant increase in VO_2 after the intervention. While there were no differences in gray or white matter volume in younger adults after 6 months, the aerobic training group showed increased gray and white matter volume in prefrontal and temporal cortices compared to the toning control group, regions that have been previously associated with age-related degeneration. Neuropsychological tests were not administered after the intervention, therefore we cannot draw conclusions about the relationship between volumetric changes and potential changes in cognition.

A number of studies have also focused on the relationship between physical activity, hippocampal volume, and memory. Higher aerobic fitness has been associated with larger hippocampal volume in nondemented older adults, and larger hippocampal volumes are thought to mediate improvements in spatial memory (Erickson et al., 2009). Research further suggests that self-reported levels of physical activity may help preserve hippocampal volume in individuals at genetic risk of Alzheimer's disease (Smith et al., 2014), and that exercise training increases hippocampal volume in healthy older adults (Erickson et al., 2011). The latter study involved a 1-year walking intervention program designed to achieve target heart rate, contrasted against a stretching and toning control program. The control group demonstrated a 1.4% decline in hippocampal volume over the 1-year interval, consistent with typical age-related hippocampal shrinkage, whereas the aerobic intervention group showed increased hippocampal volume by 2%. In addition, changes in anterior hippocampal volume were associated with changes in serum BDNF levels. Interestingly, both groups significantly improved performance on a spatial memory task between baseline and post-intervention. While changes in BDNF did not correlate with improvements in spatial memory for either group, hippocampal volume significantly correlated with the improvement in memory in the exercise intervention group.

On the other hand, Ruscheweyh et al. (2011) did not observe exercise-related changes in hippocampus volume in healthy older adults. The 6-month mild-to-moderate aerobic exercise intervention was related to increases in gray matter volume in the cingulate and prefrontal cortex (derived by VBM), which partly overlapped with the findings reported by Colcombe et al. (2006). Moreover, increased physical activity was associated with improved episodic memory scores, and showed a trend toward significance for peripheral BDNF levels. Since the effectiveness of the exercise intervention was not assessed with VO_2 measures, it remains unclear whether the changes in gray matter volume and episodic memory recall stem from improved aerobic fitness. It should also be stated that changes in gray matter volume in the parahippocampus and the hippocampus would have been difficult to detect with the method utilized in this study: voxel-based morphometry using MRI at 1.5 T.

In order to obtain high functional and structural resolution of the hippocampus, Maass et al. (2015) used 3 T MRI for perfusion imaging and 7 T MRI for structural imaging. Forty sedentary healthy older adults (60–77 years) were randomly allocated to training on a treadmill, or to progressive muscle relaxation/stretching (90 min per week for both groups). The exercise intervention significantly improved aerobic fitness levels (VO_2 measures), which was positively related to changes in hippocampal perfusion and hippocampal head volume. Fitness and perfusion changes were also positively associated with changes in recognition memory and early recall for configural spatial object memory, but not with delayed recall or verbal memory. The authors also observed that exercise-related improvement in perfusion declined with age, suggesting that younger adults may show greater improvements in vascular hippocampal plasticity after physical exercise. The same research group published another paper based on this data (Maass et al., 2016) in which they explored whether the observed improvements were related to changes in growth factors (i.e., peripheral BDNF, IGF, VEGF, and PDGF). Aerobic exercise did not affect levels of growth factors in the blood, with the exception of a significant decrease in IGF-II levels. Fitness-related benefits in perfusion of the hippocampus and hippocampal volume were also not related to growth factors. These findings indicate that increased perfusion may have enhanced neural function via mechanisms other than angiogenesis, neurogenesis, or synaptogenesis. Interestingly, changes in IGF-1, hippocampal volume, and verbal recall performance were related independent of exercise. It was suggested that other factors such as nutrition or glucose metabolism may have affected IGF-1 levels. Given the small sample size and relatively short intervention period, it is also possible that there was insufficient power to detect effects of exercise-induced vascular hippocampal plasticity in relation to growth factors.

A 5-year prospective study of changes in brain structure in older adults reported that maintenance of brain volume was associated with total physical activity (indicated by accelerometry), and that white matter atrophy was associated with sedentary behavior, independent of lifestyle physical activity (Arnardottir et al., 2016). These findings suggest that it is worth investigating whether physical activity interventions can help weaken the association between brain atrophy and sedentary behavior. Presently, a limited amount of evidence suggests that exercise interventions can induce structural changes in frontal and medial temporal cortices, areas that show the greatest age-related losses in structural integrity. It is difficult to ascertain the extent to which these changes relate to potential improvements (or preservation) of cognitive function, let alone predict whether any benefits would remain present after the end of an intervention period.

5.2.4.2 Functional Imaging

Is cardiovascular fitness in older adults associated with a more plastic and efficient brain, which translates to better performance on cognitive tasks? Colcombe et al. (2004) investigated whether this is the case in an fMRI study of attentional control, as measured by the Erikson flanker paradigm. The results demonstrated that highly fit subjects (cross-sectional design) and aerobically trained subjects (a 6-month exercise intervention) showed greater task-related activity in prefrontal and parietal regions that have been previously associated with spatial selection and inhibitory control, compared to low-fit or stretching and toning control subjects, respectively. Both groups (fit and trained), also responded more efficiently to conflicting cues presented in the flanker paradigm (i.e., they showed less interference) compared to the control groups. Thus it is not surprising that high aerobic fitness was also associated with a lower amount of activity in the anterior cingulate cortex (ACC), an area that tends to be active during response conflict. The benefit of the next exercise intervention that will be presented is a 2-year follow-up that focused on brain activation during a cognitive task. Specifically, the 1-year RCT for sedentary elders (titled LIFE-p), which involved either physical activity or successful aging health education, demonstrated that (1) physical activity significantly improved scores on a standardized measure of lower extremity performance compared to health education (Pahor et al., 2006), (2) group differences in cognitive function were not significant after the intervention, however, improvements in physical function correlated with improvements in cognitive scores, and (Williamson et al., 2009), (3) physical activity participants that remained active 2 years after completing the LIFE-p program performed significantly better on a test of psychomotor performance and displayed higher brain activation within brain regions important for processing speed (left dorsolateral prefrontal, posterior parietal, and anterior cingulate cortices) compared to regions active in the health education group. In addition, greater physical activity levels were related to greater fMRI activation in the inferior frontal gyrus 2 years after the intervention ended (Rosano et al., 2010).

Unlike task-related brain activation studies, research on functional connectivity between regions during rest (or low cognitive demand) does not suffer from certain limitations such as individual differences in performance on behavioral tasks, strategy, etc. For example, Voss et al. (2010) conducted a 1-year-long intervention in which older adults were assigned to an aerobic walking program, or a control group that joined a flexibility, toning, and balancing program. After the intervention, the walking group displayed increased functional connectivity between regions supporting the default mode network and the frontal executive network. This was interpreted as an increase in task-independent differentiation between the executive and the default mode networks. Changes in connectivity in the default mode network were positively related to executive function performance. However, the walking group only showed a trend toward a significant improvement in executive function performance compared to the control group.

Interestingly, the control group showed increased connectivity in regions of the default mode network at 6 months, and increased connectivity in the fronto-parietal network at 12 months. Increased connectivity in the DMN could be explained by the fact that this network tends to be active when an individual is focused inward to the self, which might have been stimulated by the yoga-like training. These findings suggest that exercise interventions can influence functional connectivity in the aging brain. The extent to which these changes influence cognitive functions remains debatable. The former statement was supported by the results of a study published by Burdette et al. (2010): older adults who participated in a 4-month exercise intervention showed increased hippocampal cerebral blood flow, and improved connectivity in the hippocampus compared to a control group. The exercise group also showed interconnectivity between the hippocampus and the anterior cingulate gyrus that was not present in the control group. However, the authors only analyzed that data of 11 participants, and did not relate changes in functional brain networks to measures of cognitive function. Voss and colleagues published another study in 2013 in which they examined the effect of a 1-year aerobic training intervention on growth factor levels and functional connectivity in healthy older adults. Increased connectivity in the temporal lobe, specifically between the bilateral parahippocampus and the bilateral middle temporal gyrus, was related to increased growth factors in the aerobic group, but not in the nonaerobic control group (Voss et al., 2013). Finally, Voss et al. (2015) observed that in healthy older adults, higher cardiorespiratory fitness is related to greater functional connectivity in a number of brain networks—the effect was strongest for regions within the default mode network. The results also suggested that the positive relationship between cardiorespiratory fitness and functional connectivity may be independent of habitual physical activity, highlighting the fact that substantial individual differences exist in the responsiveness of cardiorespiratory fitness to training. These differences can be partially explained by genetics and by the fact that regular exercise can be achieved without reducing sedentary behavior. As mentioned previously, sedentary behavior has been related to brain atrophy in older adults (Arnardottir et al., 2016). Since aging is associated with decreased functional connectivity in DMN, frontoparietal, and fronto-executive networks, along with a reduction in growth factor levels, findings that exercise can, to a certain extent, counteract these changes present an important step toward understanding the neuroprotective effects of exercise on declining cognitive function.

To conclude, while there is little doubt that voluntary exercise in rodents results in enhanced spatial learning and hippocampal plasticity, the relationship between physical activity and cognition appears to be far more complex in humans. The majority of cross-sectional and longitudinal studies suggest that cardiovascular fitness is positively associated with performance on tests of cognitive function in adulthood. Metaanalyses of observational studies suggest that regular physical activity can alleviate and/or help delay cognitive decline and dementia, even in individuals that are genetically susceptible to them. Less conclusive evidence can be derived from intervention studies, most of which are based on aerobic exercise programs. For older adults, two metaanalyses concluded that aerobic intervention studies positively affect cognitive functions, such as attention, processing speed, executive function, and memory (Hindin and Zelinski, 2012; Smith et al., 2010). In contrast, two metaanalyses reported that there is insufficient evidence to support this claim (Angevaren et al., 2008; Snowden et al., 2011). Balance and flexibility training, multicomponent training, and resistance training interventions have also been associated with improvement in cognitive function, however, more research is required to verify these findings. Neuroimaging research suggests that exercise interventions induce functional and structural changes in frontal and medial temporal lobes, areas that are susceptible to age-related atrophy. There is only limited evidence to suggest that these neuroprotective effects translate to improved cognitive function. In order to directly compare exercise intervention studies, procedures and outcome measures should be harmonized in a way that enables maximum efficacy. Moreover, it is unclear which type of exercise program provides the greatest cognitive benefits and to what extent these changes can be explained by mental and social components of exercise. In fact, Moreau et al. (2015) reported that exercise that included cognitive challenges resulted in greater improvements in spatial ability and working memory scores compared to aerobic exercise that did not require top down control. Despite some inconsistencies, at least 150 min of moderate intensity physical activity per week seems to be one of the best methods currently available to promote brain health and counteract age-related cognitive decline.

5.3 MEDITATION

Meditation originates from Vedic, Buddhist, and Taoist traditions, and typically involves attentional practice aimed at achieving a wakeful and tranquil state. Different types of meditation include transcendental meditation, mindfulness, zen meditation, vihangam yoga, and tai chi, to name a few. The two most widely researched meditation techniques are transcendental meditation and mindfulness (Walsh and Shapiro, 2006). Transcendental meditation was introduced by Maharishi Mahesh Yogi, and involves the use of a mantra that quiets the mind until it achieves a calm and silent state of transcendental consciousness—a process called transcending (Forem, 2012). Mindfulness meditation emerged from Buddhist traditions, and involves nonjudgmental attention to experiences in the present moment (Tang et al., 2015). Vipassana meditation, dzogchen, and zen are examples of traditional techniques that inspired the development of mindfulness-based approaches such as integrative body–mind training (IBMT) and mindfulness-based stress reduction (MBSR).

Many researchers agree that there are at least three categories of meditation techniques: focused attention, open monitoring, and loving-kindness or compassion meditation (Hofmann et al., 2011; Lutz et al., 2008). Focused attention involves directing and sustaining attention to a selected object (e.g., breathing or a mantra), detecting distractors, and shifting attention back to the object. At advanced stages, regulative skills are exercised less frequently and sustained focus becomes "effortless." In contrast, open monitoring does not involve focusing on an object, but entails nonreactive monitoring from moment to moment. The goal is to gain awareness of cognitive and emotional interpretations of endogenous and exogenous stimuli. Loving-kindness and compassion meditations focus on fostering unconditional kindness to all people and cultivating compassionate attitudes, thereby promoting altruistic behavior. These techniques are often used interchangeably within one type of meditation. While the three categories fail to encompass the complexity and unique components of various types of meditation, they can help synthesize heterogeneous research on the psychological and neurophysiological effects of meditation.

Most of the studies that will be discussed in this section evaluate the effectiveness of meditation training programs for beginners, contrasted against an active or no-treatment control group. However, research based on expert meditation practitioners may also help explain what happens in the brain during contemplative practice. For example, Carter et al. (2005) conducted a study supported by His Holiness the Dalai Lama that involved 76 Tibetan Buddhist monks. The monks varied in age (18–63) and years of meditation training (5–64), and had no experience with psychophysical testing. In order to examine the effects of meditation on perceptual rivalry with minimal disruption to the monks, the testing took place at, or near, their mountain retreats in India. In the first part of the study, binocular rivalry was induced with the help of display goggles: horizontal gratings were presented to the right eye and vertical gratings were presented to the left eye. Instead of seeing a blend of the two images, participants usually report perceptual alternations between the images. Binocular rivalry was assessed in 23 monks during (or in some cases after) two types of meditation: compassion and one-point (focused attention). As predicted, the former did not affect change in rivalry rate. The results for the latter were striking: most monks were able to control their perception rivalry during and after one-point meditation (i.e., they reported increases in perceptual dominance durations). Three monks even reported complete perceptual stability throughout the 5-min meditation period. In the second part of the study, motion induced blindness was examined—a phenomenon in which small objects "disappear" when they are surrounded by a moving pattern. In total, 61 nonmeditators (age range: 21–50 years), who were also naïve to psychophysical testing, were recruited to serve as controls. The results showed that the monks were able to maintain the disappearance of the objects for a significantly longer time than the control participants. This study was the first study to demonstrate that meditation can alter the normal fluctuations in conscious state induced by perceptual rivalry.

The following two subsections will focus on the effects of meditation on cognition and neuroplasticity. These are by no means the only dependent measures reported in the literature: anxiety, wellbeing, emotion regulation, positive and negative emotions, stress, self-realization, and interpersonal relationships have also been examined. For example, a metaanalysis of 21 psychological effects of meditation on healthy adults (number of studies = 163) showed the largest mean-weighted effect sizes for positive changes in relationships (interpersonal: $r = 0.44$), state anxiety ($r = 0.37$), negative emotions ($r = 0.34$), and trait anxiety ($r = 0.32$), medium beneficial effects for measures of attention ($r = 0.28$) and stress ($r = 0.27$), and the lowest effects for measures of intelligence ($r = 0.22$), learning and memory ($r = 0.21$), negative personality traits ($r = 0.18$), and emotion regulation ($r = 0.17$). However, the methodological quality of the studies varied—for example, several studies did not report whether the participants were randomly allocated to a meditation or control group. The results also showed different effect sizes with respect to publication outlet (articles vs. book chapters), and with respect to type of meditation, for example, transcendental versus mindfulness meditation (Sedlmeier et al., 2012). Given that relaxation control groups and no-treatment controls barely differed, the authors were able to conclude that meditation is more than just a relaxation technique and that it has a substantial effect on psychological variables, indicated by a medium-sized global effect. Finally, the results suggested that meditation affects negative emotional variables more strongly than cognitive variables.

A more recent metaanalysis of RCTs with active control groups found that meditation can reduce negative dimensions of psychological stress, and that mindfulness meditation in particular is associated with moderate improvements in anxiety, depression, and pain, and small improvements in stress/distress and mental-health related quality of life. There was low or insufficient evidence for the effectiveness of any of the meditation programs on positive mood, attention, substance abuse, eating habits, sleep, and weight (Goyal et al., 2014b). This systematic review triggered a debate about the use of RCTs with active controls. On the one side, Walach et al. (2014) claimed that randomization diminishes some of the crucial components of meditation: conscious choice and active engagement, and that long-term cohort studies provide valuable information that cannot be produced by RCTs. In response, Goyal et al. (2014a) argued that while cohort studies can help establish outcomes of a risk (or risk reduction) factor, active controls in RCT studies control for nonspecific effects of time, attention, and expectations for improvement—analogous to the use of placebos in pharmaceutical trials. However, even RCTs suffer from biases such as lack of blinding of outcome assessment, high attrition, lack of allocation concealment, and lack of intention-to-treat analysis (Goyal et al., 2014b). There is no doubt that meditation research would benefit from more rigorous studies, both RCTs with active control groups and long-term cohort studies.

There has been an exponential increase in the number of neuroimaging studies of meditation, most of which focus on mindfulness-based approaches and transcendental meditation. Meditation has been associated with altered sensory, cognitive, and neurophysiological effects, that is, state effects, and with trait effects that refer to lasting effects that transcend the practice period. A systematic review of EEG studies of state and trait effects included studies with different experimental designs: meditation compared to rest or control periods, pre- versus postmeditation effects, and meditators versus controls. The results revealed increased theta band power and decreased frequency in the alpha band, increased alpha–theta range coherence, along with various gamma band effects; however, no clear consensus could be reached. Some studies also suggested that meditation increases attentional resources and alters the P300 component (Cahn and Polich, 2006). A recent metaanalysis of MRI studies revealed that meditation is associated with activation in brain areas involved in processing self-relevant information (e.g., the precuneus), self-regulation, focused problem solving, adaptive behavior (e.g., the anterior cingulate), in interoception (e.g., the insula) and in reorienting attention (e.g., the angular gyrus). Hence, functional and structural brain changes may underlie the general effects of meditation observed in everyday life (Boccia et al., 2015).

Indeed, there is accumulating evidence for structural differences between the brains of meditators and nonmeditators. In comparison to matched controls, meditation practitioners show increased cortical thickness (Grant et al., 2010; Kang et al., 2013; Lazar et al., 2005), gray matter volume (Hernández et al., 2016; Leung et al., 2013; Luders et al., 2009; Pagnoni and Cekic, 2007), and gray matter density (Hölzel et al., 2008; Vestergaard-Poulsen et al., 2009) in several brain areas, including but not limited to the ventrolateral prefrontal cortex, rostrolateral prefrontal cortex, inferior temporal gyrus, insular cortex, anterior cingulate cortex, fusiform gyrus, and the hippocampus. Greater structural connectivity (fractional anisotropy values) in meditators compared to controls has been reported throughout the entire brain within projection, commissural, and association pathways (Luders et al., 2011), corpus callosum (Luders et al., 2012), and regions of interest associated with mindfulness—thalamus, insula, amygdala, hippocampus (Laneri et al., 2015). Due to the cross-sectional nature of these studies, causal relations between meditation and changes in brain morphometry cannot be drawn. Studies that investigate pre–post changes in brain structure associated with meditation interventions can help explain these relations, however, few such studies have been conducted. For example, Hölzel et al. (2011) reported that an MBSR intervention led to increases in gray matter density in the posterior cingulate cortex, the temporoparietal junction, and the cerebellum compared to a wait-list control group. In addition, increased fractional anisotropy in the anterior cingulate has been reported after only 11 h of integrative mind–body training (Tang et al., 2010). Future research on meditation-related changes in brain function and structure should employ longitudinal designs, particularly randomized controlled designs with active control groups and long-term follow-up.

5.3.1 Effects of Meditation Training on Intelligence

Even though meditation-induced changes in cognitive function appear to be secondary to its beneficial effect on negative affect, it is worth investigating whether regular meditative practice improves general cognitive ability. To our knowledge, there are only a handful of studies that examine the effect of meditation training on measures of intelligence. First, research based on adolescents and young adults will be presented, followed by studies that involved middle-aged and older adults.

Two longitudinal studies published in 1981 and 1986 reported improved performance on IQ tests among university students that joined a transcendental meditation (TM) and TM-Sidhi program (Aron, 1981; Dillbeck et al., 1986). However, neither study used a control group therefore it is possible that the increases in IQ scores reflected general effects of education or even test-retest effects. For this reason, Cranson et al. (1991) conducted a 2-year longitudinal study that investigated the effect of TM and TM-Sidhi program on Cattell's culture fair intelligence test (CFIT) and Hick's reaction time, contrasted against a nonmeditating control group. It should be noted that the students were not randomly assigned to the experimental and control group; 45 students enrolled in a twice daily TM and TM-Sidhi program, while 55 students served as a no-treatment control group. The experimental group demonstrated significant improvement on the CFIT ($f = 2.79$, $p < 0.005$) and choice reaction time ($t = 9.10$, $p < 0.0001$) after 2 years of meditation compared to the control group. Baseline IQ scores did not differ significantly between the two groups, nor did potential confounds related to performance on IQ measures (e.g., age, education level, father's income) account for the results. Nevertheless, it is possible that some of these effects stemmed from nonspecific effects of the meditation program, such as social support, placebo and expectancy effects, demand characteristics, and experimenter bias. An active control group could have accounted for some of these effects.

So and Orme-Johnson (2001) conducted three randomized experiments on 362 high school students at two different schools in Taiwan. In the first study, male and female students were given a lecture on TM and among those who expressed interest, half were randomly assigned to learn the TM technique, whereas the others were allocated to a waitlist control group that napped while the experimental group meditated. Those who were not interested in learning the meditation technique were assigned to a no-interest group. In the second study, female students were allocated to a TM group, a no-treatment control group, or to a contemplation meditation control group. In the third study, male students were assigned either to a TM group or to a no-treatment control group. The experimental group practiced the technique for 15–20 min twice a day for 6 months (studies 1 and 2) or for 12 months (study 3). Before and after the intervention, the students completed the following tests: Cattell's Culture Fair Intelligence Test (CFIT), inspection time (IT), constructive thinking inventory (CTI), group embedded figures test (GEFT), test for creative thinking-drawing production (TCT-DP), and state-trait anxiety inventory (STAI). Importantly, the students, teachers, and test-administrators were not told that these tests were part of the meditation study, but believed that they were part of a standard school evaluation program. Average effect sizes across the three experiments for the TM groups compared to the control groups showed post-intervention improvement on all tests: creativity ($ES_{TCT\text{-}DP}$ = 0.77), practical intelligence (ES_{CTI} = 0.62), field independence (ES_{GEFT} = 0.58), state and trait anxiety (ES_{STAI} = 0.53, and ES_{STAI} = 0.52, respectively), inspection time (ES_{IT} = 0.39), and fluid intelligence (ES_{CFIT} = 0.34). The authors hypothesized that TM contributed to these effects by inducing "restful alertness" that is supposedly characterized by decreased metabolism, heart rate, and respiration rate. Interestingly, the results of the second study demonstrated that contemplation meditation increased performance on inspection time and field independence, but not on other measures. Given that the first study included a napping control group, which serves as a better control than no treatment, the results of study 1 should be examined separately. Namely, the TM group improved significantly on 6 out of 7 tests, particularly on measures of field independence, inspection time, and creativity. However, there was no significant difference between TM and napping in fluid intelligence scores obtained on the CFIT.

In a more recent RCT, 6 weeks of integrative body–mind training[d] significantly improved attention and academic performance of 13–18 year old students compared to a relaxation training control group (Tang et al., 2014). Moreover, only the integrative mind-body training students showed a significant increase in Raven scores after training. These results were not supported by a study that used older participants and a shorter intervention period (Paholpak et al., 2012). Specifically, medical students were randomly allocated to a breathing meditation group (N = 30) or to a no-intervention control group (N = 28). The experimental group meditated for 20 min/day for 4 weeks—a relatively short period. After the intervention, the groups did not differ with respect to psychiatric symptoms, memory, intelligence, and academic achievement. All in all, the evidence is mixed thus it remains unclear whether meditation improves general cognitive ability in adolescents and young adults. Nevertheless, the idea that meditation could be introduced in schools in order to reduce stress and positively affect mental health, perhaps even boost academic performance, has become increasingly popular in recent years. The outcomes of meditation interventions in schools will be discussed in Section 5.3.2.1, albeit limited to studies that measured changes in cognitive performance.

Hereafter, the discussion will focus on the effects of meditation on measures of intelligence in middle-aged and older adults. Paralleled by experts claiming that regular exercise can help reduce age-related cognitive decline, scientists (mainly those who practice meditation themselves) have proposed that meditation might have a similar protective effect. Dr. Sanford Nidich, a professor at Maharishi University of Management, was one of the first researchers to study the physiological and psychological effects of TM in the 1970s. He has since published a number of papers, some of which focused on intellectual

d. A type of meditation that consists of body relaxation, mental imagery, and mindfulness training.

development. In one of his studies, cognitive abilities of older adults who had been practicing TM for over 10 years were compared to those of nonmeditating elderly adults (Nidich et al., 2005). Both groups solved tests that are part of the Woodcock-Johnson tests of cognitive ability, and gave blood samples that were analyzed for lipid peroxide.[e] Older adults practicing TM ($N = 18$) demonstrated higher levels of fluid intelligence, long-term memory, speed of processing ability, and verbal intelligence compared to elderly controls ($N = 23$). Lower levels of lipid peroxide were related to higher scores on measures of long-term memory and fluid intelligence. Despite the fact that the groups did not differ in terms of age, gender distribution, or level of education, a number of other lifestyle factors such as exercise habits, diet, and cognitive engagement could have mediated the results.

The effects of meditation on age-related decline in cognitive performance was also investigated by Gard et al. (2014), who recruited 16 experienced Kripalu yoga practitioners (average age = 49.4 years), 16 insight meditation practitioners (average age = 54.1 years), and 15 controls who had no experience with yoga or meditation (average age = 52.9 years). The groups were matched for gender, education, race, and handedness, and did not significantly differ based on current number of hours spent exercising and engaged in cognitive activities. Cognitive performance was assessed with five scales, including Raven's advanced progressive matrices (a measure of fluid intelligence), and the American version of the national adult reading test (a measure of verbal intelligence). Scores on the latter did not significantly differ between groups, indicating that they were well matched. In addition, resting state functional brain connectivity was examined with graph-based analyses. The results showed that yoga practitioners and meditators considered as a single group had a lower rate of decline of fluid intelligence than controls (permutation test; $p = 0.04$), which was mostly driven by the difference between yoga practitioners and controls (permutation test; $p = 0.06$). There was no significant difference in rate of decline of fluid intelligence between yoga practitioners and meditators. In addition, resting state functional brain networks of yoga practitioners and meditators combined were more resilient to simulated damage than those of controls. The two groups also had significantly shorter characteristic path length, greater global efficiency, and greater small-worldness than controls (Gard et al., 2014). These findings provide a first step toward understanding the effects of meditation and yoga on age-related decline in fluid intelligence that will hopefully be followed by more studies—as of now, information in the published literature is limited.

The next study that will be described did not measure meditation-related changes in intelligence scores, nevertheless, it can help explain the mechanisms that underlie the protective effects of meditation against cognitive decline.

e. Biomarkers of lipid peroxidation can be used to assess oxidative stress status, disease state, and effects of drugs and antioxidants (Niki, 2014)

Luders et al. (2016) used machine learning to determine anatomical correlates of age in meditators and control subjects. The results showed that at the age of 50, the brains of meditators were estimated to be 7.5 years younger than those of controls. According to the authors, dendritogenesis, synaptogenesis, myelinogenesis, or even neurogenesis might underlie meditation-related changes in brain tissue. Furthermore, Innes and Selfe (2014) suggested that meditation could help delay or prevent cognitive decline by (1) reducing activation and reactivity of the sympathoadrenal system and the HPA axis, thereby alleviating the effects of stress, (2) shifting the autonomic nervous system balance from primarily sympathetic to parasympathetic, leading to positive changes in mood, sleep, and energy state, and (3) stimulating telomerase activity, which could help preserve immune function and potentially reduce neuronal loss.

5.3.2 Effects of Meditation Training on Attention, Cognitive Inhibition, and Working Memory

5.3.2.1 *School-Based Interventions*

Albeit meditation is not generally included in the school curriculum in the Western world, its use is on the rise. Since meditation should only be introduced in schools if it is associated with positive student outcomes, adequate research is needed to help evaluate the effectiveness of meditation interventions in schools (Waters et al., 2015). Waters et al. (2015) conducted a systematic review of studies published in peer-reviewed journals that report the results of school-based meditation interventions. Fifteen studies met their criteria, only two of which investigated the relationship between meditation and performance on standardized academic tests. Namely, Nidich and Nidich (1989) investigated the effect of a 1-year-long TM intervention on academic achievement in third to seventh graders ($N = 42$). A within-group comparison showed significantly increased scores on Iowa tests of basic skills and Iowa tests of educational development for all students. Effect sizes were not reported. The second study utilized a quasiexperimental design to examine the effect of a 3-month (twice daily) transcendental meditation intervention on academic achievement in an urban public middle school (Nidich et al., 2011). Participants included 125 meditating students and 64 nonmeditating control students, who were below proficiency level in either Mathematics or English at baseline on the California Standards Test, and a matched-control subgroup of 100 students (50 meditating and 50 nonmeditating), who were also below proficiency in Mathematics and English. Results for the first group of participants showed significant improvements in Mathematics ($ES = 0.82$) and English scores ($ES = 0.44$) on the California standards test, compared to controls. Similar results were reported for the matched-control subgroup in which the two groups were equated at baseline. The results of the systematic review also indicated that meditation induces small improvements in wellbeing (metaanalysis of nine studies) and social competence (metaanalysis of six studies). The authors proposed a conceptual model in which meditation increases student success (wellbeing, social competence, and academic achievement) via two pathways: by increasing cognitive functioning, and by increasing emotional regulation (Waters et al., 2015). Given that many of these studies evaluated the meditation program in the pilot stage, it is not known whether the benefits sustained over time. As this is often the case in meditation research, many studies did not use a randomized design.

Three RCTs published after the systematic analysis provided mixed results for the efficacy of school-based meditation programs. In the first RCT, classroom-based mindfulness meditation was compared to an active control condition (nondidactive, experiential activities) in 6th grade children. The results of the 6-week intervention revealed that (1) both groups produced equally large decreases in clinical symptoms and attention problems, (2) meditators showed greater improvements over control students on measures of affect, but the differences were not statistically significant, and (3) meditators showed larger improvements than controls on measures of suicidal ideation (Britton et al., 2014). The second study was a pilot that tested cognitive effects of mindfulness practice (Wimmer et al., 2016). Thirty-four students (average age = 10.8 years) were allocated randomly to a mindfulness intervention group ($N = 16$) or to a concentration training group ($N = 8$). A passive control group was also included that consisted of 10 students from a parallel class in the same school. Before and after the 18-week intervention (25 training sessions), the students completed a number of cognitive tests. While performance on some tests did not show any group differences (reversible figures task, WCST), mindfulness training benefited cognitive inhibition and data-driven information processing. The third RCT investigated the effect of mindfulness meditation on working memory capacity in 198 adolescents aged from 12 to 15 years attending school in a lower-SES neighborhood (Quach et al., 2016). After completing baseline assessment, the participants were randomly allocated to mindfulness meditation ($N = 54$), hatha yoga ($N = 65$), or a waitlist control condition ($N = 53$). The interventions groups had two 45-min sessions per week for 4 weeks that were delivered during their physical education class periods. The students were also encouraged to meditate/practice yoga at home on a daily basis and to keep practice logs. The waitlist control group attended regular PE classes. Working memory capacity was measured with an automated operation span task, which involves performing a memory task while answering mathematics equation questions. Results demonstrated a significant increase in WM capacity for participants in the meditation group ($p < 0.001$), whereas the hatha yoga and waitlist control groups did not show significant changes. In addition, all three groups exhibited significantly reduced stress and anxiety after the intervention, suggesting that mindfulness meditation and hatha yoga were not superior in their ability to reduce stress and anxiety. The results of this study suggest that even short mindfulness meditation interventions can improve working memory capacity in adolescents. These findings need to be replicated in more diverse populations; it would also be interesting to investigate whether changes in working memory capacity affect overall academic performance.

To conclude, school-based meditation interventions differ substantially in duration and intensity of training, type of meditation, inclusion of control groups, research design, age of participants, and use of dependent variables. Methodological differences can, to a certain extent, explain the disparate findings. While there appears to be some preliminary evidence that school-based meditation positively affects certain cognitive functions, the enthusiasm of introducing meditation programs in schools surpasses empirical support for the efficacy of these programs.

5.3.2.2 Community-Based Interventions

Research on the effects of meditation training on specific cognitive functions such as attention, cognitive flexibility, and WM can help explain the relationship between meditative practice and general cognitive ability.

A major component of meditation is mental training of attention, hence it is reasonable to assume that practice of meditation improves attentional function. The question is whether mental training of certain attentional skills enhances performance on novel tasks that do not require meditation. In order to investigate this, Slagter et al. (2007) examined the effects of 3 months of intensive Vipassana meditation on the distribution of limited attentional resources in meditation practitioners. The control group consisted of novices that received a 1-hour meditation class and were asked to meditate daily for 1 week preceding the two EEG sessions. In each EEG session, the participants solved an attentional blink task in which they had to identify two targets amongst a rapid stream of distracters. When the second target is presented within half a second of the first one, it is often not detected due to competition for limited attentional resource—this has been termed the attentional-blink deficit. The practitioner group showed a significant reduction in attentional-blink magnitude compared to the novice group. Moreover, intensive meditation reduced brain-resource allocation to the first target, as indicated by a reduction in P3b amplitude in response to the first target. Importantly, participants that showed the greatest reduction in P3b amplitude demonstrated the greatest reduction in attentional blink. These results suggest that meditation training can increase control over the distribution of limited brain resources. An analysis of the same EEG data published at a later date revealed that meditation training was associated with enhanced theta phase locking after successfully detected targets, which enabled more rapid processing of new target information (Slagter et al., 2009).

Jha et al. (2007) investigated the effects of meditation on different components of attention: conflict monitoring, orienting, and alerting. Three groups of individuals were recruited for the study: (1) experienced practitioners ($N = 17$) that attended a 1-month-long intensive mindfulness retreat, (2) meditation novices ($N = 17$) that participated in an 8-week mindfulness-based stress reduction program, and (3) nonmeditators that did not receive any treatment ($N = 17$). The subjects in all three groups participated in two testing sessions (baseline and posttraining), in which they solved the attention network test. The test involves determining the direction of an arrow (pointing left or right) flanked by four other arrows. As predicted, baseline performance of experienced mindfulness practitioners was better than that of the other two groups (i.e., they showed improved conflict monitoring). However, no significant differences in conflict monitoring were observed among the three groups in the second testing session—a paradox given that one would expect to find a difference in relation to the control group. After the retreat, the meditation practitioners demonstrated improved altering performance compared to the mindfulness-based stress reduction program and control participants. In contrast, the mindfulness-based stress-reduction training resulted in improved orienting relative to the retreat and control groups. These findings must be interpreted with caution; at the very least, they suggest that the relationship between mindfulness training and attentional subsystems is much more complex than previously thought.

Another study that examined the effects of meditation training on performance on the attention network test reported a significantly larger improvement in conflict scores after 5 days of integrative mind-body training than a relaxation control group (Tang et al., 2007). However, there were no differences between the two groups on either alerting or orienting efficiency scores. Elliott et al. (2014) used a modified version of the attention network test to demonstrate that an intensive week-long meditation retreat can improve the efficiency of executive attention and decouple the executive and alerting networks. In addition, MacLean et al. (2010) reported that a 3-month meditation training retreat improved performance on a visual discrimination task that was maintained after a 5-month follow up. It should be noted that self-selection might present a problem for these types of studies: people who believe in the benefits of meditation are more likely to apply for a meditation retreat and report that they benefited from one (Goyal et al., 2014a,b). For example, the participants in the study by Elliott et al. (2014) paid between $600 and $1095 for a 7 day retreat that could have provided further incentive for increasing test effort after the training ended. As stressed by Davidson and Kaszniak (2015), one of the major limitations of meditation research is that double-blind placebo-controlled designs cannot be implemented because the participants that are assigned to meditation training are aware of the nature of the intervention. Active control conditions that match the meditation condition in all nonspecific factors and blinding procedures represent the best solutions to this problem.

Jensen et al. (2012) also observed that previous meditation studies did not control for attentional effort or the effects of stress reduction. Attentional effort refers to the allocation of additional attentional resources in response to deteriorating attentional performance on a task with the goal of improving said performance. In light of these issues, Jensen et al. (2012) conducted a blinded, randomized study in which they investigated the effects of mindfulness-based stress reduction on attention. Three groups (16 participants per group; all were meditation novices) that were balanced for age, sex, marital status, education, and perceived stress, were allocated to one of the following groups: mindfulness-based stress reduction, nonmindfulness stress reduction, or an inactive control group. Attentional effort was manipulated by offering a financial incentive to half of the participants in the inactive control group. Pre- and post-intervention testing (intervention duration = 8 weeks) involved five tests of attention, questionnaires, and measurement of saliva cortisol levels in response to awakening. The latter two measurements confirmed that the mindfulness-based stress reduction intervention was effective in reducing stress. As intended, the nonmindfulness program also reduced stress but did not affect mindfulness. Mindfulness-based stress reduction improved the ability to maintain a selective attentional focus significantly more than in any other group. While the groups did not significantly differ in WM capacity in the two testing sessions, only the participants in the mindfulness-based stress reduction group showed a significant improvement in working memory capacity after the intervention. On the other hand, simply increasing test effort, and the nonmindfulness stress reduction program, produced even larger effects than the mindfulness-based approach on measures of temporal orienting, sustained readiness to react, and attentional set shifting. For example, the controls that received a financial incentive showed great improvement in attentional set shifting and even outperformed the mindfulness group. In general, raw reaction time scores were particularly vulnerable to attentional effort. The results of this study substantiate the importance of controlling for attentional effort and nonspecific reductions of stress.

Given that meditation has been equated with attention training, it is surprising that there are relatively few behavioral studies that show solid empirical evidence in favor of meditation-related increases in attentional functions. Converging evidence from functional and structural neuroimaging studies suggests that meditation practice is associated with (1) activations in areas supporting attentional functions (for a review, see Tomasino et al., 2013), and (2) changes in brain morphology in the same networks (Fox et al., 2014). The metaanalysis performed by Fox et al. (2014) did not show statistically significant correlations between structural differences and total meditation practice time, and other behavioral measures, suggesting that further research is needed to understand the effects of long-term meditative practice on attentional function.

There is evidence to suggest that meditation optimizes cognitive control (Colzato et al., 2016) and improves the ability to suppress an automatic response (Moore and Malinowski, 2009). A review of mindfulness meditation-related benefits to executive functioning revealed consistent benefits for inhibition, but more variable advantages for set shifting and WM updating (Gallant, 2016). The latter has been explored by a limited number of pilot studies (Jha et al., 2010; Mrazek et al., 2013; van Vugt and Jha, 2011). Jha et al. (2010) conducted a quasi-experiment to examine whether 8 weeks of mindfulness training might be beneficial to US Marine Corps reservists preparing for deployment. One military cohort received mindfulness training ($N = 31$) during the high-stress predeployment period, whereas the other did not ($N = 17$)—a civilian control group was also recruited ($N = 12$). WM capacity was assessed with Ospan, a task that requires solving math problems while trying to remember a sequence of unrelated letters. Participants also completed a questionnaire for positive and negative affect. The results indicated that WM capacity deteriorated over time within the military cohort that did not receive mindfulness training, whereas these scores remained stable in the civilian cohort. A similar pattern of depletion in WM capacity was observed in the military cohort that received mindfulness training, but spent little time engaging in mindfulness training exercises. In contrast, the high practice mindfulness training group showed modest improvements in WM capacity, which corresponded with reduction in negative affect. Jha conducted another study a year later (van Vugt and Jha, 2011), in which an intensive 1-month mindfulness training resulted in faster and less variable response times on a delayed-recognition WM task compared to age- and education-matched controls who received no training. Finally, a RCT that investigated the effects of a 2-week mindfulness training program in undergraduate students reported significantly increased reading-comprehension scores and working memory capacity compared to a nutrition training control program (Mrazek et al., 2013).

Meditation research has suffered from low methodological quality in the past. It is well known that many researchers are avid meditators, and this may lead to confirmation bias. Placebo effects, demand characteristics, nonspecific effects of meditation, and publication bias present further problems for meditation research. Research findings on the psychological benefits of meditation have been variable, which can be partially explained by differences in research design and type of meditation studied. Nevertheless, the advantages of meditation have never been more publicized; the benefits of mindfulness-based approaches are particularly prominent in mainstream media, and are often presented from an uncritical point of view. That being said, new studies are emerging that uphold methodological standards, many of which employ neuroimaging techniques to further our understanding of the effects of meditation on cognitive functions. While there is substantial evidence for meditation-related reduction in psychological stress and anxiety and some evidence for improvement in various components of attention, no conclusions can be drawn with respect to its effect on measures of intelligence. Preliminary findings suggest that meditation interventions can positively affect cognitive functioning, including fluid intelligence, however, further research is needed to determine the nature of this relationship and to identify potential moderators and mediators.

5.4 NUTRITION AND COGNITIVE-ENHANCING DRUGS

The relationship between health and intelligence is complex: on the one hand, lower intelligence presents a risk factor for poor health even after controlling for socioeconomic status (Der et al., 2009; Wraw et al., 2015), and on the other, health affects intelligence. For example, cardiovascular disease and increased levels of inflammatory markers have been associated with reduced cognitive abilities (Karlsson et al., 2010; Qiu and Fratiglioni, 2015). In most cases, causal inference is not possible; illness can affect cognitive functioning or vice versa, or both can be influenced by common factors such as shared genetic etiology (Hagenaars et al., 2016). Higher intelligence predicts longer lifespan, which has been attributed to longer education, higher socioeconomic position, better health management, and access to safer environments (Whalley and Deary, 2001), and to common genetic factors influencing both intelligence and health (Arden et al., 2016). Higher intelligence may also provide a reserve capacity, which acts like a buffer against age-related cognitive decline (Stern, 2009). Adhering to a healthy lifestyle, which involves a balanced diet, regular exercise, maintaining an optimal weight, drinking alcohol in moderation, and not smoking, is associated with a lower risk of mortality (Loef and Walach, 2012). One line of research on the nutrition-intelligence relation is based on prenatal and early life influences, specifically on breastfeeding and pre- and postnatal nutrient intake. Since nutrient supplementation does not appear to enhance cognitive ability in well-nourished, healthy adults, a second line of research investigates the effects of nutrient supplementation on cognitive preservation in the elderly. These topics will be discussed in Sections 5.4.1.1 and 5.4.1.2, respectively. On the other side of the coin, there is growing public and academic interest in pharmacological cognitive enhancement, examples of which include methylphenidate (e.g., Ritalin), mixed amphetamine salts (e.g., Adderall), and modafinil (e.g., Provigil) (Schelle et al., 2014). The medical safety and ethics of pharmacological cognitive enhancement in healthy individuals will be discussed in Section 5.4.2.

5.4.1 Nutrition

Undernutrition underlies 3.1 million deaths annually in children younger than 5 years. Providing nutrition interventions to prevent maternal and child undernutrition, and micronutrient deficiencies has great public health significance. An analysis of nutrition interventions showed that deaths in children younger than 5 years could be reduced by 15% if populations could access the following evidence-based interventions: folic acid supplementation or fortification, maternal balanced energy protein supplementation, maternal calcium supplementation, multiple micronutrient supplementation in pregnancy, promotion of breast feeding, appropriate complementary feeding, vitamin A and preventive zinc supplementation. These interventions could also lead to improvements in cognitive and socioemotional development (Bhutta et al., 2013).

5.4.1.1 Breastfeeding

Does breastfeeding increase the child's intelligence? This has been debated extensively by experts over the years. The mechanisms that are thought to underlie improvements in intellectual development are related to the chemical properties of breastmilk. Not only is breastmilk considered optimal food for infants, it contains long-chain polyunsaturated fatty acids such as docosahexaenoic acid (DHA) and arachidonic acid (AA), which are important for retina and cortical brain development (Brenna et al., 2007). Until recently, DHA and AA were absent from infant formula. In addition, breastfeeding may augment bonding between mother and child, which could contribute to the child's intellectual development (Horta et al., 2007).

Researchers have also explored whether children's variants of genes modify breastfeeding effects on cognition (e.g., Morales et al., 2011). This type of research presents an advantage over observational studies, as the data tend to be more resilient to confounding by social and lifestyle factors. The FADS gene family is involved in the desaturation of fatty acids to form the more complex long chain polyunsaturated fatty acids such as DHA. Certain polymorphisms in the FADS2 gene appear to be less efficient at converting precursors to form products, thereby modifying the effect of breastfeeding on child IQ (Steer et al., 2010). However, some studies failed to detect a FADS2-breastfeeding interaction with IQ (Groen-Blokhuis et al., 2013; Martin et al., 2011). The challenge for future research is to clarify the role of long-chain polyunsaturated fatty acids for the breastfeeding-IQ relation.

RCTs present the "gold standard" for causal inference in medicine and psychology, however, it would be unethical to randomly allocate subjects not to receive breastmilk (Horta and Victora, 2013). Early studies that investigated the effects of breastfeeding on health and developmental outcomes did not consider this unethical. In a study by Lucas and Morley (1994), preterm infants were randomly assigned to receive donated banked breastmilk or formula. A follow-up at age 13–16 years demonstrated that adolescents who had been randomized to banked breastmilk had a lower concentration of C-reactive protein, a marker of inflammatory process associated with atherosclerosis. Together with measurements of blood lipoproteins, these findings suggested that breastmilk feeding has a major beneficial effect on long-term cardiovascular health (Singhal et al., 2004). Research has changed since then: studies examine whether there is a significant association between breastfeeding and an outcome of interest. Well-designed prospective cohort studies can produce powerful results. A cohort of infants is selected at birth or at the time of infant feeding, and followed until outcome measures are assessed—usually at least 2 years later, with some extending into adulthood. The home environments of breastfed infants may differ from the home environments of nonbreastfed infants, particularly with respect to parent–child interactions. Women who breastfeed are more likely to provide a cognitively stimulating environment than those who do not breastfeed (Jacobson et al., 1999). Many studies published before 2002 did not assess and control for important confounding factors such as home environment and socioeconomic status (SES) (Jain et al., 2002). In addition, Der et al. (2006) identified maternal IQ as a potential confounder that had been overlooked in many previous studies. In light of these issues, and due to the sheer amount of studies dedicated to exploring the relationship between breastfeeding and intellectual development, the results of systematic reviews published after the year 2006 will be presented. Specifically, reviews of well-controlled studies that assessed the relation between breastfeeding duration of full-term infants and standardized measures of general intelligence will be discussed.

Let's start with the research of Der et al. (2006) that involved a prospective study, siblings pairs analysis, and a metaanalysis. Data from the US national longitudinal survey of youth were used for the prospective study; the final sample included 5475 children. Breastfeeding was assessed by asking mothers how old their child was when they stopped breastfeeding. In most cases, the interviews took place within one year of the child's birth. Children's cognitive ability was measured with The Peabody Individual Achievement Test (PIAT) between the ages of 5 and 14 years. Child demographics, their home environment, and maternal characteristics including cognitive ability were used as control variables. Random effects models revealed that the unadjusted effects of breastfeeding on mental ability corresponded to an increase of 4.1 to 4.7 points, which is comparable with the effects reported in other studies (Anderson et al., 1999). However, adjusting for maternal IQ reduced this advantage by 71%–75%, and controlling for mother's education by 34%–42%. When all of the confounders were included in the model, the effect of breastfeeding was small (0.52 points) and nonsignificant. An analysis of sibling pairs that did not match on breastfeeding status or duration of breastfeeding confirmed these findings: no advantage of breastfeeding on cognitive ability was observed. Finally, the results of a metaanalysis of 12 studies that controlled for parental IQ found no significant effect when relevant confounders were taken into account. The results highlighted another important issue in this field of research: studies with fewer controls and smaller sample sizes tend to show the largest effects. Similar conclusions were put forward in a systematic review of 84 studies (Walfisch et al., 2013): much of the reported effect of breastfeeding on children's intelligence can be explained by maternal socioeconomic class and IQ.

On the other hand, systematic reviews conducted by Bernardo Lessa Horta and colleagues (Horta et al., 2015; Horta and Victora, 2013) demonstrate significant effects of breastfeeding on intelligence even net of maternal IQ. In a review of studies published between 2006 and 2011 that examined the long-term effects of breastfeeding ($N = 13$), breastfed subjects displayed higher performance in intelligence tests (mean difference: 3.45 points; 95% confidence interval: 1.92; 4.98). The pooled effect from studies with control of maternal IQ was equal to 2.19 (95% CI: 0.89; 3.50). These results suggest that while the effects of breastfeeding on IQ might not be large, they cannot entirely be explained by maternal IQ and other confounding variables (Horta and Victora, 2013). A second systematic review that included well-controlled studies published between 2011 and 2014 ($N = 17$) reported similar results: breastfeeding was associated with better performance on tests of intelligence in childhood and adolescence (mean difference: 3.44 points; 95% confidence interval: 2.30; 4.58), even after controlling for maternal IQ (mean difference: 2.62 points; 95% confidence interval: 1.25; 3.98) (Horta et al., 2015).

Kanazawa and Satoshi (2015) also challenged the view that breastfeeding has no independent causal effect on children's IQ when maternal IQ is accounted for. The following argument was put forward: assuming that breastfeeding improves infant nutrition and health, and that infant nutrition and health support intellectual development, breastfeeding should increase childhood and adult intelligence. The author further suggested that the reason some studies failed to uncover the effect of breastfeeding on intelligence after accounting for maternal IQ lies in the use of only one IQ test to measure intelligence, hence they could not perform a factor analysis to determine the general intelligence factor. In order to test this hypothesis, Kanazawa and Satoshi measured the effect of breastfeeding on intelligence in two consecutive generations of British respondents. Data from the national child development study in the United Kingdom were used to assess this effect. Intelligence was measured with multiple cognitive tests, which enabled the extraction of the latent general intelligence factor. The results demonstrated that breastfeeding for at least 1 month was significantly associated with child intelligence, even net of parental IQ. The analysis of the second generation of respondents showed that each month of breastfeeding was associated with an increase of 0.16 IQ points and that 2 years of breastfeeding may potentially translate to an increase of 3.86 IQ points (note that only 1% of the infants were breastfeed for 24 months or longer). One of the limitations of this study was that there was no data available with respect to mother's attachment or commitment to their children thus these variables were not controlled for. In addition, parental IQ of the first generation was not measured in the original study, so it had to be estimated via education and social class.

Kramer et al. (2008) adopted a different approach to study the effect of breastfeeding on child cognitive development. Instead of conducting an observational study, a cluster-randomized trial was used (the analyses were based on intention to treat). Maternity hospitals and surrounding clinics in Belarus were randomly assigned to receive an intervention that promoted breastfeeding exclusivity and duration among those mothers who had already decided to initiate breastfeeding. The control maternity hospitals and clinics continued with their standard practices and policies. The children were followed-up at age 6.5 ($N = 13{,}889$), which included administration of the Wechsler Abbreviated Scales of Intelligence (WASI). The breastfeeding promotion intervention was successful: exclusive breastfeeding at 3 months of age increased significantly in the experimental group, as did the prevalence of any breastfeeding up to 12 months. On average, the experimental group scored higher on all of the WASI measures, with cluster-adjusted mean differences (95% confidence intervals) of +7.5 (+0.8 to +14.3) for verbal IQ, +2.9 (−3.3 to +9.1) for performance IQ, and +5.9 (−1.0 to +12.8) for full-scale IQ. The wide confidence intervals reflect the differences observed in the mean WASI scores among the polyclinics, thus the magnitude of these effects remains unclear. Nevertheless, the authors concluded that prolonged and exclusive breastfeeding improves children's cognitive development. The study by Kramer et al. (2008) has been cited by many metaanalyses as evidence that the effects of breastfeeding on IQ are causal.

A recent review of the health benefits of breastfeeding reported that high-income countries have shorter breastfeeding duration than do low-income and middle-income countries. Even in the latter countries only 37% of infants younger than 6 months are exclusively breastfeed (Victora et al., 2016). Their metaanalysis indicated that breastfeeding is associated with protection against child infections and malocclusion, increases in intelligence, and probable reductions in overweight and diabetes. No significant associations were reported for allergic disorders, blood pressure, or cholesterol. Longer breastfeeding was also related to an increase in tooth decay. One aspect that has not been mentioned so far is the benefit of breastfeeding for women. Victora et al. (2016) found that breastfeeding gave protection against breast cancer, improved birth spacing, and might even protect against ovarian cancer, and type 2 diabetes. The causal role of breastfeeding in improved intellectual development remains debatable. Likewise, it is not known which combination of duration and intensity (e.g., exclusive or not exclusive) of breastfeeding is more likely to elicit improvements in IQ scores (Colson et al., 2016). Nevertheless, there is no doubt that breastfeeding is associated with a wide range of health benefits both for the infant and mother therefore it should be promoted worldwide.

5.4.1.2 Dietary Intake

While all nutrients are important for early brain development, macronutrients such as protein and long-chain polyunsaturated fatty acids, and micronutrients like iron, zinc, and iodine have profound effects on brain growth and function, particularly from the time of conception through 2–3 years of age. An adverse environmental event such as nutritional deficiency can be particularly detrimental during sensitive periods, namely, epochs when the brain is more vulnerable to environmental factors (Cusick and Georgieff, 2016). An extensive review on the role of key nutrients in neurocognitive development is out of the scope of this section, hence we will limit the discussion to pre- and postnatal exposure to nutrients that have been specifically associated with early childhood cognitive development. These include but are not limited to iron, iodine, zinc, vitamins such as folic acid, and omega-3 fatty acids. Understandably, most nutrition intervention studies focus on deficient populations, the results of which may not directly translate to healthy populations (Saint and Frick, 2015). It is important to note that supplementation of most nutrients beyond a sufficiency state does not grant an additional neurodevelopmental advantage (Cusick and Georgieff, 2012).

Iron deficiency and iron deficiency anemia are common in young children and can negatively affect psychomotor development and cognitive function (Wang et al., 2013). Lukowski et al. (2010) reported that chronic, severe iron deficiency in infancy results in long-term deficits in executive function and recognition memory (up to 19 years of age). Fetal brain malformations can also result from severe gestational zinc (Adamo and Oteiza, 2010) and iodine deficiencies (Pearce, 2012). Deficiencies in iron, iodine, zinc, and vitamin A present great public health problems in developing countries. In contrast, vitamin C, D, and B deficiencies have declined in the recent decades, but may still present a problem in some areas and during certain seasons (Müller and Krawinkel, 2005). There is evidence that providing micronutrient supplementation to children who are at risk of developing these deficiencies (or are micronutrient deficient) positively affects cognitive development. For example, a double-blind, placebo-controlled study that involved Thai primary schoolchildren demonstrated that daily supplementation of iron resulted in improvement in school performance and in nonverbal intelligence scores compared to once-weekly iron consumption or placebo (Sungthong et al., 2004). The effectiveness of nutrition interventions on cognitive function is often assessed by comparing multiple micronutrient supplementation with single nutrient and/or placebo supplementation. This approach was adopted by Solon et al. (2003), who opted for a placebo-controlled design. The results demonstrated that consumption of a multiple-micronutrient-fortified beverage for 16 weeks significantly improved total cognitive, nonverbal ability, and verbal ability in iron- and/or iodine-deficient Filipino schoolchildren.

However, not all multiple micronutrition intervention studies report improvements in general cognitive ability. Warthon-Medina et al. (2015a) conducted a RCT that provided supplements to 902 children aged 6–17 months, who lived in the province of Lima, Peru. The multiple micronutrient group received iron, zinc, folic acid, vitamin C, and vitamin A, whereas the other group received iron supplementation only. Two years later, a subsample of children ($N = 184$) was assessed on a range of cognitive tests, including the Wechsler preschool and primary scale of intelligence. The results showed no significant differences between the iron and the multiple micronutrition groups for most cognitive tests, except on a verbal subtest of the intelligence test: the girls in the multiple micronutrition group showed an advantage over the girls in the iron group. The authors concluded that multiple micronutrition supplements had no long-term additional effects on cognitive function compared to iron alone. A more rigorous research design was adopted by Murray-Kolb et al. (2012), who conducted a double-blind, placebo-controlled study in which they investigated the effects of micronutrient supplementation on intellectual and motor function in school-aged Nepalese children. The children in this study ($N = 694$) represent a subsample of children whose mothers participated in micronutrient supplementation (iron plus folic acid, only zinc, iron plus folic acid and zinc, or placebo); age at supplementation was 12–35 months and age of testing was 7–9 years of age. No beneficial effect of either iron plus folic acid or zinc supplementation on intellectual, executive, or motor function was reported. As stated by Cusick and Georgieff (2012), not only the type of nutrition supplementation but also the timing is key, as one should aim to apply a nutritional intervention inside the window of critical development. The study by Murray-Kolb et al. (2012) might have provided the supplementation too late—after the period of peak central nervous system development.

Despite some inconsistencies, systematic reviews and metaanalyses of nutrition supplementation studies in developing countries suggest that (1) multiple micronutrient supplementation during pregnancy positively affects infants' birth weight, height, and cognitive development by 2 years of age (Shrimpton et al., 2009), and (2) multiple micronutrition supplementation improves cognitive performance in children aged 5–15 years (Khor and Misra, 2012). A recent systematic review that evaluated the effects of micronutrition interventions among children aged 4–18 years in both developing and developed countries supported these findings. Micronutrition supplementation improved performance on tests of fluid intelligence, particularly among those who were iron- or iodine-deficient at baseline. The effects on other cognitive domains remain ambiguous (Lam and Lawlis, in press).

There is even less consistent evidence with respect to the effects of micronutrient supplementation on cognitive ability among healthy children. For example, a metaanalysis that focused on zinc intervention randomized controlled trials reported nonsignificant effects of zinc supplementation on cognitive functioning in apparently healthy children (Warthon-Medina et al., 2015b). The data pointed toward small improvements in executive function and motor development. In contrast, a double-blind, placebo-controlled study that examined the effect of vitamin-mineral supplementation on the intelligence of American schoolchildren reported that the supplementation raised the nonverbal intelligence of some groups of children by 2–3 points. However, most children did not show any improvements, presumably because they were already adequately nourished (Schoenthaler et al., 2000). Similar conclusions were drawn in a review by Benton (2001): 10 out of 13 studies reported improvements in nonverbal measures of intelligence following micronutrient supplementation (contrasted against placebo supplements)—importantly, the evidence suggested that only the minority of children whose diets provide low amounts of micronutrients respond to supplementation. A more recent systematic review of randomized controlled trials demonstrated that multiple micronutrient supplementation resulted in a small improvement in fluid but not crystallized intelligence in healthy children (Eilander et al., 2010). Collectively, these findings suggest that multiple micronutrient supplementation can enhance fluid intelligence in children with micronutrient-poor diets.

Observational studies have reported that pregnant women who consume more oily fish, which are rich in omega-3 fatty acids, have children with better neurodevelopmental outcomes, that is, the children are more likely to score higher on tests of intelligence (Hibbeln et al., 2007; Mendez et al., 2009). Since seafood is also a source of neurotoxic contaminants, such as methylmercury, it is important to examine the potential risks and benefits of seafood consumption to the developing fetus (Mendez et al., 2009). Oken et al. (2008) demonstrated that higher maternal fish intake was associated with better child cognitive test performance at age 3, and higher mercury levels with poorer test scores. On the other hand, maternal fish consumption during pregnancy was not associated with harmful or beneficial effects on children's cognition in mid-childhood, suggesting that prenatal exposure to nutrients and neurotoxic contaminants might be obscured by childhood experiences (Oken et al., 2016). Since there are a number of complex confounding factors that affect early childhood development causality cannot be established; RCTs are needed to determine whether prenatal exposure to omega-3 fatty acids benefits early childhood cognitive development. A metaanalysis of RCTs reported no effect of omega-3 supplementation during pregnancy on IQ or Developmental Standard Score, except in the 2–5 year age group, in which omega-3 groups showed an advantage in Developmental Standard Scores. However, the latter finding might have been confounded due to high risk of bias (Gould et al., 2013). While it remains to be determined whether prenatal omega-3 supplementation specifically affects general cognitive ability, there is no doubt that consuming food that is rich in omega-3 fatty acids benefits neurodevelopment. The US Food and Drug Administration advises to "eat 8 to 12 ounces of a variety of fish each week from choices that are lower in mercury. The nutritional value of fish is important during growth and development before birth, in early infancy for breastfed infants, and in childhood" (Fish: What Pregnant Women and Parents Should Know, 2014).

It is unlikely that adding supplements to a diet, but not changing other lifestyle behaviors, could affect general cognitive ability during adulthood. That being said, there is preliminary evidence that supplementation with a combination of huperzine A, vinpocetine, acetyl-L-carnitine, *R. rosea* and alpha-lipoic acid for 28 days improves performance on Raven's advanced progressive matrices compared to baseline in healthy adults, an effect that was not observed in the placebo group (Stough et al., 2011). The majority of research on the efficacy of these supplements involves patients with neurological disease, for example, huperzine A appears to have beneficial effects on improvements in cognitive function in subjects with Alzheimer's disease (for a review, see Yang et al., 2013). In addition, there is correlational evidence that balanced nutrition may promote successful cognitive aging for older adults. Namely, frequent consumption of fruits and vegetables, fish, and omega-3 rich oils may decrease the risk of dementia and Alzheimer disease, especially among ApoE ε4[f] noncarriers (Barberger-Gateau et al., 2013). This finding was to some extent supported by a systematic analysis of randomized controlled studies that provided omega-3 fatty acids to older adults. Three groups of subjects were compared: healthy subjects, subjects with cognitive impairment but no dementia, and subjects with Alzheimer's disease. Supplementation with omega-3 fatty acids

f. The APOE ε4 gene is important for making the protein apolipoprotein E that regulates cholesterol, triglyceride and phospholipid transport and metabolism (Chouinard-Watkins and Plourde, 2014).

did not significantly affect cognitive performance in healthy subjects or subjects with Alzheimer's disease. In contrast, the supplementation benefited immediate recall, attention, and processing speed in individuals who displayed cognitive impairment. The results of another metaanalysis of trials that provided at least 3 months of long-chain n-3 polyunsaturated fatty acids to children, adults, and the elderly confirmed these findings. Compared with placebo, long-chain fatty acids improved cognitive development in infants, but did not enhance cognitive function in children, adults, or the elderly (Jiao et al., 2014).

A number of other dietary supplements have been investigated in relation to preservation of cognitive function in the elderly. Positive effects on cognitive performance have been reported for folic acid (Durga et al., 2007), folic acid plus vitamin B12 (Walker et al., 2012), cocoa flavanol (Brickman et al., 2014; Mastroiacovo et al., 2015), and gingko biloba (Ong Lai Teik et al., 2016), to name a few. The benefits of dietary supplements are often subject to sensationalism in the media, hence it is important to note that most of this research requires further work and replication, and that nutrient supplementation should not be used as a replacement for a balanced diet. Indeed, there is little evidence that multivitamin and multimineral supplementation improves cognitive ability in normally aging adults (e.g., McNeill et al., 2007). However, we cannot exclude the possibility of positive effects in those at higher risk of nutritional deficiency.

A brief note on animal research: animal models have been of pivotal importance in advancing knowledge of the effects of nutrients on brain function. Dietary factors can influence brain processes by controlling neurotransmitter pathways, synaptic transmission, membrane fluidity, and signal-transduction pathways. In addition, gut hormones that are produced by the brain, or that can enter the brain, can also affect cognitive ability (for a review, see Gómez-Pinilla, 2008). While many animal nutrition studies have been supported by research involving humans, this is not always the case. Resveratrol, a polyphenol in red grapes, berries, and peanuts, preserves cognitive function in aging mice (Oomen et al., 2009), yet these findings were not supported in a human trial (Kennedy et al., 2010). This may be partially attributed to the fact that it is difficult to disentangle the effects of specific nutrients on brain function in the human diet. Particularly, the interactions among and between micronutrients and macronutrients are poorly understood. Nevertheless, there is promising evidence that the combined action of specific diets and exercise can activate molecular systems that are involved in synaptic plasticity (Gómez-Pinilla, 2008).

5.4.2 Cognitive-Enhancing Drugs

Pharmacological cognitive enhancement refers to the use of drugs for the purpose of improving cognitive functions by healthy individuals (Franke et al., 2014). Several creative terms have been coined to describe this phenomenon, ranging from neuroenhancement to "doping the mind" and "cosmetic neurology." The drugs themselves are often described in sensationalist terms in the media, as "smart drugs" and "brain steroids"—yet many media articles fail to mention the risks and side effects associated with the consumption of these drugs (Partridge et al., 2011). One last term that needs to be mentioned is *nootropic* (from Noos, mind; tropein, toward), which was originally used to describe the pharmacology of piracetam[g] (Giurgea, 1982), but is often applied more broadly as a synonym for neuroenhancement (e.g., nootropic drugs = cognitive enhancers). The most common pharmaceutical cognitive enhancers are methylphenidate (Ritalin), modafinil (Provigil or Nuvigil), and mixed amphetamine salts (Adderall), either purchased illegally or used off-label (Smith and Farah, 2011). Modafinil is a medication for sleep disorders, such as narcolepsy, and is prescribed for the treatment of cognitive and behavioral symptoms of dementia. Ritalin and Adderall are typically prescribed for the treatment of ADHD. Even though these stimulants can help manage ADHD symptoms, little is known about the risks and benefits of their use by healthy adults. Moreover, while stimulant drugs can improve episodic memory in adults diagnosed with ADHD, there is no evidence that this translates to improved long-term academic performance (Advokat and Scheithauer, 2013). As (re)stated by Rose (2002): "there is a fine medical and ethical line between correcting deficits and improving on normality."

The most widely used cognitive enhancer is caffeine in the form of coffee, caffeinated drinks, and tablets (Franke et al., 2014). Some do not consider caffeine as a pure cognitive enhancer since most of its cognitive enhancing properties stem from indirect effects: increases in arousal, mood, and concentration (Nehlig, 2010). Nicotine can also affect cognitive function including learning, memory, and attention. However, studies in which nicotine was administered to subjects (usually in the form of transdermal skin patches) reported both improvements and impairments in cognitive function: studies of nicotine in normal nonsmokers tend to show deleterious effects, whereas studies that tend to show improvements involve smokers deprived of cigarettes or clinical populations (e.g., patients with Alzheimer's disease, Parkinson's disease, ADHD, and schizophrenia) (Newhouse et al., 2004). Niemegeers et al. (2014) conducted an interesting double-blind placebo-controlled study in which they examined the effects of oromucosal nicotine spray on several cognitive measures in healthy elderly and young volunteers who did not smoke. The participants took part in a screening visit, three treatment periods (separated by washout periods), and a follow-up examination. The authors predicted that the elderly would benefit more from nicotine compared to young adults due to decreases in cognitive function that are associated with normal aging. The results did not support this: healthy elderly subjects showed impaired cognitive performance on measures of information-processing speed and psychomotor function compared to young adults. Nicotine did not affect cognitive function in young adults and had either no effects or deleterious effects in the elderly. An exploratory analysis

g. A cyclic derivative of the neurotransmitter GABA (Corazza et al., 2014).

suggested that participants in both groups with lower baseline functioning may have benefited from nicotine administration. As stated by the authors, these results are preliminary and need to be replicated. It should also be noted that funding for this study was provided by Janssen Pharmaceutica N.V., who were involved in the preparation of the protocol but not in the analysis or writing of the manuscript.

Husain and Mehta (2011) reviewed the effects of drugs used for cognitive enhancement in neurological conditions and in healthy people. Table 5.1 summarizes the effects of these drugs.[h] The authors concluded that the overall effects of drugs appear to be modest both in healthy individuals and patient groups, and that more significant effects can be observed in subgroups whose baseline performance is poorest, or in individuals with a particular genotype.

Healthy students and professionals alike are using cognitive enhancers in order to gain a competitive advantage over their peers. But how prevalent is this practice? The largest nationwide study of the use of prescription stimulants (Ritalin, Dexedrine, or Adderall) for nonmedical purposes among American college students reported the following prevalence: 6.9% (lifetime), 4.1% (past year), and 2.1% (past month). The use of prescription stimulants was more likely in colleges with competitive admissions criteria, among students who were members of a fraternity or sorority, and among males compared to females (McCabe et al., 2005). Smith and Farah (2011) reviewed the literature on prevalence rates among college students in the United States, and concluded that this practice is commonplace. More precise quantitative conclusions were not drawn due to the high variability of estimates between studies. Differences in prevalence data are likely to stem from different definitions of nonmedical use, sampling methods, and respondent demographics. In the United Kingdom, lifetime prevalence of pharmacological cognitive enhancement (modafinil, methylphenidate, or Adderall) was under 10%, whereas past regular and current users of these substances made up between 0.3% and 4% of the survey population. A total of 67% of the students surveyed were either unaware of or uninterested in cognitive enhancement. Older male students were more likely to use cognitive enhancers. The most common cognitive enhancer was caffeine delivered in tablets or energy drinks, whereas modafinil was the most commonly used pharmaceutical cognitive enhancer. Less than 6% of students had used methylphenidate or Adderall as a cognitive enhancer at least once. Even though these estimates are lower than the authors predicted they would be, there is concern regarding the relatively high interest in pharmaceutical cognitive enhancement among students (Singh et al., 2014).

Prevalence data prompt further questions: are students' beliefs about the efficacy of these drugs supported by research? Do these substances really work as "smart pills"? It is possible that at least some of the cognitive benefits stem from placebo effects. Other possibilities include inflated perceptions about the

h. Reprinted from Husain and Mehta (2011), with permission from Elsevier

TABLE 5.1 Summary of the Effects of Drugs Frequently Used as Cognitive Enhancers

Cognitive enhancer	Neuromodulatory mechanism	Cognitive functions improved	Known brain systems most affected	Currently recommended clinical use
Methylphenidate, amphetamine	Dopamine and noradrenaline reuptake inhibitors	Response inhibition, working memory, attention, vigilance	Frontoparietal attentional systems, striatum, default mode networks	ADHD, wake-promoting agent
Caffeine	Nonselective adenosine receptor antagonist	Vigilance, working memory, incidental learning	Frontal lobe attentional systems	—
Nicotine	Nicotinic cholinergic receptor agonist	Working memory, episodic memory, attention	Fronto-parietal attentional systems, medial temporal lobe, default mode networks	—
Modafinil	Unknown, but effects on dopamine, noradrenaline and orexin systems proposed	Working memory, episodic memory, attention	Frontal lobe attentional systems	Wake-promoting agent
Atomoxetine, reboxetine	Noradrenaline reuptake inhibitors	Response inhibition, working memory, attention	Frontoparietal attentional systems	ADHD, depression
Donepezil, galantamine, rivastigmine (AChEI)	Blocks enzymatic breakdown of acetylcholine	Episodic memory, attention	Frontal lobe attentional systems	Alzheimer's disease, PDD, DLB
Memantine	Noncompetitive, low-affinity, open channel blocker of the NMDA receptor	Episodic memory, attention	Frontal and parietal lobe	Alzheimer's disease

Source: Reprinted from Husain, M., Mehta, M.A., 2011.Cognitive enhancement by drugs in health and disease. Trends Cogn. Sci., 15(1), 28–36, with permission from Elsevier.

quality/quantity of work accomplished and enhanced wakefulness or motivation that improves the work but is not related to improved cognitive abilities (Smith and Farah, 2011). Systematic research is needed to examine the effects of prescription stimulants on cognitive functions in healthy adults. Due to obvious ethical reasons, such studies are limited to examining single-dose effects rather than long-term effects, and often rely on small samples. The systematic review of pharmacological cognitive enhancement by Smith and Farah (2011) reported mixed findings: studies have reported improvements in consolidation of long-term declarative memory, but the patterns of evidence for executive function and cognitive control are less clear. Over a third of findings reported no effect on cognitive function in healthy adults.

Similar results were reported by Ilieva, Hook, and Farah in a systematic review of double-blind placebo-controlled trials that investigated the effects of methlyphenidate and amphetamine on cognitive functions in healthy adults. Namely, small effects of these drugs were found for executive function and memory. An analysis limited to high-quality studies also showed a small but significant enhancement of inhibitory control and short-term episodic memory, along with medium sized effects for delayed episodic memory (*n.b.,* the latter might have been affected by publication bias). Overall, the small effects suggest that the use of these drugs would not deliver a practically significant performance advantage in healthy people, corroborating the view that the media has inflated the benefits of pharmacological cognitive enhancement. Nevertheless it is possible that (1) there are individual differences in response to stimulants, (2) a small effect can still make a difference between passing or failing a test, (3) these drugs improve cognitive functions that were not examined in the metaanalysis, (4) cognitive enhancement would be different under detrimental conditions such as sleep deprivation, and (5) drugs affect cognitive performance via changes in affective states (Ilieva et al., 2015).

Pharmacological cognitive enhancement presents a topic of great interest in the media and in the academic literature. The three most common concerns that have been discussed are *medical safety, coercion*, and *fairness* (Schelle et al., 2014). The former focuses on the tradeoffs between benefits and risks. Even though a prescription drug is regarded as safe and effective for therapeutic purposes, there is no guaranty that this will hold true for unregulated off-label use. Moreover, the risk-benefit ratio that might be acceptable to patients (e.g., those who suffer from neurodegenerative disorders) may not be acceptable to healthy individuals seeking enhancement. Since the long-term effects of pharmacological cognitive enhancement on the developing brain are unclear, unregulated use of cognitive enhancers by children and adolescents represent a great concern among clinicians (Greely et al., 2008; Schermer et al., 2009). Methylphenidate (Ritalin) and amphetamine amplify the activity of dopamine, which can make cognitive tasks more interesting and rewarding, however, this can lead to compulsive use and addiction in vulnerable people (Volkow and Swanson, 2008). The second concern, *coercion*, refers to implicit and direct pressures that can arise from the availability of pharmacological cognitive enhancement. The perception that other students or colleagues who take cognitive enhancers are more productive or perform better on cognitive tasks might lead to the belief that one must take these substances in order to keep up (Maslen et al., 2014). Finally, concerns regarding *fairness* have been divided into three subthemes: (1) the potential to exacerbate existing socio-economic inequality—the rich and informed are likely to have greater access to cognitive enhancers, (2) the issue whether pharmacological cognitive enhancement to enhance academic performance constitutes cheating, and (3) the assumption that high performance can be achieved with less effort due to pharmacological cognitive enhancement and whether such performance can be viewed as authentic (Forlini and Racine, 2012; Schelle et al., 2014).

Views on pharmacological cognitive enhancement range from favorable and optimistic (Greely et al., 2008) to skeptical (e.g., Lynch et al., 2011). Research suggests that certain drugs can improve performance on specific cognitive tasks in the short term, whereas the long-term effects remain unknown. In order to truly qualify as a "smart drug," (1) an agent should produce capabilities that transfer to a broad range of cognitive activities, and (2) this transfer should last for a substantial period of time (Hunt and Jaeggi, 2013). In addition, the drug should be safe for use and devoid of side effects. Needless to say, the drugs that are available currently do not match these criteria. It is likely that new generations of cognitive enhancers will be developed in the future thus it is important to continue to investigate the risks (and benefits) of pharmacological cognitive enhancement, and to decide whether the current regulatory structures are adequate for the control of this type of drug use, or whether new policies and laws need to be written (Farah et al., 2004).

In conclusion, nutrition plays an important role in human neurocognitive functioning. Research shows that breastfeeding duration positively correlates with intelligence (for a systematic review, see Horta et al., 2015) and that malnutrition in infancy is negatively associated with measures of IQ (Venables and Raine, 2016). Micronutrients, such as iron, iodine, zinc, and various vitamins, as well as macronutrients, including omega-3 fatty acids and protein, play a crucial role in early brain development. Intervention studies suggest that multiple micronutrient supplementation to infants and children who have micronutrient deficiencies can improve fluid intelligence. The effects on other cognitive domains, and in healthy children remain ambiguous (Lam and Lawlis, in press). There is little evidence to suggest that nutrient supplementation enhances cognitive ability in well-nourished, healthy adults, however, supplementation of certain nutrients can help ameliorate cognitive decline in the elderly, particularly in individuals who are at risk of nutrient deficiency, and/or in individuals with age-related neurological disease. Similarly, the greatest benefit from pharmaceutical cognitive enhancement has been observed in individuals whose baseline performance is poorest, in individuals with a particular genotype, and in patient groups. Prescription stimulants can produce short-term improvements in memory and executive function (Ilieva et al., 2015; Smith and Farah, 2011) in healthy individuals, however, the reported effects are small and may not be of practical significance. Illegal and off-label long-term use of cognitive enhancers can present significant health risks to the user, and raises a number of ethical issues.

REFERENCES

Abbott, R.D., White, L.R., Ross, G.W., Masaki, K.H., Curb, J.D., Petrovitch, H., 2004. Walking and dementia in physically capable elderly men. JAMA 292 (12), 1447–1453.

Abdelnour, A.F., Huppert, T., 2009. Real-time imaging of human brain function by near-infrared spectroscopy using an adaptive general linear model. NeuroImage 46 (1), 133–143.

Aberg, M.A.I., Pedersen, N.L., Torén, K., Svartengren, M., Bäckstrand, B., Johnsson, T., Kuhn, H.G., 2009. Cardiovascular fitness is associated with cognition in young adulthood. Proc. Natl. Acad. Sci. USA 106 (49), 20906–20911.

Adamo, A.M., Oteiza, P.I., 2010. Zinc deficiency and neurodevelopment: the case of neurons. BioFactors 36 (2).

Adlard, P.A., Perreau, V.M., Engesser-Cesar, C., Cotman, C.W., 2004. The timecourse of induction of brain-derived neurotrophic factor mRNA and protein in the rat hippocampus following voluntary exercise. Neurosci. Lett. 363 (1), 43–48.

Advokat, C., Scheithauer, M., 2013. Attention-deficit hyperactivity disorder (ADHD) stimulant medications as cognitive enhancers. Front. Neurosci. 7, 82.

Ainslie, P.N., Cotter, J.D., George, K.P., Lucas, S., Murrell, C., Shave, R., Atkinson, G., 2008. Elevation in cerebral blood flow velocity with aerobic fitness throughout healthy human ageing. J. Physiol. 586 (16), 4005–4010.

Anderson, J.W., Johnstone, B.M., Remley, D.T., 1999. Breast-feeding and cognitive development: a meta-analysis. Am. J. Clin. Nutr. 70 (4), 525–535, http://www.ncbi.nlm.nih.gov/pubmed/10500022.

Angevaren, M., Aufdemkampe, G., Verhaar, H.J.J., Aleman, A., Vanhees, L., 2008. Physical activity and enhanced fitness to improve cognitive function in older people without known cognitive impairment. Cochrane Database Syst. Rev. (2), CD005381.

Arden, R., Luciano, M., Deary, I.J., Reynolds, C.A., Pedersen, N.L., Plassman, B.L., Visscher, P.M., 2016. The association between intelligence and lifespan is mostly genetic. Int. J. Epidemiol. 45 (1), 178–185.

Arnardottir, N.Y., Koster, A., Domelen, D.R., Van, Brychta, R.J., Caserotti, P., Eiriksdottir, G., Sveinsson, T., 2016. Association of change in brain structure to objectively measured physical activity and sedentary behavior in older adults: Age, Gene/Environment Susceptibility-Reykjavik Study. Behav. Brain Res. 296, 118–124.

Arns, M., de Ridder, S., Strehl, U., Breteler, M., Coenen, A., 2009. Efficacy of neurofeedback treatment in ADHD: the effects on inattention, impulsivity and hyperactivity: a meta-analysis. Clin. EEG Neurosci. 40 (3), 180–189.

Arns, M., Heinrich, H., Strehl, U., 2014. Evaluation of neurofeedback in ADHD: the long and winding road. Biol. Psychol. 95, 108–115.

Aron, A., 1981. The transcendental meditation program in the college curriculum: a 4-year longitudinal study of effects on cognitive and affective functioning. Coll. Stud. J. 15 (2), 140–146.

Awh, E., Vogel, E.K., 2015. Attention: feedback focuses a wandering mind. Nat. Neurosci. 18 (3), 327–328.

Babiloni, C., Del Percio, C., Lizio, R., Marzano, N., Infarinato, F., Soricelli, A., Rossini, P.M., 2014. Cortical sources of resting state electroencephalographic alpha rhythms deteriorate across time in subjects with amnesic mild cognitive impairment. Neurobiol. Aging 35 (1), 130–142.

Barberger-Gateau, P., Lambert, J.-C., Féart, C., Pérès, K., Ritchie, K., Dartigues, J.-F., Alpérovitch, A., 2013. From genetics to dietetics: the contribution of epidemiology to understanding Alzheimer's disease. J. Alzheimer Dis. 33 (Suppl. 1), S457–S463.

Barnes, D.E., Yaffe, K., Satariano, W.A., Tager, I.B., 2003. A longitudinal study of cardiorespiratory fitness and cognitive function in healthy older adults. J. Am. Geriatr. Soc. 51 (4), 459–465.

Becerra, J., Fernández, T., Roca-Stappung, M., Díaz-Comas, L., Galán, L., Bosch, J., Harmony, T., 2012. Neurofeedback in healthy elderly human subjects with electroencephalographic risk for cognitive disorder. J. Alzheimer Dis. 28 (2), 357–367.

Belsky, D.W., Caspi, A., Israel, S., Blumenthal, J.A., Poulton, R., Moffitt, T.E., 2015. Cardiorespiratory fitness and cognitive function in midlife: neuroprotection or neuroselection? Ann. Neurol. 77 (4), 607–617.

Benton, D., 2001. Micro-nutrient supplementation and the intelligence of children. Neurosci. Biobehav. Rev. 25 (4), 297–309, http://www.ncbi.nlm.nih.gov/pubmed/11445136.

Bhutta, Z.A., Das, J.K., Rizvi, A., Gaffey, M.F., Walker, N., Horton, S., Black, R.E., 2013. Evidence-based interventions for improvement of maternal and child nutrition: what can be done and at what cost? Lancet 382 (9890), 452–477.

Black, J.E., Isaacs, K.R., Anderson, B.J., Alcantara, A.A., Greenough, W.T., 1990. Learning causes synaptogenesis, whereas motor activity causes angiogenesis, in cerebellar cortex of adult rats. Proc. Natl. Acad. Sci. USA 87 (14), 5568–5572, http://www.ncbi.nlm.nih.gov/pubmed/1695380.

Blondell, S.J., Hammersley-Mather, R., Veerman, J.L., 2014. Does physical activity prevent cognitive decline and dementia? A systematic review and meta-analysis of longitudinal studies. BMC Public Health 14 (1), 510.

Boccia, M., Piccardi, L., Guariglia, P., Boccia, M., Piccardi, L., Guariglia, P., 2015. The meditative mind: a comprehensive meta-analysis of MRI studies. BioMed Res. Int. 2015, 1–11.

Booth, J.N., Leary, S.D., Joinson, C., Ness, A.R., Tomporowski, P.D., Boyle, J.M., Reilly, J.J., 2014. Associations between objectively measured physical activity and academic attainment in adolescents from a UK cohort. Br. J. Sports Med. 48 (3), 265–270.

Boyke, J., Driemeyer, J., Gaser, C., Büchel, C., May, A., 2008. Training-induced brain structure changes in the elderly. J. Neurosci. 28 (28), 7031–7035.

Bray, S., Shimojo, S., O'Doherty, J.P., 2007. Direct instrumental conditioning of neural activity using functional magnetic resonance imaging-derived reward feedback. J. Neurosci. 27 (28), 7498–7507.

Brenna, J.T., Varamini, B., Jensen, R.G., Diersen-Schade, D.A., Boettcher, J.A., Arterburn, L.M., 2007. Docosahexaenoic and arachidonic acid concentrations in human breast milk worldwide. Am. J. Clin. Nutr. 85 (6), 1457–1464, http://www.ncbi.nlm.nih.gov/pubmed/17556680.

Brickman, A.M., Khan, U.A., Provenzano, F.A., Yeung, L.-K., Suzuki, W., Schroeter, H., Small, S.A., 2014. Enhancing dentate gyrus function with dietary flavanols improves cognition in older adults. Nat. Neurosci. 17 (12), 1798–1803.

Britton, W.B., Lepp, N.E., Niles, H.F., Rocha, T., Fisher, N.E., Gold, J.S., 2014. A randomized controlled pilot trial of classroom-based mindfulness meditation compared to an active control condition in sixth-grade children. J. Sch. Psychol. 52 (3), 263–278.

Brouwer, R.M., Hedman, A.M., van Haren, N.E.M., Schnack, H.G., Brans, R.G.H., Smit, D.J.A., Hulshoff Pol, H.E., 2014. Heritability of brain volume change and its relation to intelligence. NeuroImage 100, 676–683.

Brown, A.K., Liu-Ambrose, T., Tate, R., Lord, S.R., 2009. The effect of group-based exercise on cognitive performance and mood in seniors residing in intermediate care and self-care retirement facilities: a randomised controlled trial. Br. J. Sports Med. 43 (8), 608–614.

Brown, A.D., McMorris, C.A., Longman, R.S., Leigh, R., Hill, M.D., Friedenreich, C.M., Poulin, M.J., 2010. Effects of cardiorespiratory fitness and cerebral blood flow on cognitive outcomes in older women. Neurobiol. Aging 31 (12), 2047–2057.

Bugg, J.M., Head, D., 2011. Exercise moderates age-related atrophy of the medial temporal lobe. Neurobiol. Aging 32 (3), 506–514.

Bugg, J.M., Zook, N.A., DeLosh, E.L., Davalos, D.B., Davis, H.P., 2006. Age differences in fluid intelligence: contributions of general slowing and frontal decline. Brain Cogn. 62 (1), 9–16.

Burdette, J.H., Laurienti, P.J., Espeland, M.A., Morgan, A., Telesford, Q., Vechlekar, C.D., Rejeski, W.J., 2010. Using network science to evaluate exercise-associated brain changes in older adults. Front. Aging Neurosci. 2, 23.

Cahn, B.R., Polich, J., 2006. Meditation states and traits: EEG, ERP, and neuroimaging studies. Psychol. Bull. 132 (2), 180–211.

Carro, E., Nuñez, A., Busiguina, S., Torres-Aleman, I., 2000. Circulating insulin-like growth factor I mediates effects of exercise on the brain. J. Neurosci. 20 (8), 2926–2933, http://www.ncbi.nlm.nih.gov/pubmed/10751445.

Carter, O.L., Presti, D.E., Callistemon, C., Ungerer, Y., Liu, G.B., Pettigrew, J.D., 2005. Meditation alters perceptual rivalry in Tibetan Buddhist monks. Curr Biol. 15 (11), R412–R413.

Cassilhas, R.C., Viana, V.A.R., Grassmann, V., Santos, R.T., Santos, R.F., Tufik, S., Mello, M.T., 2007. The impact of resistance exercise on the cognitive function of the elderly. Med. Sci. Sports Exerc. 39 (8), 1401–1407.

Cavanagh, J.F., Frank, M.J., 2014. Frontal theta as a mechanism for cognitive control. Trends Cogn. Sci. 18 (8), 414–421.

Chang, M., Jonsson, P.V., Snaedal, J., Bjornsson, S., Saczynski, J.S., Aspelund, T., Launer, L.J., 2010. The effect of midlife physical activity on cognitive function among older adults: AGES—Reykjavik Study. J. Gerontol. A Biol. Sci. Med. Sci. 65 (12), 1369–1374.

Chang, Y.K., Labban, J.D., Gapin, J.I., Etnier, J.L., 2012. The effects of acute exercise on cognitive performance: a meta-analysis. Brain Res. 1453, 87–101.

Chouinard-Watkins, R., Plourde, M., 2014. Fatty acid metabolism in carriers of apolipoprotein e epsilon 4 allele: is it contributing to higher risk of cognitive decline and coronary heart disease? Nutrients 6 (10), 4452–4471.

Clark, P.J., Brzezinska, W.J., Puchalski, E.K., Krone, D.A., Rhodes, J.S., 2009. Functional analysis of neurovascular adaptations to exercise in the dentate gyrus of young adult mice associated with cognitive gain. Hippocampus 19 (10), 937–950.

Colcombe, S., Kramer, A.F., 2003. Fitness effects on the cognitive function of older adults: a meta-analytic study. Psychol. Sci. 14 (2), 125–130, http://www.ncbi.nlm.nih.gov/pubmed/12661673.

Colcombe, S.J., Kramer, A.F., Erickson, K.I., Scalf, P., McAuley, E., Cohen, N.J., Elavsky, S., 2004. Cardiovascular fitness, cortical plasticity, and aging. Proc. Natl. Acad. Sci. USA 101 (9), 3316–3321.

Colcombe, S.J., Erickson, K.I., Scalf, P.E., Kim, J.S., Prakash, R., McAuley, E., Kramer, A.F., 2006. Aerobic exercise training increases brain volume in aging humans. Biol. Sci. Med. Sci. 61A (11), 1166–1170.

Colom, R., Burgaleta, M., Román, F.J., Karama, S., Alvarez-Linera, J., Abad, F.J., Haier, R.J., 2013. Neuroanatomic overlap between intelligence and cognitive factors: morphometry methods provide support for the key role of the frontal lobes. NeuroImage 72, 143–152.

Colson, A., Cooke, R., Lutter, R., 2016. How does breastfeeding affect IQ? Applying the classical model of structured expert judgment. RFF, 16–28, https://ssrn.com/abstract=2849605 or http://dx.doi.org/10.2139/ssrn.2849605.

Colzato, L.S., van der Wel, P., Sellaro, R., Hommel, B., 2016. A single bout of meditation biases cognitive control but not attentional focusing: evidence from the global–local task. Conscious. Cogn. 39, 1–7.

Corazza, O., Bersani, F.S., Brunoro, R., Valeriani, G., Martinotti, G., Schifano, F., 2014. The diffusion of performance and image-enhancing drugs (PIEDs) on the internet: the abuse of the cognitive enhancer Piracetam. Subst. Use Misuse 49 (14), 1849–1856.

Cotman, C.W., Berchtold, N.C., 2002. Exercise: a behavioral intervention to enhance brain health and plasticity. Trends Neurosci. 25 (6), 295–301, http://www.ncbi.nlm.nih.gov/pubmed/12086747.

Cotman, C.W., Berchtold, N.C., Christie, L.-A., 2007. Exercise builds brain health: key roles of growth factor cascades and inflammation. Trends Neurosci. 30 (9), 464–472.

Cranson, R.W., Orme-Johnson, D.W., Gackenbach, J., Dillbeck, M.C., Jones, C.H., Alexander, C.N., 1991. Transcendental meditation and improved performance on intelligence-related measures: a longitudinal study. Pers. Individ. Dif. 12 (10), 1105–1116.

Cusick, S.E., Georgieff, M.K., 2012. Nutrient supplementation and neurodevelopment: timing is the key. Arch. Pediatr. Adolesc. Med. 166 (5), 481–482.

Cusick, S.E., Georgieff, M.K., 2016. The role of nutrition in brain development: the golden opportunity of the first 1000 days. J. Pediatr. 175, 16–21.

Daley, A.J., Ryan, J., 2000. Academic performance and participation in physical activity by secondary school adolescents. Percept. Motor Skills 91 (2), 531–534.

Davidson, R.J., Kaszniak, A.W., 2015. Conceptual and methodological issues in research on mindfulness and meditation. Am. Psychol. 70 (7), 581–592.

de Ruiter, M.A., Oosterlaan, J., Schouten-van Meeteren, A.Y.N., Maurice-Stam, H., van Vuurden, D.G., Gidding, C., Grootenhuis, M.A., 2016. Neurofeedback ineffective in paediatric brain tumour survivors: results of a double-blind randomised placebo-controlled trial. Eur. J. Cancer 64, 62–73.

deBettencourt, M.T., Cohen, J.D., Lee, R.F., Norman, K.A., Turk-Browne, N.B., 2015. Closed-loop training of attention with real-time brain imaging. Nat. Neurosci. 18 (3), 470–475.

Der, G., Batty, G.D., Deary, I.J., 2006. Effect of breast feeding on intelligence in children: prospective study, sibling pairs analysis, and meta-analysis. BMJ 333 (7575).

Der, G., Batty, G.D., Deary, I.J., 2009. The association between IQ in adolescence and a range of health outcomes at 40 in the 1979 US National Longitudinal Study of Youth. Intelligence 37 (6), 573–580.

Dik, M.G., Pluijm, S.M., Jonker, .C., Deeg, D.J., Lomecky, M.Z., Lips, P., 2003. Insulin-like growth factor I (IGF-I) and cognitive decline in older persons. Neurobiol. Aging 24 (4), 573–581.

Dillbeck, M.C., Assimakis, P.D., Raimondi, D., Orme-Johnson, D.W., Rowe, R., 1986. Longitudinal effects of the transcendental meditation and TM-Sidhi program on cognitive ability and cognitive style. Percept. Motor Skills 62 (3), 731–738.

Ding, Y.-H., Li, J., Zhou, Y., Rafols, J.A., Clark, J.C., Ding, Y., 2006. Cerebral angiogenesis and expression of angiogenic factors in aging rats after exercise. Curr. Neurovasc. Res. 3 (1), 15–23, http://www.ncbi.nlm.nih.gov/pubmed/16472122.

Duffy, S.N., Craddock, K.J., Abel, T., Nguyen, P.V., 2001. Environmental enrichment modifies the PKA-dependence of hippocampal LTP and improves hippocampus-dependent memory. Learn. Mem. 8 (1), 26–34.

Durga, J., van Boxtel, M.P., Schouten, E.G., Kok, F.J., Jolles, J., Katan, M.B., Verhoef, P., 2007. Effect of 3-year folic acid supplementation on cognitive function in older adults in the FACIT trial: a randomised, double blind, controlled trial. Lancet 369 (9557), 208–216.

Eadie, B.D., Redila, V.A., Christie, B.R., 2005. Voluntary exercise alters the cytoarchitecture of the adult dentate gyrus by increasing cellular proliferation, dendritic complexity, and spine density. J. Comp. Neurol. 486 (1), 39–47.

Egan, M.F., Kojima, M., Callicott, J.H., Goldberg, T.E., Kolachana, B.S., Bertolino, A., Weinberger, D.R., 2003. The BDNF val66met polymorphism affects activity-dependent secretion of BDNF and human memory and hippocampal function. Cell 112 (2), 257–269.

Egner, T., Gruzelier, J.H., 2004. EEG Biofeedback of low beta band components: frequency-specific effects on variables of attention and event-related brain potentials. Clin. Neurophysiol. 115 (1), 131–139.

Eilander, A., Gera, T., Sachdev, H.S., Transler, C., van der Knaap, H.C., Kok, F.J., Osendarp, S.J., 2010. Multiple micronutrient supplementation for improving cognitive performance in children: systematic review of randomized controlled trials. Am. J. Clin. Nutr. 91 (1), 115–130.

Elliott, J.C., Wallace, B.A., Giesbrecht, B., 2014. A week-long meditation retreat decouples behavioral measures of the alerting and executive attention networks. Front. Hum. Neurosci. 8, 69.

Engelbregt, H.J., Keeser, D., van Eijk, L., Suiker, E.M., Eichhorn, D., Karch, S., Pogarell, O., 2016. Short and long-term effects of sham-controlled prefrontal EEG-neurofeedback training in healthy subjects. Clin. Neurophysiol. 127 (4), 1931–1937.

Enriquez-Geppert, S., Huster, R.J., Herrmann, C.S., 2013a. Boosting brain functions: Improving executive functions with behavioral training, neurostimulation, and neurofeedback. Int. J. Psychophysiol. 88 (1), 1–16.

Enriquez-Geppert, S., Huster, R.J., Scharfenort, R., Mokom, Z.N., Vosskuhl, J., Figge, C., Herrmann, C.S., 2013b. The morphology of midcingulate cortex predicts frontal-midline theta neurofeedback success. Front. Hum. Neurosci. 7, 453.

Enriquez-Geppert, S., Huster, R.J., Figge, C., Herrmann, C.S., 2014a. Self-regulation of frontal-midline theta facilitates memory updating and mental set shifting. Front. Behav. Neurosci. 8, 420.

Enriquez-Geppert, S., Huster, R.J., Scharfenort, R., Mokom, Z.N., Zimmermann, J., Herrmann, C.S., 2014b. Modulation of frontal-midline theta by neurofeedback. Biol. Psychol. 95, 59–69.

Erickson, K.I., Prakash, R.S., Voss, M.W., Chaddock, L., Hu, L., Morris, K.S., Kramer, A.F., 2009. Aerobic fitness is associated with hippocampal volume in elderly humans. Hippocampus 19 (10), 1030–1039.

Erickson, K.I., Voss, M.W., Prakash, R.S., Basak, C., Szabo, A., Chaddock, L., Kramer, A.F., 2011. Exercise training increases size of hippocampus and improves memory. Proc. Natl. Acad. Sci. USA 108 (7), 3017–3022.

Erickson, K.I., Banducci, S.E., Weinstein, A.M., Macdonald, A.W., Ferrell, R.E., Halder, I., Manuck, S.B., 2013. The brain-derived neurotrophic factor Val66Met polymorphism moderates an effect of physical activity on working memory performance. Psychol. Sci. 24 (9), 1770–1779.

Eriksson, P.S., Perfilieva, E., Björk-Eriksson, T., Alborn, A.M., Nordborg, C., Peterson, D.A., Gage, F.H., 1998. Neurogenesis in the adult human hippocampus. Nat. Med. 4 (11), 1313–1317.

Esteban-Cornejo, I., Tejero-Gonzalez, C.M., Sallis, J.F., Veiga, O.L., 2015. Physical activity and cognition in adolescents: a systematic review. J. Sci. Med. Sport 18 (5), 534–539.

Etnier, J.L., Nowell, P.M., Landers, D.M., Sibley, B.A., 2006. A meta-regression to examine the relationship between aerobic fitness and cognitive performance. Brain Res. Rev. 52 (1), 119–130.

Evans, J.R., Abarbanel, A., 1999. Introduction to quantitative EEG and neurofeedback. Elsevier, New York, NY.

Fabel, K., Fabel, K., Tam, B., Kaufer, D., Baiker, A., Simmons, N., Palmer, T.D., 2003. VEGF is necessary for exercise-induced adult hippocampal neurogenesis. Eur. J. Neurosci. 18 (10), 2803–2812.

Farah, M.J., Illes, J., Cook-Deegan, R., Gardner, H., Kandel, E., King, P., Wolpe, P.R., 2004. Science and society: neurocognitive enhancement: what can we do and what should we do? Nat. Rev. Neurosci. 5 (5), 421–425.

Farmer, J., Zhao, X., van Praag, H., Wodtke, K., Gage, F., Christie, B., 2004. Effects of voluntary exercise on synaptic plasticity and gene expression in the dentate gyrus of adult male Sprague–Dawley rats in vivo. Neuroscience 124 (1), 71–79.

Farrer, L.A., 1997. Effects of age, sex, and ethnicity on the association between apolipoprotein E genotype and Alzheimer disease. JAMA 278 (16), 1349.

Ferrari, M., Quaresima, V., 2012. A brief review on the history of human functional near-infrared spectroscopy (fNIRS) development and fields of application. NeuroImage 63 (2), 921–935.

Fish: What Pregnant Women and Parents Should Know, 2014. http://www.fda.gov/Food/FoodborneIllnessContaminants/Metals/ucm393070.htm

Fjell, A.M., Walhovd, K.B., 2010. Structural brain changes in aging: courses, causes and cognitive consequences. Rev. Neurosci. 21 (3), 187–221, http://www.ncbi.nlm.nih.gov/pubmed/20879692.

Florin, E., Bock, E., Baillet, S., 2014. Targeted reinforcement of neural oscillatory activity with real-time neuroimaging feedback. NeuroImage 88, 54–60.

Forem, J., 2012. Transcendental meditation: the essential teachings of Maharishi Mahesh Yogi. Hay House, London, UK.

Forlini, C., Racine, E., 2012. Added stakeholders, added value(s) to the cognitive enhancement debate: are academic discourse and professional policies sidestepping values of stakeholders? AJOB Primary Res. 3 (1), 33–47.

Forte, R., Boreham, C.A.G., Leite, J.C., De Vito, G., Brennan, L., Gibney, E.R., Pesce, C., 2013. Enhancing cognitive functioning in the elderly: multicomponent vs resistance training. Clin. Interv. Aging 8, 19–27.

Fox, K.C.R., Nijeboer, S., Dixon, M.L., Floman, J.L., Ellamil, M., Rumak, S.P., Christoff, K., 2014. Is meditation associated with altered brain structure? A systematic review and meta-analysis of morphometric neuroimaging in meditation practitioners. Neurosci. Biobehav. Rev. 43, 48–73.

Franke, A.G., Bagusat, C., Rust, S., Engel, A., Lieb, K., 2014. Substances used and prevalence rates of pharmacological cognitive enhancement among healthy subjects. Eur. Arch. Psychiatry Clin. Neurosci. 264 (S1), 83–90.

Gallant, S.N., 2016. Mindfulness meditation practice and executive functioning breaking down the benefit. Conscious. Cogn. 40, 116–130.

Gard, T., Taquet, M., Dixit, R., Hölzel, B.K., de Montjoye, Y.-A., Brach, N., Lazar, S.W., 2014. Fluid intelligence and brain functional organization in aging yoga and meditation practitioners. Front. Aging Neurosci. 6, 76.

Gertz, K., Priller, J., Kronenberg, G., Fink, K.B., Winter, B., Schröck, H., Endres, M., 2006. Physical activity improves long-term stroke outcome via endothelial nitric oxide synthase-dependent augmentation of neovascularization and cerebral blood flow. Circ. Res. 99 (10), 1132–1140.

Gill, S.S., Seitz, D.P., 2015. Lifestyles and cognitive health: what older individuals can do to optimize cognitive outcomes. JAMA 314 (8), 774–775.

Giurgea, C.E., 1982. The nootropic concept and its prospective implications. Drug Dev. Res. 2 (5), 441–446.

Gómez-Pinilla, F., 2008. Brain foods: the effects of nutrients on brain function. Nat. Rev. Neurosci. 9 (7), 568–578.

Gondoh, Y., Sensui, H., Kinomura, S., Fukuda, H., Fujimoto, T., Masud, M., Takekura, H., 2009. Effects of aerobic exercise training on brain structure and psychological well-being in young adults. J. Sports Med. Phys. Fitness 49 (2), 129–135, http://www.ncbi.nlm.nih.gov/pubmed/19528889.

Gothe, N.P., McAuley, E., 2015. Yoga and cognition. Psychosom. Med. 77 (7), 784–797.
Gould, J.F., Smithers, L.G., Makrides, M., 2013. The effect of maternal omega-3 (n-3) LCPUFA supplementation during pregnancy on early childhood cognitive and visual development: a systematic review and meta-analysis of randomized controlled trials. Am. J. Clin. Nutr. 97 (3), 531–544.
Goyal, M., Bass, E.B., Haythornthwaite, J.A., 2014a. Meditation intervention reviews—reply. JAMA Intern. Med. 174 (7), 1195.
Goyal, M., Singh, S., Sibinga, E.M.S., Gould, N.F., Rowland-Seymour, A., Sharma, R., Haythornthwaite, J.A., 2014b. Meditation programs for psychological stress and well-being. JAMA Intern. Med. 174 (3), 357.
Grant, J.A., Courtemanche, J., Duerden, E.G., Duncan, G.H., Rainville, P., 2010. Cortical thickness and pain sensitivity in zen meditators. Emotion 10 (1), 43–53.
Greely, H., Campbell, P., Sahakian, B., Harris, J., Kessler, R.C., 2008. Towards responsible use of cognitive-enhancing drugs by the healthy. Nature 456 (7223), 702–705.
Groen-Blokhuis, M.M., Franic´, S., van Beijsterveldt, C.E.M., de Geus, E., Bartels, M., Davies, G.E., Boomsma, D.I., 2013. A prospective study of the effects of breastfeeding and FADS2 polymorphisms on cognition and hyperactivity/attention problems. Am. J. Med. Genet. B Neuropsych. Genet. 162 (5), 457–465.
Grundy, S.M., Cleeman, J.I., Daniels, S.R., Donato, K.A., Eckel, R.H., Franklin, B.A., Costa, F., 2005. Diagnosis and management of the metabolic syndrome: an American Heart Association/National Heart, Lung, and Blood Institute scientific statement. Circulation 112 (17), 2735–2752.
Gruzelier, J.H., 2014. EEG-neurofeedback for optimising performance. I: a review of cognitive and affective outcome in healthy participants. Neurosci. Biobehav. Rev. 44, 124–141.
Guiney, H., Lucas, S.J., Cotter, J.D., Machado, L., 2015. Evidence cerebral blood-flow regulation mediates exercise-cognition links in healthy young adults. Neuropsychology 29 (1), 1–9.
Hagenaars, S.P., Harris, S.E., Davies, G., Hill, W.D., Liewald, D.C.M., Ritchie, S.J., Deary, I.J., 2016. Shared genetic aetiology between cognitive functions and physical and mental health in UK Biobank (N = 112 151) and 24 GWAS consortia. Mol. Psychiatry 21 (11), 1624–1632.
Haier, R.J., Siegel, B.V., MacLachlan, A., Soderling, E., Lottenberg, S., Buchsbaum, M.S., 1992. Regional glucose metabolic changes after learning a complex visuospatial/motor task: a positron emission tomographic study. Brain Res. 570 (1), 134–143.
Halder, S., Varkuti, B., Bogdan, M., Kübler, A., Rosenstiel, W., Sitaram, R., Birbaumer, N., 2013. Prediction of brain-computer interface aptitude from individual brain structure. Front. Hum. Neurosci. 7, 105.
Hanslmayr, S., Sauseng, P., Doppelmayr, M., Schabus, M., Klimesch, W., 2005. Increasing Individual upper alpha power by neurofeedback improves cognitive performance in human subjects. Appl. Psychophysiol. Biofeedback 30 (1), 1–10.
Head, D., Bugg, J.M., Goate, A.M., Fagan, A.M., Mintun, M.A., Benzinger, T., Morris, J.C., 2012. Exercise engagement as a moderator of the effects of APOE genotype on amyloid deposition. Arch. Neurol. 69 (5), 636–643.
Henderson, A., Jorm, A., Korten, A., Christensen, H., Jacomb, P., Easteal, S., Mackinnon, A., 1995. Apolipoprotein E allele ∈4, dementia, and cognitive decline in a population sample. Lancet 346 (8987), 1387–1390.
Hernández, S.E., Suero, J., Barros, A., González-Mora, J.L., Rubia, K., Giedd, J., Schandry, R., 2016. Increased grey matter associated with long-term Sahaja Yoga meditation: a voxel-based morphometry study. PLoS One 11 (3), e0150757.
Herting, M.M., Nagel, B.J., 2012. Aerobic fitness relates to learning on a virtual Morris Water Task and hippocampal volume in adolescents. Behav. Brain Res. 233 (2), 517–525.

Hibbeln, J.R., Davis, J.M., Steer, C., Emmett, P., Rogers, I., Williams, C., Golding, J., 2007. Maternal seafood consumption in pregnancy and neurodevelopmental outcomes in childhood (ALSPAC study): an observational cohort study. Lancet 369 (9561), 578–585.

Hillman, C.H., Erickson, K.I., Kramer, A.F., 2008. Be smart, exercise your heart: exercise effects on brain and cognition. Nat. Rev. Neurosci. 9 (1), 58–65.

Hindin, S.B., Zelinski, E.M., 2012. Extended practice and aerobic exercise interventions benefit untrained cognitive outcomes in older adults: a meta-analysis. J. Am. Geriatr. Soc. 60 (1), 136–141.

Ho, A.J., Raji, C.A., Becker, J.T., Lopez, O.L., Kuller, L.H., Hua, X., Thompson, P.M., 2011. The effects of physical activity, education, and body mass index on the aging brain. Hum. Brain Map. 32 (9), 1371–1382.

Hofmann, S.G., Grossman, P., Hinton, D.E., 2011. Loving-kindness and compassion meditation: potential for psychological interventions. Clin. Psychol. Rev. 31 (7), 1126–1132.

Holmes, M.M., Galea, L.A.M., Mistlberger, R.E., Kempermann, G., 2004. Adult hippocampal neurogenesis and voluntary running activity: circadian and dose-dependent effects. J. Neurosci. Res. 76 (2), 216–222.

Hölzel, B.K., Ott, U., Gard, T., Hempel, H., Weygandt, M., Morgen, K., Vaitl, D., 2008. Investigation of mindfulness meditation practitioners with voxel-based morphometry. Social Cogn. Affect. Neurosci. 3 (1), 55–61.

Hölzel, B.K., Carmody, J., Vangel, M., Congleton, C., Yerramsetti, S.M., Gard, T., Lazar, S.W., 2011. Mindfulness practice leads to increases in regional brain gray matter density. Psychiatry Res. Neuroimaging 191 (1), 36–43.

Horta, B., Victora, C.G., 2013. Long-term effects of breastfeeding—a systematic review. World Health Organization, Geneva, Switzerland.

Horta, B., Bahl, R., Martines, J., Victora, C.G., 2007. Evidence on the long-term effects of breastfeeding: systematic reviews and meta-analysis. World Health Organization, Geneva, Switzerland.

Horta, B., Loret de Mola, C., Victora, C.G., 2015. Breastfeeding and intelligence: a systematic review and meta-analysis. Acta Paediatr. 104 (S467), 14–19.

Hosseini, S.M.H., Pritchard-Berman, M., Sosa, N., Ceja, A., Kesler, S.R., 2016. Task-based neurofeedback training: a novel approach toward training executive functions. NeuroImage 134, 153–159.

Huang, T., Larsen, K.T., Ried-Larsen, M., Møller, N.C., Andersen, L.B., 2014. The effects of physical activity and exercise on brain-derived neurotrophic factor in healthy humans: a review. Scand. J. Med. Scie. Sports 24 (1), 1–10.

Hunt, E., Jaeggi, S., 2013. Challenges for research on intelligence. J. Intell. 1 (1), 36–54.

Husain, M., Mehta, M.A., 2011. Cognitive enhancement by drugs in health and disease. Trends Cogn. Sci. 15 (1), 28–36.

Ilieva, I.P., Hook, C.J., Farah, M.J., 2015. Prescription stimulants' effects on healthy inhibitory control, working memory, and episodic memory: a meta-analysis. J. Cogn. Neurosci. 27 (6), 1069–1089.

Innes, K.E., Selfe, T.K., 2014. Meditation as a therapeutic intervention for adults at risk for Alzheimer's disease—potential benefits and underlying mechanisms. Front. Psychiatry 5, 40.

Jacobson, S.W., Chiodo, L.M., Jacobson, J.L., 1999. Breastfeeding effects on intelligence quotient in 4- and 11-year-old children. Pediatrics 103 (5.).

Jain, A., Concato, J., Leventhal, J.M., 2002. How good is the evidence linking breastfeeding and intelligence? Pediatrics 109 (6), 1044–1053, http://www.ncbi.nlm.nih.gov/pubmed/12042541.

Jensen, O., Tesche, C.D., 2002. Frontal theta activity in humans increases with memory load in a working memory task. Eur. J. Neurosci. 15 (8), 1395–1399, http://www.ncbi.nlm.nih.gov/pubmed/11994134.

Jensen, C.G., Vangkilde, S., Frokjaer, V., Hasselbalch, S.G., 2012. Mindfulness training affects attention—or is it attentional effort? J. Exp. Psychol. Gen. 141 (1), 106–123.

Jha, A.P., Krompinger, J., Baime, M.J., 2007. Mindfulness training modifies subsystems of attention. Cogn. Affect. Behav. Neurosci. 7 (2), 109–119.

Jha, A.P., Stanley, E.A., Kiyonaga, A., Wong, L., Gelfand, L., 2010. Examining the protective effects of mindfulness training on working memory capacity and affective experience. Emotion 10 (1), 54–64.

Jiao, J., Li, Q., Chu, J., Zeng, W., Yang, M., Zhu, S., 2014. Effect of n-3 PUFA supplementation on cognitive function throughout the life span from infancy to old age: a systematic review and meta-analysis of randomized controlled trials. Am. J. Clin. Nutr. 100 (6), 1422–1436.

Kanazawa, S., Satoshi, 2015. Breastfeeding is positively associated with child intelligence even net of parental IQ. Dev. Psychol. 51 (12), 1683–1689.

Kang, D.-H., Jo, H.J., Jung, W.H., Kim, S.H., Jung, Y.-H., Choi, C.-H., Kwon, J.S., 2013. The effect of meditation on brain structure: cortical thickness mapping and diffusion tensor imaging. Social Cogn. Affect. Neurosci. 8 (1), 27–33.

Karlsson, H., Ahlborg, B., Dalman, C., Hemmingsson, T., 2010. Association between erythrocyte sedimentation rate and IQ in Swedish males aged 18-20. Brain Behav. Immun. 24 (6), 868–873.

Karp, A., Paillard-Borg, S., Wang, H.-X., Silverstein, M., Winblad, B., Fratiglioni, L., 2006. Mental, physical and social components in leisure activities equally contribute to decrease dementia risk. Dement. Geriatr. Cogn. Disord. 21 (2), 65–73.

Keizer, A.W., Verment, R.S., Hommel, B., 2010a. Enhancing cognitive control through neurofeedback: a role of gamma-band activity in managing episodic retrieval. NeuroImage 49 (4), 3404–3413.

Keizer, A.W., Verschoor, M., Verment, R.S., Hommel, B., 2010b. The effect of gamma enhancing neurofeedback on the control of feature bindings and intelligence measures. Int J. Psychophysiol. 75 (1), 25–32.

Kelly, M.E., Loughrey, D., Lawlor, B.A., Robertson, I.H., Walsh, C., Brennan, S., 2014. The impact of exercise on the cognitive functioning of healthy older adults: a systematic review and meta-analysis. Ageing Res. Rev. 16, 12–31.

Kennedy, D.O., Wightman, E.L., Reay, J.L., Lietz, G., Okello, E.J., Wilde, A., Haskell, C.F., 2010. Effects of resveratrol on cerebral blood flow variables and cognitive performance in humans: a double-blind, placebo-controlled, crossover investigation. Am. J. Clin. Nutr. 91 (6), 1590–1597.

Khor, G.L., Misra, S., 2012. Micronutrient interventions on cognitive performance of children aged 5–15 years in developing countries. Asia Pacific J. Clin. Nutr. 21 (4), 476–486, http://www.ncbi.nlm.nih.gov/pubmed/23017305.

Killgore, W.D.S., Olson, E.A., Weber, M., 2013. Physical exercise habits correlate with gray matter volume of the hippocampus in healthy adult humans. Sci. Rep. 3, 3457.

Kimura, K., Obuchi, S., Arai, T., Nagasawa, H., Shiba, Y., Watanabe, S., Kojima, M., 2010. The influence of short-term strength training on health-related quality of life and executive cognitive function. J. Physiol. Anthropol. 29 (3), 95–101, http://www.ncbi.nlm.nih.gov/pubmed/20558967.

Klimesch, W., 2012. Alpha-band oscillations, attention, and controlled access to stored information. Trends Cogn. Sci. 16 (12), 606–617.

Koush, Y., Rosa, M.J., Robineau, F., Heinen, K.W., Rieger, S., Weiskopf, N., Scharnowski, F., 2013. Connectivity-based neurofeedback: dynamic causal modeling for real-time fMRI. NeuroImage 81, 422–430.

Kramer, A.F., Erickson, K.I., 2007. Capitalizing on cortical plasticity: influence of physical activity on cognition and brain function. Trends Cogn. Sci. 11 (8), 342–348.

Kramer, M.S., Aboud, F., Mironova, E., Vanilovich, I., Platt, R.W., Matush, L., AS, R., 2008. Breastfeeding and child cognitive development. Arch. Gen. Psychiatry 65 (5), 578.

Lam, L.F., Lawlis, T.R. Feeding the brain—the effects of micronutrient interventions on cognitive performance among school-aged children: a systematic review of randomized controlled trials. Clin. Nutr. (in press).

Laneri, D., Schuster, V., Dietsche, B., Jansen, A., Ott, U., Sommer, J., 2015. Effects of long-term mindfulness meditation on brain's white matter microstructure and its aging. Front. Aging Neurosci. 7, 254.

Larson, E.B., 2006. Exercise is associated with reduced risk for incident dementia among persons 65 years of age and older. Ann. Intern. Med. 144 (2), 73.

Laurin, D., Verreault, R., Lindsay, J., MacPherson, K., Rockwood, K., 2001. Physical activity and risk of cognitive impairment and dementia in elderly persons. Arch. Neurol. 58 (3), 498–504.

Lazar, S.W., Kerr, C.E., Wasserman, R.H., Gray, J.R., Greve, D.N., Treadway, M.T., Fischl, B., 2005. Meditation experience is associated with increased cortical thickness. Neuroreport 16 (17), 1893–1897, http://www.ncbi.nlm.nih.gov/pubmed/16272874.

Leung, M.-K., Chan, C.C.H., Yin, J., Lee, C.-F., So, K.-F., Lee, T.M.C., 2013. Increased gray matter volume in the right angular and posterior parahippocampal gyri in loving-kindness meditators. Social Cogn. Affect. Neurosci. 8 (1), 34–39.

Li, H., Liang, A., Guan, F., Fan, R., Chi, L., Yang, B., 2013. Regular treadmill running improves spatial learning and memory performance in young mice through increased hippocampal neurogenesis and decreased stress. Brain Res. 1531, 1–8.

Lisman, J., 2010. Working memory: the importance of theta and gamma oscillations. Curr. Biol. 20 (11), R490–R492.

Liu-Ambrose, T., Nagamatsu, L.S., Graf, P., Beattie, B.L., Ashe, M.C., Handy, T.C., 2010. Resistance training and executive functions: a 12-month randomized controlled trial. Arch. Intern. Med. 170 (2), 170–178.

Liu-Ambrose, T., Nagamatsu, L.S., Voss, M.W., Khan, K.M., Handy, T.C., 2012. Resistance training and functional plasticity of the aging brain: a 12-month randomized controlled trial. Neurobiol. Aging 33 (8), 1690–1698.

Loef, M., Walach, H., 2012. The combined effects of healthy lifestyle behaviors on all cause mortality: a systematic review and meta-analysis. Prevent. Med. 55 (3), 163–170.

Lubar, J.F., Swartwood, M.O., Swartwood, J.N., O'Donnell, P.H., 1995. Evaluation of the effectiveness of EEG neurofeedback training for ADHD in a clinical setting as measured by changes in T.O.V.A. scores, behavioral ratings, and WISC-R performance. Biofeedback Self Regul. 20 (1), 83–99.

Lucas, A., Morley, R., 1994. Does early nutrition in infants born before term programme later blood pressure? BMJ 309 (6950), 304–309.

Luders, E., Toga, A.W., Lepore, N., Gaser, C., 2009. The underlying anatomical correlates of long-term meditation: larger hippocampal and frontal volumes of gray matter. NeuroImage 45 (3), 672–678.

Luders, E., Clark, K., Narr, K.L., Toga, A.W., 2011. Enhanced brain connectivity in long-term meditation practitioners. NeuroImage 57 (4), 1308–1316.

Luders, E., Phillips, O.R., Clark, K., Kurth, F., Toga, A.W., Narr, K.L., 2012. Bridging the hemispheres in meditation: thicker callosal regions and enhanced fractional anisotropy (FA) in long-term practitioners. NeuroImage 61 (1), 181–187.

Luders, E., Cherbuin, N., Gaser, C., 2016. Estimating brain age using high-resolution pattern recognition: younger brains in long-term meditation practitioners. NeuroImage 134, 508–513.

Lukowski, A.F., Koss, M., Burden, M.J., Jonides, J., Nelson, C.A., Kaciroti, N., Lozoff, B., 2010. Iron deficiency in infancy and neurocognitive functioning at 19 years: evidence of long-term deficits in executive function and recognition memory. Nutr. Neurosci. 13 (2), 54–70.

Lutz, A., Slagter, H.A., Dunne, J.D., Davidson, R.J., 2008. Attention regulation and monitoring in meditation. Trends Cogn. Sci. 12 (4), 163–169.

Lynch, G., Palmer, L.C., Gall, C.M., 2011. The likelihood of cognitive enhancement. Pharmacol. Biochem. Behav. 99 (2), 116–129.

Maass, A., Düzel, S., Goerke, M., Becke, A., Sobieray, U., Neumann, K., Düzel, E., 2015. Vascular hippocampal plasticity after aerobic exercise in older adults. Mol. Psychiatry 20 (5), 585–593.

Maass, A., Düzel, S., Brigadski, T., Goerke, M., Becke, A., Sobieray, U., Düzel, E., 2016. Relationships of peripheral IGF-1, VEGF and BDNF levels to exercise-related changes in memory, hippocampal perfusion and volumes in older adults. NeuroImage 131, 142–154.

MacLean, K.A., Ferrer, E., Aichele, S.R., Bridwell, D.A., Zanesco, A.P., Jacobs, T.L., Saron, C.D., 2010. Intensive meditation training improves perceptual discrimination and sustained attention. Psychol. Sci. 21 (6), 829–839.

Markowska, A., Mooney, M., Sonntag, W., 1998. Insulin-like growth factor-1 ameliorates age-related behavioral deficits. Neuroscience 87 (3), 559–569.

Marlatt, M.W., Potter, M.C., Lucassen, P.J., van Praag, H., 2012. Running throughout middle-age improves memory function, hippocampal neurogenesis, and BDNF levels in female C57BL/6J mice. Dev. Neurobiol. 72 (6), 943–952.

Martin, N.W., Benyamin, B., Hansell, N.K., Montgomery, G.W., Martin, N.G., Wright, M.J., Bates, T.C., 2011. Cognitive function in adolescence: testing for interactions between breast-feeding and FADS2 polymorphisms. J. Am. Acad. Child Adolesc. Psychiatry 50 (1.), 55.e4-62.e4.

Maslen, H., FaulmÃ¼ller, N., Savulescu, J., 2014. Pharmacological cognitive enhancement: how neuroscientific research could advance ethical debate. Front. Syst. Neurosci. 8, 107.

Masley, S., Roetzheim, R., Gualtieri, T., 2009. Aerobic exercise enhances cognitive flexibility. J. Clin. Psychol. Med. Set. 16 (2), 186–193.

Mastroiacovo, D., Kwik-Uribe, C., Grassi, D., Necozione, S., Raffaele, A., Pistacchio, L., Desideri, G., 2015. Cocoa flavanol consumption improves cognitive function, blood pressure control, and metabolic profile in elderly subjects: the Cocoa, Cognition, and Aging (CoCoA) Study--a randomized controlled trial. Am. J. Clin. Nutr. 101 (3), 538–548.

McCabe, S.E., Knight, J.R., Teter, C.J., Wechsler, H., 2005. Non-medical use of prescription stimulants among US college students: prevalence and correlates from a national survey. Addiction 100 (1), 96–106.

McEwen, B.S., Morrison, J.H., 2013. The brain on stress: vulnerability and plasticity of the prefrontal cortex over the life course. Neuron 79 (1), 16–29.

McNeill, G., Avenell, A., Campbell, M.K., Cook, J.A., Hannaford, P.C., Kilonzo, M.M., Rees, L., 2007. Effect of multivitamin and multimineral supplementation on cognitive function in men and women aged 65 years and over: a randomised controlled trial. Nutr. J. 6 (1), 10.

Mellinger, J., Schalk, G., Braun, C., Preissl, H., Rosenstiel, W., Birbaumer, N., Kübler, A., 2007. An MEG-based brain–computer interface (BCI). NeuroImage 36 (3), 581–593.

Mendez, M.A., Torrent, M., Julvez, J., Ribas-Fitó, N., Kogevinas, M., Sunyer, J., 2009. Maternal fish and other seafood intakes during pregnancy and child neurodevelopment at age 4 years. Public Health Nutr. 12 (10), 1702–1710.

Mitchell, D.J., McNaughton, N., Flanagan, D., Kirk, I.J., 2008. Frontal-midline theta from the perspective of hippocampal theta. Prog. Neurobiol. 86 (3), 156–185.

Moore, A., Malinowski, P., 2009. Meditation, mindfulness and cognitive flexibility. Conscious Cogn. 18 (1), 176–186.

Morales, E., Bustamante, M., Gonzalez, J.R., Guxens, M., Torrent, M., Mendez, M., Sunyer, J., 2011. Genetic Variants of the FADS gene cluster and ELOVL gene family, colostrums LC-PUFA levels, breastfeeding, and child cognition. PLoS One 6 (2), e17181.

Moreau, D., Morrison, A.B., Conway, A.R.A., 2015. An ecological approach to cognitive enhancement: complex motor training. Acta Psychol. 157, 44–55.

Mrazek, M.D., Franklin, M.S., Phillips, D.T., Baird, B., Schooler, J.W., 2013. Mindfulness training improves working memory capacity and GRE performance while reducing mind wandering. Psychol. Sci. 24 (5), 776–781.

Müller, O., Krawinkel, M., 2005. Malnutrition and health in developing countries. Can. Med. Assoc. J. 173 (3), 279–286.

Murray-Kolb, L.E., Khatry, S.K., Katz, J., Schaefer, B.A., Cole, P.M., LeClerq, S.C., Christian, P., 2012. Preschool micronutrient supplementation effects on intellectual and motor function in school-aged Nepalese children. Arch. Pediatr. Adolesc. Med. 166 (5), 404–410.

Murrell, C.J., Cotter, J.D., Thomas, K.N., Lucas, S.J.E., Williams, M.J.A., Ainslie, P.N., 2013. Cerebral blood flow and cerebrovascular reactivity at rest and during sub-maximal exercise: effect of age and 12-week exercise training. Age 35 (3), 905–920.

Mustroph, M.L., Chen, S., Desai, S.C., Cay, E.B., DeYoung, E.K., Rhodes, J.S., 2012. Aerobic exercise is the critical variable in an enriched environment that increases hippocampal neurogenesis and water maze learning in male C57BL/6J mice. Neuroscience 219, 62–71.

Nan, W., Rodrigues, J.P., Ma, J., Qu, X., Wan, F., Mak, P.-I., Rosa, A., 2012. Individual alpha neurofeedback training effect on short term memory. Int. J. Psychophysiol. 86 (1), 83–87.

Naseer, N., Hong, K.-S., 2015. fNIRS-based brain-computer interfaces: a review. Front. Hum. Neurosci. 9, 3.

Naseer, N., Hong, M.J., Hong, K.-S., 2014. Online binary decision decoding using functional near-infrared spectroscopy for the development of brain-computer interface. Exp. Brain Res. 232 (2), 555–564.

Nehlig, A., 2010. Is caffeine a cognitive enhancer? J. Alzheimer Dis. 20 (S1), 85–94.

Nelson, M.E., Rejeski, W.J., Blair, S.N., Duncan, P.W., Judge, J.O., King, A.C., Castaneda-Sceppa, C., 2007. Physical activity and public health in older adults: recommendation from the American College of Sports Medicine and the American Heart Association. Med. Sci. Sports Exerc. 39 (8), 1435–1445.

Newhouse, P.A., Potter, A., Singh, A., 2004. Effects of nicotinic stimulation on cognitive performance. Curr. Opin. Pharmacol. 4 (1), 36–46.

Ngandu, T., Lehtisalo, J., Solomon, A., Levälahti, E., Ahtiluoto, S., Antikainen, R., Kivipelto, M., 2015. A 2 year multidomain intervention of diet, exercise, cognitive training, and vascular risk monitoring versus control to prevent cognitive decline in at-risk elderly people (FINGER): a randomised controlled trial. Lancet 385 (9984), 2255–2263.

Nidich, S., Mjasiri, S., Nidich, R., Rainforth, M., Grant, J., Valosek, L., Zigler, R.L., 2011. Academic achievement and transcendental meditation: a study with at-risk urban middle school students. Education 131 (3), 556–564.

Nidich, S.I., Nidich, R.J., 1989. Increased academic achievement at Maharishi School of the Age of Enlightenment: a replication study. Education 109 (3), 302–304.

Nidich, S.I., Schneider, R.H., Nidich, R.J., Foster, G., Sharma, H., Salerno, J., et al., 2005. Effect of the transcendental meditation program on intellectual development in community-dwelling older adults. J. Social Behav. Personal. 17 (1), 217.

Niemegeers, P., Dumont, G.J.H., Quisenaerts, C., Morrens, M., Boonzaier, J., Fransen, E., Sabbe, B.G.C., 2014. The effects of nicotine on cognition are dependent on baseline performance. Eur. Neuropsychopharmacol. 24 (7), 1015–1023.

Niki, E., 2014. Biomarkers of lipid peroxidation in clinical material. Biochim. Biophys. Acta 1840 (2), 809–817.

Niv, S., 2013. Clinical efficacy and potential mechanisms of neurofeedback. Pers. Individ. Dif. 54 (6), 676–686.

Okazaki, Y.O., Horschig, J.M., Luther, L., Oostenveld, R., Murakami, I., Jensen, O., 2015. Real-time MEG neurofeedback training of posterior alpha activity modulates subsequent visual detection performance. NeuroImage 107, 323–332.

Oken, E., Radesky, J.S., Wright, R.O., Bellinger, D.C., Amarasiriwardena, C.J., Kleinman, K.P., Gillman, M.W., 2008. Maternal fish intake during pregnancy, blood mercury levels, and child cognition at age 3 years in a US cohort. Am. J. Epidemiol. 167 (10), 1171–1181.

Oken, E., Rifas-Shiman, S.L., Amarasiriwardena, C., Jayawardene, I., Bellinger, D.C., Hibbeln, J.R., Gillman, M.W., 2016. Maternal prenatal fish consumption and cognition in mid childhood: mercury, fatty acids, and selenium. Neurotoxicol. Teratol., http://dx.doi.org/10.1016/j.clnu.2016.06.013.

Ong Lai Teik, D., Lee, X.S., Lim, C.J., Low, C.M., Muslima, M., Aquili, L., Inzlicht, M., 2016. Ginseng and ginkgo biloba effects on cognition as modulated by cardiovascular reactivity: a randomised trial. PLoS One 11 (3), e0150447.

Oomen, C., Farkas, E., Roman, V., Van Der Beek, E.M., Luiten, P.G., Meerlo, P., 2009. Resveratrol preserves cerebrovascular density and cognitive function in aging mice. Front. Aging Neurosci. 1, 4.

Orlando, P.C., Rivera, R.O., 2004. Neurofeedback for elementary students with identified learning problems. J. Neurother. 8 (2), 5–19.

Pagnoni, G., Cekic, M., 2007. Age effects on gray matter volume and attentional performance in Zen meditation. Neurobiol. Aging 28, 1623–1627.

Paholpak, S., Piyavhatkul, N., Rangseekajee, P., Krisanaprakornkit, T., Arunpongpaisal, S., Pajanasoontorn, N., Unprai, P., 2012. Breathing meditation by medical students at Khon Kaen University: effect on psychiatric symptoms, memory, intelligence and academic achievement. J. Med. Assoc. Thai. 95 (3), 461–469, http://www.ncbi.nlm.nih.gov/pubmed/22550848.

Pahor, M., Blair, S.N., Espeland, M., Fielding, R., Gill, T.M., Guralnik, J.M., Studenski, S., 2006. Effects of a physical activity intervention on measures of physical performance: results of the lifestyle interventions and independence for Elders Pilot (LIFE-P) study. J. Gerontol. A Biol. Sci. Med. Sci. 61 (11), 1157–1165, http://www.safetylit.org/citations/index.php?fuseaction=citations.viewdetails&citationIds%5B%5D=citjournalarticle_376414_38.

Partridge, B.J., Bell, S.K., Lucke, J.C., Yeates, S., Hall, W.D., 2011. Smart drugs as common as coffee: media hype about neuroenhancement. PLoS One 6 (11), e28416.

Patel, N.K., Newstead, A.H., Ferrer, R.L., 2012. The effects of yoga on physical functioning and health related quality of life in older adults: a systematic review and meta-analysis. J. Altern. Complement. Med. 18 (10), 902–917.

Pearce, E.N., 2012. Effects of iodine deficiency in pregnancy. J. Trace Elem. Med. Biol. 26 (2), 131–133.

Pedersen, B.K., 2006. The anti-inflammatory effect of exercise: its role in diabetes and cardiovascular disease control. Essays Biochem. 42, 105–117.

Pereira, A.C., Huddleston, D.E., Brickman, A.M., Sosunov, A.A., Hen, R., McKhann, G.M., Small, S.A., 2007. An in vivo correlate of exercise-induced neurogenesis in the adult dentate gyrus. Proc. Natl. Acad. Sci. USA 104 (13), 5638–5643.

Perry, V.H., 2004. The influence of systemic inflammation on inflammation in the brain: implications for chronic neurodegenerative disease. Brain Behav. Immun. 18 (5), 407–413.

Physical Activity Guidelines for Americans, 2008. Available from: www.health.gov/paguidelines

Podewils, L.J., Guallar, E., Kuller, L.H., Fried, L.P., Lopez, O.L., Carlson, M., Lyketsos, C.G., 2005. Physical activity, APOE genotype, and dementia risk: findings from the Cardiovascular Health Cognition Study. Am. J. Epidemiol. 161 (7), 639–651.

Prakash, R.S., Voss, M.W., Erickson, K.I., Kramer, A.F., 2015. Physical activity and cognitive vitality. Annu. Rev. Psychol. 66, 769–797.

Prince, S.A., Adamo, K.B., Hamel, M.E., Hardt, J., Connor Gorber, S., Tremblay, M., 2008. A comparison of direct versus self-report measures for assessing physical activity in adults: a systematic review. Int. J. Behav. Nutr. Phys. Activ. 5, 56.

Prince, M., Bryce, R., Albanese, E., Wimo, A., Ribeiro, W., Ferri, C.P., 2013. The global prevalence of dementia: a systematic review and metaanalysis. Alzheimer Dement. 9 (1.), 63.e2-75.e2.

Qiu, C., Fratiglioni, L., 2015. A major role for cardiovascular burden in age-related cognitive decline. Nat. Rev. Cardiol. 12 (5), 267–277.

Quach, D., Jastrowski Mano, K.E., Alexander, K., 2016. A Randomized controlled trial examining the effect of mindfulness meditation on working memory capacity in adolescents. J. Adolesc. Health 58 (5), 489–496.

Raz, N., Lindenberger, U., Rodrigue, K.M., Kennedy, K.M., Head, D., Williamson, A., Acker, J.D., 2005. Regional brain changes in aging healthy adults: general trends, individual differences and modifiers. Cereb. Cortex 15 (11), 1676–1689.

Reiner, M., Rozengurt, R., Barnea, A., 2014. Better than sleep: theta neurofeedback training accelerates memory consolidation. Biol. Psychol. 95, 45–53.

Ritchie, S.J., Booth, T., Valdés Hernández, M., del, C., Corley, J., Maniega, S.M., Gow, A.J., Deary, I.J., 2015. Beyond a bigger brain: multivariable structural brain imaging and intelligence. Intelligence 51, 47–56.

Rogala, J., Jurewicz, K., Paluch, K., Kublik, E., Cetnarski, R., Wróbel, A., 2016. The do's and don'ts of neurofeedback training: a review of the controlled studies using healthy adults. Front. Hum. Neurosci. 10, 301.

Rosano, C., Venkatraman, V.K., Guralnik, J., Newman, A.B., Glynn, N.W., Launer, L., Aizenstein, H., 2010. Psychomotor speed and functional brain MRI 2 years after completing a physical activity treatment. J. Gerontol. A Biol. Sci. Med. Sci. 65A (6), 639–647.

Rose, S.P.R., 2002. Smart Drugs: do they work? Are they ethical? Will they be legal? Nat. Rev. Neurosci. 3 (12), 975–979.

Rovio, S., Kåreholt, I., Helkala, E.-L., Viitanen, M., Winblad, B., Tuomilehto, J., Kivipelto, M., 2005. Leisure-time physical activity at midlife and the risk of dementia and Alzheimer's disease. Lancet Neurol. 4 (11), 705–711.

Ruiz, J.R., Ortega, F.B., Castillo, R., Martín-Matillas, M., Kwak, L., Vicente-Rodríguez, G., Moreno, L.A., 2010. Physical activity, fitness, weight status, and cognitive performance in adolescents. J. Pediatr. 157 (6.), 917–922.

Ruscheweyh, R., Willemer, C., Krüger, K., Duning, T., Warnecke, T., Sommer, J., Flöel, A., 2011. Physical activity and memory functions: an interventional study. Neurobiol. Aging 32 (7), 1304–1319.

Ryan, S.M., Nolan, Y.M., 2016. Neuroinflammation negatively affects adult hippocampal neurogenesis and cognition: can exercise compensate? Neurosci. Biobehav. Rev. 61, 121–131.

Saint, S.E., Frick, J.E., 2015. Prenatal supplementation and its effects on early childhood cognitive outcome. Dietary supplements in health promotion. Taylor and Francis Group, Boca Raton, FL.

Scharnowski, F., Weiskopf, N., 2015. Cognitive enhancement through real-time fMRI neurofeedback. Curr. Opin. Behav. Sci. 4, 122–127.

Schelle, K.J., FaulmÃ¼ller, N., Caviola, L., Hewstone, M., 2014. Attitudes toward pharmacological cognitive enhancement: a review. Front. Syst. Neurosci. 8, 53.

Schermer, M., Bolt, I., de Jongh, R., Olivier, B., 2009. The future of psychopharmacological enhancements: expectations and policies. Neuroethics 2 (2), 75–87.

Schiepers, O.J.G., Harris, S.E., Gow, A.J., Pattie, A., Brett, C.E., Starr, J.M., Deary, I.J., 2012. APOE E4 status predicts age-related cognitive decline in the ninth decade: longitudinal follow-up of the Lothian Birth Cohort 1921. Mol. Psychiatry 17 (3), 315–324.

Schoenthaler, S.J., Bier, I.D., Young, K., Nichols, D., Jansenns, S., 2000. The effect of vitamin-mineral supplementation on the intelligence of American schoolchildren: a randomized, double-blind placebo-controlled trial. J. Altern. Complement. Med. 6 (1), 19–29.

Schuit, A.J., Feskens, E.J., Launer, L.J., Kromhout, D., 2001. Physical activity and cognitive decline, the role of the apolipoprotein e4 allele. Med. Sci. Sports Exerc. 33 (5), 772–777, http://www.ncbi.nlm.nih.gov/pubmed/11323547.

Sedlmeier, P., Eberth, J., Schwarz, M., Zimmermann, D., Haarig, F., Jaeger, S., Kunze, S., 2012. The psychological effects of meditation: a meta-analysis. Psychol. Bull. 138 (6), 1139–1171.

Shen, J., Zhang, G., Yao, L., Zhao, X., 2015. Real-time fMRI training-induced changes in regional connectivity mediating verbal working memory behavioral performance. Neuroscience 289, 144–152.

Sherwood, M.S., Weisend, M.P., Kane, J.H., Parker, J.G., 2016. Combining real-time fMRI neurofeedback training of the DLPFC with n-back practice results in neuroplastic effects confined to the neurofeedback target region. Front. Behav. Neurosci. 10, 138.

Shrimpton, R., Huffman, S.L., Zehner, E.R., Darnton-Hill, I., Dalmiya, N., 2009. Multiple micronutrient supplementation during pregnancy in developing-country settings: policy and program implications of the results of a meta-analysis. Food N Bull. 30 (4 Suppl.), S556–S573.

Singh, I., Bard, I., Jackson, J., Pustovrh, T., Mali, F., Smith, M., Reips, U., 2014. Robust resilience and substantial interest: a survey of pharmacological cognitive enhancement among university students in the UK and Ireland. PLoS One 9 (10), e105969.

Singhal, A., Cole, T.J., Fewtrell, M., Lucas, A., 2004. Breastmilk feeding and lipoprotein profile in adolescents born preterm: follow-up of a prospective randomised study. Lancet 363 (9421), 1571–1578.

Singh-Manoux, A., Hillsdon, M., Brunner, E., Marmot, M., 2005. Effects of physical activity on cognitive functioning in middle age: evidence from the Whitehall II prospective cohort study. Am. J. Public Health 95 (12), 2252–2258.

Slagter, H.A., Lutz, A., Greischar, L.L., Francis, A.D., Nieuwenhuis, S., Davis, J.M., Davidson, R.J., 2007. Mental training affects distribution of limited brain resources. PLoS Biol. 5 (6), e138.

Slagter, H.A., Lutz, A., Greischar, L.L., Nieuwenhuis, S., Davidson, R.J., 2009. Theta phase synchrony and conscious target perception: impact of intensive mental training. J. Cogn. Neurosci. 21 (8), 1536–1549.

Smith, M.E., Farah, M.J., 2011. Are prescription stimulants smart pills? The epidemiology and cognitive neuroscience of prescription stimulant use by normal healthy individuals. Psychol. Bull. 137 (5), 717–741.

Smith, P.J., Blumenthal, J.A., Hoffman, B.M., Cooper, H., Strauman, T.A., Welsh-Bohmer, K., Sherwood, A., 2010. Aerobic exercise and neurocognitive performance: a meta-analytic review of randomized controlled trials. Psychosom. Med. 72 (3), 239–252.

Smith, J.C., Nielson, K.A., Woodard, J.L., Seidenberg, M., Durgerian, S., Hazlett, K.E., Rao, S.M., 2014. Physical activity reduces hippocampal atrophy in elders at genetic risk for Alzheimer's disease. Front. Aging Neurosci. 6, 61.

Snowden, M., Steinman, L., Mochan, K., Grodstein, F., Prohaska, T.R., Thurman, D.J., Anderson, L.A., 2011. Effect of exercise on cognitive performance in community-dwelling older adults: review of intervention trials and recommendations for public health practice and research. J. Am. Geriatr. Soc. 59 (4), 704–716.

So, K.-T., Orme-Johnson, D.W., 2001. Three randomized experiments on the longitudinal effects of the transcendental meditation technique on cognition. Intelligence 29 (5), 419–440.

Sofi, F., Valecchi, D., Bacci, D., Abbate, R., Gensini, G.F., Casini, A., Macchi, C., 2011. Physical activity and risk of cognitive decline: a meta-analysis of prospective studies. J. Intern. Med. 269 (1), 107–117.

Solon, F.S., Sarol, J.N., Bernardo, A.B.I., Solon, J.A.A., Mehansho, H., Sanchez-Fermin, L.E., Juhlin, K.D., 2003. Effect of a multiple-micronutrient-fortified fruit powder beverage on the nutrition status, physical fitness, and cognitive performance of schoolchildren in the Philippines. Food Nutr. Bull. 24 (4 Suppl.), S129–S140, http://www.ncbi.nlm.nih.gov/pubmed/17016955.

Staufenbiel, S.M., Brouwer, A.-M., Keizer, A.W., van Wouwe, N.C., 2014. Effect of beta and gamma neurofeedback on memory and intelligence in the elderly. Biol. Psychol. 95, 74–85.

Steer, C.D., Davey Smith, G., Emmett, P.M., Hibbeln, J.R., Golding, J., 2010. FADS2 polymorphisms modify the effect of breastfeeding on child IQ. PLoS One 5 (7), e11570.

Stern, Y., 2009. Cognitive reserve. Neuropsychologia 47 (10), 2015–2028.

Stomrud, E., Hansson, O., Minthon, L., Blennow, K., Rosén, I., Londos, E., 2010. Slowing of EEG correlates with CSF biomarkers and reduced cognitive speed in elderly with normal cognition over 4 years. Neurobiol. Aging 31 (2), 215–223.

Stough, C., Camfield, D., Kure, C., Tarasuik, J., Downey, L., Lloyd, J., Reynolds, J., 2011. Improving general intelligence with a nutrient-based pharmacological intervention. Intelligence 39 (2), 100–107.

Stroth, S., Hille, K., Spitzer, M., Reinhardt, R., 2009. Aerobic endurance exercise benefits memory and affect in young adults. Neuropsychol. Rehabil. 19 (2), 223–243.

Sturman, M.T., Morris, M.C., Mendes de Leon, C.F., Bienias, J.L., Wilson, R.S., Evans, D.A., 2005. Physical activity, cognitive activity, and cognitive decline in a biracial community population. Arch. Neurol. 62 (11), 1750–1754.

Sulzer, J., Haller, S., Scharnowski, F., Weiskopf, N., Birbaumer, N., Blefari, M.L., Sitaram, R., 2013. Real-time fMRI neurofeedback: progress and challenges. NeuroImage 76, 386–399.

Sungthong, R., Mo-suwan, L., Chongsuvivatwong, V., Geater, A.F., 2004. Once-weekly and 5-days a week iron supplementation differentially affect cognitive function but not school performance in Thai children. J. Nutr. 134 (9), 2349–2354, http://www.ncbi.nlm.nih.gov/pubmed/15333727.

Surmeli, T., Ertem, A., 2010. Post WISC-R and TOVA improvement with QEEG guided neurofeedback training in mentally retarded: a clinical case series of behavioral problems. Clin. EEG Neurosci. 41 (1), 32–41.

Swain, R., Harris, A., Wiener, E., Dutka, M., Morris, H., Theien, B., Greenough, W., 2003. Prolonged exercise induces angiogenesis and increases cerebral blood volume in primary motor cortex of the rat. Neuroscience 117 (4), 1037–1046.

Szuhany, K., Bugatti, M., Otto, M., 2015. A meta-analytic review of the effects of exercise on brain-derived neurotrophic factor. J. Psychiatric Res. 60, 56–64, http://www.sciencedirect.com/science/article/pii/S0022395614002933.

Tang, Y.-Y., Ma, Y., Wang, J., Fan, Y., Feng, S., Lu, Q., Posner, M.I., 2007. Short-term meditation training improves attention and self-regulation. Proc. Natl. Acad. Sci. USA 104 (43), 17152–17156.

Tang, Y.-Y., Lu, Q., Geng, X., Stein, E.A., Yang, Y., Posner, M.I., 2010. Short-term meditation induces white matter changes in the anterior cingulate. Proc. Natl. Acad. Sci. USA 107 (35), 15649–15652.

Tang, Y.Y., Tang, R., Jiang, C., Posner, M.I., 2014. Short-term meditation intervention improves self-regulation and academic performance. J. Child Adolesc. Behav. 2 (4), 154.

Tang, Y.-Y., Hölzel, B.K., Posner, M.I., 2015. The neuroscience of mindfulness meditation. Nat. Rev. Neurosci. 16 (4), 213–225.

Taylor-Piliae, R.E., Froelicher, E.S., 2004. Effectiveness of Tai Chi exercise in improving aerobic capacity: a meta-analysis. J. Cardiovasc. Nurs. 19 (1), 48–57, http://www.ncbi.nlm.nih.gov/pubmed/14994782.

Thibault, R.T., Lifshitz, M., Raz, A., 2016. The self-regulating brain and neurofeedback: experimental science and clinical promise. Cortex 74, 247–261.

Thomas, A.G., Dennis, A., Rawlings, N.B., Stagg, C.J., Matthews, L., Morris, M., Johansen-Berg, H., 2016. Multi-modal characterization of rapid anterior hippocampal volume increase associated with aerobic exercise. NeuroImage 131, 162–170.

Tomasino, B., Fregona, S., Skrap, M., Fabbro, F., 2013. Meditation-related activations are modulated by the practices needed to obtain it and by the expertise: an ALE meta-analysis study. Front. Hum. Neurosci. 6, 346.

Trejo, J.L., Carro, E., Torres-Aleman, I., 2001. Circulating insulin-like growth factor I mediates exercise-induced increases in the number of new neurons in the adult hippocampus. J. Neurosci. 21 (5), 1628–1634, http://www.ncbi.nlm.nih.gov/pubmed/11222653.

Van der Borght, K., Havekes, R., Bos, T., Eggen, B.J.L., Van der Zee, E.A., 2007. Exercise improves memory acquisition and retrieval in the Y-maze task: relationship with hippocampal neurogenesis. Behav. Neurosci. 121 (2), 324–334.

van der Ploeg, H.P., Chey, T., Korda, R.J., Banks, E., Bauman, A., 2012. Sitting time and all-cause mortality risk in 222 497 Australian adults. Arch. Intern. Med. 172 (6), 494–500.

van Praag, H., 2009. Exercise and the brain: something to chew on. Trends Neurosci. 32 (5), 283–290.

van Praag, H., Christie, B.R., Sejnowski, T.J., Gage, F.H., 1999a. Running enhances neurogenesis, learning, and long-term potentiation in mice. Proc. Natl. Acad. Sci. US A 96 (23), 13427–13431, http://www.ncbi.nlm.nih.gov/pubmed/10557337.

van Praag, H., Kempermann, G., Gage, F.H., 1999b. Running increases cell proliferation and neurogenesis in the adult mouse dentate gyrus. Nat. Neurosci. 2 (3), 266–270.

van Praag, H., Shubert, T., Zhao, C., Gage, F.H., 2005. Exercise enhances learning and hippocampal neurogenesis in aged mice. J. Neurosci. 25 (38), 8680–8685.

van Vugt, M.K., Jha, A.P., 2011. Investigating the impact of mindfulness meditation training on working memory: a mathematical modeling approach. Cogn. Affect. Behav. Neurosci. 11 (3), 344–353.

Venables, P.H., Raine, A., 2016. The impact of malnutrition on intelligence at 3 and 11 years of age: the mediating role of temperament. Dev. Psychol. 52 (2), 205–220.

Verghese, J., Lipton, R.B., Katz, M.J., Hall, C.B., Derby, C.A., Kuslansky, G., Buschke, H., 2003. Leisure activities and the risk of dementia in the elderly. N. Engl. J. Med. 348 (25), 2508–2516.

Vernon, D.J., 2005. Can neurofeedback training enhance performance? An evaluation of the evidence with implications for future research. Appl. Psychophysiol. Biofeedback 30 (4), 347–364.

Vernon, D., Egner, T., Cooper, N., Compton, T., Neilands, C., Sheri, A., Gruzelier, J., 2003. The effect of training distinct neurofeedback protocols on aspects of cognitive performance. Int. J. Psychophysiol. 47 (1), 75–85.

Vernon, D., Dempster, T., Bazanova, O., Rutterford, N., Pasqualini, M., Andersen, S., 2009. Alpha neurofeedback training for performance enhancement: reviewing the methodology. J. Neurother. 13 (4), 214–227.

Vestergaard-Poulsen, P., van Beek, M., Skewes, J., Bjarkam, C.R., Stubberup, M., Bertelsen, J., Roepstorff, A., 2009. Long-term meditation is associated with increased gray matter density in the brain stem. NeuroReport 20 (2), 170–174.

Victora, C.G., Bahl, R., Barros, A.J.D., França, G.V.A., Horton, S., Krasevec, J., Rollins, N.C., 2016. Breastfeeding in the 21st century: epidemiology, mechanisms, and lifelong effect. Lancet 387 (10017), 475–490.

Volkow, N.D., Swanson, J.M., 2008. The action of enhancers can lead to addiction. Nature 451 (7178), 520–1520.

Voss, M.W., Prakash, R.S., Erickson, K.I., Basak, C., Chaddock, L., Kim, J.S., Kramer, A.F., 2010. Plasticity of brain networks in a randomized intervention trial of exercise training in older adults. Front. Aging Neurosci. 2, 32.

Voss, M.W., Erickson, K.I., Prakash, R.S., Chaddock, L., Kim, J.S., Alves, H., Kramer, A.F., 2013. Neurobiological markers of exercise-related brain plasticity in older adults. Brain Behav. Immun. 28, 90–99.

Voss, M.W., Weng, T.B., Burzynska, A.Z., Wong, C.N., Cooke, G.E., Clark, R., Kramer, A.F., 2015. Fitness, but not physical activity, is related to functional integrity of brain networks associated with aging. NeuroImage 131, 113–125.

Walach, H., Schmidt, S., Esch, T., 2014. Meditation intervention reviews: comments and responses. JAMA Intern. Med. 174 (7), 1193–1194.

Walfisch, A., Sermer, C., Cressman, A., Koren, G., 2013. Breast milk and cognitive development—the role of confounders: a systematic review. BMJ Open 3 (8), e003259.

Walker, J.G., Batterham, P.J., Mackinnon, A.J., Jorm, A.F., Hickie, I., Fenech, M., Christensen, H., 2012. Oral folic acid and vitamin B-12 supplementation to prevent cognitive decline in community-dwelling older adults with depressive symptoms—the Beyond Ageing Project: a randomized controlled trial. Am. J. Clin. Nutr. 95 (1), 194–203.

Walsh, R., Shapiro, S.L., 2006. The meeting of meditative disciplines and Western psychology: a mutually enriching dialogue. Am. Psychol. 61 (3), 227–239.

Wang, J.-R., Hsieh, S., 2013. Neurofeedback training improves attention and working memory performance. Clin. Neurophysiol. 124 (12), 2406–2420.

Wang, B., Zhan, S., Gong, T., Lee, L., 2013. Iron therapy for improving psychomotor development and cognitive function in children under the age of three with iron deficiency anaemia. Cochrane Database Syst. Rev. (6), CD001444.

Warthon-Medina, M., Moran, V.H., Stammers, A.-L., Dillon, S., Qualter, P., Nissensohn, M., Lowe, N.M., 2015a. Zinc intake, status and indices of cognitive function in adults and children: a systematic review and meta-analysis. Eur. J. Clin. Nutr. 69 (6), 649–661.

Warthon-Medina, M., Qualter, P., Zavaleta, N., Dillon, S., Lazarte, F., Lowe, N.M., 2015b. The long term impact of micronutrient supplementation during infancy on cognition and executive function performance in pre-school children. Nutrients 7 (8), 6606–6627.

Waters, L., Barsky, A., Ridd, A., Allen, K., 2015. Contemplative education: a systematic, evidence-based review of the effect of meditation interventions in schools. Educ. Psychol. Rev. 27 (1), 103–134.

Wayne, P.M., Walsh, J.N., Taylor-Piliae, R.E., Wells, R.E., Papp, K.V., Donovan, N.J., Yeh, G.Y., 2014. Effect of tai chi on cognitive performance in older adults: systematic review and meta-analysis. J. Am. Geriatr. Soc. 62 (1), 25–39.

Weiskopf, N., 2012. Real-time fMRI and its application to neurofeedback. NeuroImage 62 (2), 682–692.

Weuve, J., Kang, J.H., Manson, J.E., Breteler, M.M.B., Ware, J.H., Grodstein, F., 2004. Physical activity, including walking, and cognitive function in older women. JAMA 292 (12), 1454–1461.

Whalley, L.J., Deary, I.J., 2001. Longitudinal cohort study of childhood IQ and survival up to age 76. BMJ 322 (7290), 819.

Whiteman, A.S., Young, D.E., Budson, A.E., Stern, C.E., Schon, K., 2016. Entorhinal volume, aerobic fitness, and recognition memory in healthy young adults: a voxel-based morphometry study. NeuroImage 126, 229–238.

Williamson, J.D., Espeland, M., Kritchevsky, S.B., Newman, A.B., King, A.C., Pahor, M., LIFE Study Investigators, 2009. Changes in cognitive function in a randomized trial of physical activity: results of the lifestyle interventions and independence for elders pilot study. J. Gerontol. A Biol. Sci. Med. Sci. 64 (6), 688–694.

Wilson, P.W.F., Schaefer, E.J., Larson, M.G., Ordovas, J.M., 1996. Apolipoprotein E alleles and risk of coronary disease: a meta-analysis. Arterioscler. Thromb. Vasc. Biol. 16 (10), 1250–1255.

Wimmer, L., Bellingrath, S., Von Stockhausen, L., 2016. Cognitive effects of mindfulness training: results of a pilot study based on a theory driven approach. Front. Psychol. 7, 1037.

Wraw, C., Deary, I.J., Gale, C.R., Der, G., 2015. Intelligence in youth and health at age 50. Intelligence 53, 23–32.

Yang, G., Wang, Y., Tian, J., Liu, J.-P., Hanns, H., Gabriele, N., Wang, Y., 2013. Huperzine A for Alzheimer's disease: a systematic review and meta-analysis of randomized clinical trials. PLoS One 8 (9), e74916.

Young, K.D., Zotev, V., Phillips, R., Misaki, M., Yuan, H., Drevets, W.C., Furey, M., 2014. Real-time fMRI neurofeedback training of amygdala activity in patients with major depressive disorder. PLoS One 9 (2), e88785.

Zhang, G., Yao, L., Zhang, H., Long, Z., Zhao, X., Baddeley, A., Hodges, J., 2013. Improved working memory performance through self-regulation of dorsal lateral prefrontal cortex activation using real-time fMRI. PLoS One 8 (8), e73735.

Zheng, G., Liu, F., Li, S., Huang, M., Tao, J., Chen, L., 2015. Tai chi and the protection of cognitive ability: a systematic review of prospective studies in healthy adults. Am. J. Prev. Med. 49 (1), 89–97.

Zhu, N., Jacobs, D.R., Schreiner, P.J., Yaffe, K., Bryan, N., Launer, L.J., Sternfeld, B., 2014. Cardiorespiratory fitness and cognitive function in middle age: the CARDIA study. Neurology 82 (15), 1339–1346.

Zoefel, B., Huster, R.J., Herrmann, C.S., 2011. Neurofeedback training of the upper alpha frequency band in EEG improves cognitive performance. NeuroImage 54 (2), 1427–1431.

Chapter 6

Once Upon a Time We Were on the Moon

In the final chapter, authors usually summarize the previous ones, point to shortcomings in research methodology, and discuss the theoretical background of analyzed objectives and assumptions. Subsequently, improvements are suggested by proposing new experiments and hypotheses that would finally provide answers to the unresolved issues. However, to my knowledge, the proposed research is seldom carried out. Some authors also have the urge to make prophetic predictions—proposing the changes that research will bring to society and our daily lives in the future. This may sound pessimistic, but these forecasts rarely come true.

To give just one example from personal experience. In 1969, I was a teenager who stayed up late to watch the live TV broadcast of Apollo 11 spaceflight that landed the first humans on the Moon. It was in July during the summer school break, so no one in the family objected that I spent most of the time during a period of 4 days, from the start of the Saturn V rocket till the landing of the Lunar module, in front of our tiny black-and-white TV set. Because the inserts of direct coverage of the flight were rather short, although repeated several times, the broadcast was also packed with commentaries and predictions what impact the landing will have on humanity. It was speculated that the module Eagle would sink into the Moon sand, or that the importance of art will vanish only to be replaced by the appealing creativity of technological invention. People wondered what the first words of the first man on the Moon would be. Then, when Armstrong uttered the words, it was seriously estimated that, by the end of the century, we will face lunar colonies living in bases built on the Moon, and missions to orbit or even land on Mars. As stressed by Goodings et al. (2013), this was a fascinating moment that was lost quickly. However, at that time, one would probably be diagnosed insane if they had predicted that, in 2016, a mission to the Moon would be regarded as impossible, not to speak about Mars.

It is amazing that most expert models of futuristic worlds end up being proven wrong. In contrast, Jules Verne, the father of science fiction, made several correct predictions. Submarines, newscasts, skywriting—atmospheric advertisements, videoconferencing—phonotelephone, taser—a gun that delivers a

Increasing Intelligence. http://dx.doi.org/10.1016/B978-0-12-809407-5.00006-3

strong electric jolt, are just a few of his ideas that came true. Probably the most prescient anticipation of later history was his 1865 novel *From the Earth to the Moon*, in which he correctly predicted that the United States would be the country to launch the first lunar mission, that Florida would make the best launch site, and that the crew size would be three astronauts. Close to reality were also the dimensions of the space capsule and the duration of the voyage to the Moon, as well as the landing in the ocean.

Hence, a better approach to the question what direction research about increasing intelligence will take in the future is to consider Hollywood production of science fiction. I have seen four films that could be classified narrowly into the genre of increasing intelligence. The first one was Ralph Nelson's *Charly* (1968), followed by *The Lawnmower Man* (1992) staring Pierce Brosnan as Dr. Lawrence Angelo, and the more recent ones *Limitless* (2011) with Robert de Niro as the bad guy Van Loon, and the last one Luc Besson's *Lucy* (2014). The timeline of production suggests that there is an increasing interest in the topic, which could be explained as an influence of research on the motion picture industry. The common characteristic of all four plots is that the respondents who undergo the interventions—Charly, Jobe, Eddie, and Lucy—could be regarded as losers that have immense difficulties in their everyday and professional lives. On the extremes with respect to intelligence are Jobe, a simple-minded gardener and Charly, an intellectually disabled young adult who can hardly spell his own name. The interventions were mainly based on drugs either accidentally consumed as in the cases of Eddie and Lucy, or given on purpose in combination with a training based on computer games and virtual reality as in the case of Jobe. An exception is Charly who together with the lab mouse Algernon undergoes a surgical procedure that triples his IQ. Hence, if we believe in science fiction, nootropic drugs in combination with computer games will in the future be the most promising cognitive enhancers. One last word in this brief discussion of Hollywood production, what were the outcomes of the interventions? Jobe and Lucy physically disintegrate—Lucy in the space-time continuum—and Jobe escapes into the worldwide network. Both confirmed their existence by phone—Jobe makes the phones all around the globe ring, while Lucy is more modest, she sends a text message stating that she is everywhere. Eddie runs for the US senate, impressing Robert de Niro with his ability to forecast events based on his extraordinary processing speed to extrapolate multiple sensory information, as well as his girlfriend, with his Chinese language skills. For Charly, the intervention shows side effects, the lab mouse Algernon dies, and Charly returns to his state prior to surgery, happily playing with children.

The fate of the main characters—Charly, Jobe, Eddie, and Lucy—opens yet another question that I will address in the last section: do we really need more intelligence, or are we better off with less?

6.1 WHY BECOME MORE INTELLIGENT?

The main purpose of increasing intelligence advocated in this book was mainly as a tool for elucidating our understanding of the brain–intelligence relationship. Do not turn a blind eye to the fact that the main interest is actually the result—more intelligent people. What would this "more" mean for each of us and society? The first part of the question is easy to answer, as there are numerous correlations between IQ and all kinds of characteristics, from ordinary ones such as academic performance (Strenze, 2015; see Section 1.1.2), to more bizarre ones like the acceptability of genetically-modified organisms (Šorgo et al., 2012).

Some of the benefits of having a high IQ can only be realized in competition with others. Hence, if everyone would be more intelligent, this would not present an advantage for the individual. From a selfish point of view, to take full advantage of an IQ increase, it would be desirable if not all but just one individual, or a selected minority, would have the opportunity of becoming more intelligent. In contrast, an increase in IQ of the whole human population would be more beneficial for the society.

To estimate whether having a higher IQ presents an advantage for an individual, one should know what the main priorities of people are—what they value most in life. In 2011, the OECD launched the *Better Life Initiative*. The main objective of the initiative was to ask people what aspects of life matter most to them, and shape their quality of life. The OECD web application, the Better Life Index (www.oecdbetterlifeindex.org) identified 11 topics (safety, housing, jobs, income, work-life balance, life satisfaction, health, education, governance, community, and environment) as essential in the areas of material living conditions and quality of life (OECD, 2015). The findings are based on voluntary submissions of over 60,000 individuals from 184 countries, and can only be considered as indicative and not representative of the population as a whole. Life satisfaction (10.34%), health (10.32%) and education (9.86%) have been consistently the top rated topics, while governance or civic engagement (6.76%), community (8.07%), and income (8.77%) have been ranked the lowest. However, there are also considerable differences between countries. People of Cyprus selected a work-life balance as their highest priority, while the people of Ukraine and Albania valued income. Almost all of South America was most preoccupied with issues regarding education. In Venezuela, safety was the first, and in Guyana housing was a priority.

Considering just the top three ranked topics, life satisfaction, health, and education, we know that at least for the last one there is a robust positive correlation with intelligence. As already discussed in Section 1.1.2, several correlational studies between intelligence and success in life, defined as socioeconomic success (academic and job performance), have shown the highest correlations with academic performance (0.58), followed by educational attainment (0.56). The number of individuals included in these studies was 191,033 (Hunter and

Hunter, 1984; Poropat, 2009; Strenze, 2007, 2015). It could be argued that in modern societies, intelligence explains approximately 30% of success in academic and job performance.

Similar correlations with intelligence were also reported for different health indicators. Lynn and Vanhanen (2012) analyzed 24 studies showing a negative correlation between –0.34 and –0.84 for infant mortality, and a positive for life expectancy ranging between 0.51 and 0.84. Further, there is substantial empirical evidence linking cognitive ability to mortality risk (Anstey et al., 2006). In a recent study Aichele et al. (2016) showed that processing speed was among the analyzed cognitive variables the best predictor of survival. The authors explained this finding with the sensitivity of processing speed to the effects of aging that can signal variation in multiple underlying processes linked to mortality risk (e.g., cardiovascular disease, prevalence of cerebral white-matter lesions). Comparable findings were also reported by Roberts et al. (2009), who showed that processing speed predicted mortality risk more strongly than physical activity, resting heart rate, psychological distress, waist-to-hip ratio, weekly alcohol consumption, body mass index, and socioeconomic advantage. Only smoking status and blood pressure were better predictors than processing speed.

The finding that intelligent people live longer (Deary, 2008) can be also explained by the fact that they are better educated, have a higher income, and have the tendency to pursue more healthy habits in terms of sports, smoking, and dietary regime. In this explanation, intelligence is considered just as a tool to indirectly achieve a longer life expectancy. On the other hand, the cognitive reserve theory (Stern, 2002; see also Section 2.2.1) suggests that high intelligence provides a reserve capacity, a buffer against the effects of aging and disease. In this framework, intelligence is conceptualized as a functionally relevant phenotype where a cognitively optimized brain acts against aging or intercurring neurological insults. The latter hypothesis has been recently confirmed in a study by Santarnecchi et al. (2015), pointing to a direct link between intelligence and intrinsic brain properties to cope with the loss of its functional units. Moreover, it was shown that brain robustness in high IQ individuals was supported by neocortical regions associated with language and memory processing, whereas regions related to emotional processing were mostly responsible for lower IQ individuals. This pattern also suggests a more favorable prognosis for postlesional or neurodegenerative recovery.

Carol Graham (2012) in her book *The Pursuit of Happiness: an Economy of Well-being*, proposed a new science of happiness. If we believe in Gallup's global emotions report for 2016 (Gallup, 2016), the happiest people live in Paraguay, Guatemala, and Honduras, whereas people from Syria, Turkey, and Nepal have the least positive experiences. Has this anything to do with IQ? At first glance, one would expect that smart people are healthier, more successful at school, excel others in job performance, and in general, cope better with their life, hence they are happier and more satisfied. This is also the most common assumption of laypeople (Grossmann et al., 2013). However, research has provided less clear cut results for the relation between intelligence and life satisfaction, well-being, and happiness.

The Terman study of the gifted provided the first opportunity to examine the role of giftedness in shaping aspects of the self-narrative over the life course (Terman et al., 1925; Terman and Oden, 1959). Selecting the *crème de la crème* of California's students with respect to intelligence resulted in the identification of 1528 gifted pupils (11 years of age) with IQs above 140, eighty of whom had IQs above 170. Together, they became known as the "Termites," and have been studied in follow-up research over the decades, and will continue to do so until the last participant dies or withdraws from the study[a] (Holahan and Holahan, 1999). With respect to happiness, the participants in their mid-seventies had on average high levels of subjective well-being. The main variables contributing to this were self-rated health and adequacy of finances, social relationships, and positive approaches to life, represented by goals, ambitions, and behavioral involvement in activities (Holahan and Sears, 1995). In another study, it was established that knowing at a younger age (14 years or less) that one was intellectual gifted, was related to a smaller likelihood of believing that one had lived up to one's intellectual abilities at midlife, and to less favorable psychological well-being at age 80 (Holahan and Holahan, 1999).

On the other hand, Wirthwein and Rost (2011) found no difference between gifted and nongifted respondents in any of the components of subjective well-being (e.g., positive and negative affectivity, general life satisfaction, and satisfaction with life in specific domains such as work, spouse/partner, self and friends, health, and leisure). There was a small negative trend for gifted adults for the domain of leisure. Similarly, Sękowski and Siekańska (2008) found that the traits typical of the high achievers group in interpersonal relations were relative shyness and lower sociability. These traits, however, did not indicate social maladjustment or intensified problems in social contacts. In the same direction points a recent study involving German high school students (Bergold et al., 2015). No differences in life satisfaction between gifted (IQ > 120) and nongifted adolescents could be observed. In yet another study by Wigtil and Henriques (2015), performed on a sample of 3829 incoming first-year university students, contrary to expectations, it was found that the relationship between intelligence and psychological well-being was somewhat negative. The negative correlation was most pronounced for SAT Maths scores with personal growth, positive relations with others, and purpose in life.

An interesting finding was revealed in a metaanalysis by Veenhoven and Choi (2012) comparing the relation between intelligence and happiness at the microlevel of individuals with the one at the macrolevel of nations. On the

a. Howard Friedman and Leslie Martin (Friedman and Martin, 2011), who are currently involved in the research, published a book in 2011 summarizing their findings: *The Longevity Project: Surprising Discoveries for Health and Long Life from the Landmark Eight-Decade Study.*

individual level, 23 empirical studies were included that involved 10 different nations and 15,827 respondents. The findings of studies correlating earlier intelligence and current happiness as well as those between current intelligence and current happiness found almost no correlation between intelligence and happiness. In contrast, on the macro level, data for 192 nations for the second half of the 20th century showed a strong correlation between IQ and happiness in nations ($r = 0.60$). These findings suggest that the smartness of a population is more beneficial than the more selfish variant of being smarter than others. The authors concluded that intelligence adds to happiness only indirectly with its effects on society.

Using more complex structural modeling methodology, weak relations between multidimensional well-being, intelligence, and own education in a female sample (4634 women, aged 18–82) was established (Zhou et al., 2016). Similarly, in yet another study employing structural modeling, a measure of *g* showed positive correlations with life satisfaction and career success, and no correlation with career satisfaction (Amdurer et al., 2014). Furthermore, some researchers have attempted to make a distinction between trait intelligence measured with standard IQ tests and wise thinking, defined as the degree to which people use various pragmatic schemas to deal with social conflicts (Grossmann et al., 2013). The central finding was that crystallized intelligence, processing speed, and working memory showed no relationship to well-being, whereas the ability to reason wisely about social conflicts was positively associated with different aspects of well-being, such as global life satisfaction, social relationships, and positively solved social conflicts, as well as less negative affect in daily life, a lower propensity to brood, and greater longevity. An important aspect of the study was also the finding that wisdom improves into old age. The authors speculated that the increasing number of older adults may be of great value for well-being of future communities (Grossmann et al., 2013).

There are also positive correlations between intelligence and the other OECD topics identified as essential for quality of life. The Terman study showed that average salary of the participants at their midlife was twice that of the average white-collar job at the same time (Terman and Oden, 1959). For example, one of the most famous termites was Jess Oppenheimer, the writer of the classic 1950s sitcom *I Love Lucy,* a series broadcast on CBS. A study by Zagorsky (2007) further showed that, on the one hand, people with above-average IQ scores are three times as likely as below-average IQ individuals to have a high income, whereas on the other hand, people with above-average IQ are only 1.2 times as likely as individuals with below-average intelligence to have a comparatively high net worth. This would mean that being highly intelligent can sometimes increase the risk of being in financial difficulty.

Furthermore, it was shown that intelligence is positively related to ecological sustainability (Salahodjaev, 2016). This relationship was much more pronounced in high-income countries than in lower income countries. Being more intelligent also means being more physically attractive (Kanazawa, 2011;

Zebrowitz et al., 2002); having a larger bias blind spot which means identifying thinking biases in others but not in oneself (West et al., 2012), having less satisfaction with more frequent socialization with friends (Li and Kanazawa, 2016); being less religious (Lynn et al., 2009; Lynn and Vanhanen, 2012), and being more suicidal (Lynn and Vanhanen, 2012; Voracek, 2009).

All in all, confronting the pros and cons of being more intelligent, either on an individual micro level, or on the macro level of society, the pros clearly prevail. Intelligence has made us the dominant species on Earth, with no natural enemies, except ourselves. Yet, we still have not solved the problem of hunger and poverty, and often our discussions lead to devastating wars. Furthermore, we have created new problems like pollution, extinction of species, global warming, population growth, and many experts believe that we are on the edge of a global economic collapse. Can higher human intelligence solve them?

REFERENCES

Aichele, S., Rabbitt, P., Ghisletta, P., 2016. Think fast, feel fine, live long: a 29-year study of cognition, health, and survival in middle-aged and older adults. Psychol. Sci. 27 (4), 518–529.

Amdurer, E., Boyatzis, R.E., Saatcioglu, A., Smith, M.L., Taylor, S.N., 2014. Long term impact of emotional, social and cognitive intelligence competencies and GMAT on career and life satisfaction and career success. Front. Psychol. 5, 1447.

Anstey, K.J., Mack, H.A., von Sanden, C., 2006. The relationship between cognition and mortality in patients with stroke, coronary heart disease, or cancer. Eur. Psychol. 11 (3), 182–195.

Bergold, S., Wirthwein, L., Rost, D.H., Steinmayr, R., 2015. Are gifted adolescents more satisfied with their lives than their non-gifted peers? Front. Psychol. 6, 1623.

Deary, I., 2008. Why do intelligent people live longer? Nature 456 (7219), 175–176.

Friedman, H.S., Martin, L.R., 2011. The Longevity Project: surprising discoveries for health and long life from the landmark eight-decade study. Center Point Pub, Thorndike, ME.

Gallup, 2016. Gallup Global Emotions 2016. Gallup, Washington, DC

Goodings, L., Brown, S.D., Parker, M., 2013. Organising images of futures-past: remembering the Apollo moon landings. Int. J. Manag. Con. Philos. 7 (3/4), 263.

Graham, C., 2012. The pursuit of happiness: an economy of well-being. Brookings Inst. Press, Washington, DC.

Grossmann, I., Na, J., Varnum, M.E.W., Kitayama, S., Nisbett, R.E., 2013. A route to well-being: Intelligence versus wise reasoning. J. Exp. Psychol. Gen. 142 (3), 944–953, https://doi.org/10.1037/a0029560

Holahan, C.K., Holahan, C.J., 1999. Being labeled as gifted, self-appraisal, and psychological well-being: a life span developmental perspective. Int. J. Aging Hum. Dev. 48 (3), 161–173.

Holahan, C.K., Sears, R.R., 1995. The gifted group in later maturity. Stanford University Press, Stanford, CA.

Hunter, J.E., Hunter, R.F., 1984. Validity and utility of alternative predictors of job performance. Psychol. Bull. 96 (1), 72–98.

Kanazawa, S., 2011. Intelligence and physical attractiveness. Intelligence 39 (1), 7–14.

Li, N.P., Kanazawa, S., 2016. Country roads, take me home… to my friends: how intelligence, population density, and friendship affect modern happiness. Br. J. Psychol. 107 (4), 675–697.

Lynn, R., Vanhanen, T., 2012. National IQs: a review of their educational, cognitive, economic, political, demographic, sociological, epidemiological, geographic, and climatic correlates. Intelligence 40 (2), 226–234.

Lynn, R., Harvey, J., Nyborg, H., 2009. Average intelligence predicts atheism rates across 137 nations. Intelligence 37 (1), 11–15.

OECD, 2015. *How's Life? 2015.* OECD Publishing. Available from: http://www.oecd-ilibrary.org/economics/how-s-life-2015_how_life-2015-en

Poropat, A.E., 2009. A meta-analysis of the five-factor model of personality and academic performance. Psychol. Bull. 135 (2), 322–338.

Roberts, B.A., Der, G., Deary, I.J., Batty, G.D., 2009. Reaction time and established risk factors for total and cardiovascular disease mortality: comparison of effect estimates in the follow-up of a large, UK-wide, general-population based survey. Intelligence 37 (6), 561–566.

Salahodjaev, R., 2016. Does intelligence improve environmental sustainability? An empirical test: intelligence and environmental sustainability. Sust. Dev. 24 (1), 32–40.

Santarnecchi, E., Rossi, S., Rossi, A., 2015. The smarter, the stronger: intelligence level correlates with brain resilience to systematic insults. Cortex 64, 293–309.

Sękowski, A., Siekańska, M., 2008. National academic award winners over time: their family situation, education and interpersonal relations. High Ability Stud. 19 (2), 155–171.

Šorgo, A., Jaušovec, N., Jaušovec, K., Puhek, M., 2012. The influence of intelligence and emotions on the acceptability of genetically modified organisms. Electr. J. Biotechnol. 15 (1).

Stern, Y., 2002. What is cognitive reserve? Theory and research application of the reserve concept. J. Int. Neuropsychol. Soc. 8 (3), 448–460.

Strenze, T., 2007. Intelligence and socioeconomic success: a meta-analytic review of longitudinal research. Intelligence 35 (5), 401–426.

Strenze, T., 2015. *Intelligence and socioeconomic success. A study of correlations, causes and consequences* (Dissertationes Sociologicae Universitatis Tartuensis). University of Tartu press. Available from: http://dspace.ut.ee/dspace;handle/bitstream/handle/10062/47823/strenze_tarmo.pdf?sequence=5andisAllowed=y

Terman, L.M., Oden, M.H., 1959. The gifted group at mid-life: vol. 5, Genetic studies of genius. Stanford University Press, Stanford, CA.

Terman. L.M., Baldwin. B.T., Bronson. E., Devoss, J.C., Fuller, F.L., Goodenough, T.L., Kelley, M. Lima, Marshall, H., Moore, A.H., Raubenheimer, A.S. Ruch, G.M., Willoughby, R.L., Wyman, J.B., Yates, H.H., 1925. Mental and physical traits of a thousand gifted children: Vol. I . Genetic studies of genius. Stanford, CA: Stanford University Press

Veenhoven, R., Choi, Y., 2012. Does intelligence boost happiness? Smartness of all pays more than being smarter than others. Int. J. Happ. Dev. 1 (1), 5.

Voracek, M., 2009. National intelligence, suicide rate, and subjective well-being. Percept. Motor Skills 109 (3), 718–720.

West, R.F., Meserve, R.J., Stanovich, K.E., 2012. Cognitive sophistication does not attenuate the bias blind spot. J. Person. Soc. Psychol. 103 (3), 506–519.

Wigtil, C.J., Henriques, G.R., 2015. The relationship between intelligence and psychological well-being in incoming college students. Psychol. Wellbeing 5 (1).

Wirthwein, L., Rost, D.H., 2011. Giftedness and subjective well-being: a study with adults. Learn. Individ. Differ. 21 (2), 182–186.

Zagorsky, J.L., 2007. Do you have to be smart to be rich? The impact of IQ on wealth, income and financial distress. Intelligence 35 (5), 489–501.

Zebrowitz, L.A., Hall, J.A., Murphy, N.A., Rhodes, G., 2002. Looking smart and looking good: facial cues to intelligence and their origins. Pers. Soc. Psychol. Bull. 28 (2), 238–249.

Zhou, L., Lin, H., Lin, Y.-C., 2016. Education, intelligence, and well-being: evidence from a semi-parametric latent variable transformation model for multiple outcomes of mixed types. Soc. Indicat. Res. 125 (3), 1011–1033.

Index

H

I

J

K

L

S

T

U

V

W

Kolibri 360 GmbH
Ferdinand-Jühlke-Straße 7
99095 Erfurt, Deutschland
produktsicherheit@kolibri360.de